Land, Value, Community

SUNY series in Environmental Philosophy and Ethics
J. Baird Callicott and John van Buren, editors

Land, Value, Community

Callicott and Environmental Philosophy

Edited by
Wayne Ouderkirk
and
Jim Hill

State University of New York Press

Cover image: Digital Stock

Published by
State University of New York Press, Albany

Printed in the United States of America

For information, address State University of New York Press,
90 State Street, Suite 700, Albany, NY 12207.

Production by Judith Block
Marketing by Anne Valentine

Library of Congress Cataloging-in-Publication Data

Land, value, community : Callicott and environmental philosophy / edited by Wayne Ouderkirk, Jim Hill.
p. cm. — (SUNY series in environmental philosophy and ethics)
Includes bibliographical references and index.
ISBN 0-7914-5229-8 (alk. paper) — ISBN 0-7914-5230-1 (pbk. : alk. paper)
1. Callicott, J. Baird. 2. Human ecology—Philosophy. I. Ouderkirk, Wayne. II. Hill, Jim, 1937– III. Series.

GF21 .L35 2002
304.2—dc21

2001031198

10 9 8 7 6 5 4 3 2 1

Contents

Acknowledgments

The following selections were either previously published or are based on earlier works and are reprinted here, in whole or in part, with permission of the authors and original publishers: "Ecological Morality and Nonmoral Sentiments" by Ernest Partridge originally appeared in *Environmental Ethics* 18 (1996): 149–163. In her "Biocentrism, Biological Science, and Ethical Theory," Kristin Shrader-Frechette develops ideas from her "Biological Holism and the Evolution of Ethics," in *Between the Species* 6 (1990): 185–192. Wendy Donner's "Callicott on Intrinsic Value and Moral Standing in Environmental Ethics" is one section of her earlier essay, "Inherent Value and Moral Standing in Environmental Change," which appeared in *Earthly Goods: Environmental Change and Social Justice*, edited by Fen Osler Hampson and Judith Reppy (Ithaca, N.Y.: Cornell University Press, 1996), 57–65. "Epistemology and Environmental Values" is an abbreviated version of Bryan Norton's essay of the same title, originally in *The Monist*, Special Issue on Intrinsic Value in Nature, 75 (1992): 208–226. Peter Wenz's essay, "Minimal, Moderate and Extreme Moral Pluralism," appeared in longer form in *Environmental Ethics* 15 (1993): 149–163. Andrew Light's essay is reprinted from: "Callicott and Naess on Pluralism," *Inquiry* 39, no. 2 (June 1996): 273–294, by permission of Scandinavian University Press, Oslo, Norway. And "Callicott's Last Stand" by Lee Hester, Dennis McPherson, Annie Booth, and Jim Cheney, although originally written for this volume, appeared earlier, in different form, in *Environmental Ethics* 22 (2000): 273–290.

Introduction: Callicott and Environmental Philosophy

Wayne Ouderkirk

Over the last twenty-five years, environmental philosophy has exploded into a vigorous and important area of research and writing. At first a form of applied ethics, it has rapidly become a matrix of ethical, metaphysical, epistemological, social, and political speculation, with an array of special problems or issues, several major theoretical models or paradigms, and the other fundamentals—journals, conferences, graduate programs—that mark a philosophically significant area of study.[1] This vigor is partial testimony to its importance because the creation of an environmentally benign, beneficial worldview, which would include a defensible and practicable environmental ethic, is clearly a major necessity resulting from our continuing and deepening environmental crisis.

J. Baird Callicott has been, and continues to be, one of the central figures in the development of environmental philosophy. To say that he has helped set the terms of the discussion, that he has developed one of the central theoretical models in the field, the land ethic, and that his work has provoked reactions and reflections that have both clarified other models and opened new avenues for continued work is no exaggeration.

This book examines environmental philosophy by analyzing Callicott's views critically. There are several reasons for this approach. First, one cannot discuss the field without considering Callicott's views. And the reverse is also true: If one wants to examine Callicott's views, there is no escaping a discussion of the larger field. He is that important a figure. Third, because he has been such a force, his theory warrants extended examination and analysis. Finally, by presenting his critics' evaluations of his theories, their own preferred ideas for future work, along with Callicott's response to those ideas, we can get a partial picture of some of the next important developments in the field. Not that there is here a crystal ball, but certainly that potent mixture—Callicott and his critics—will be at the center of whatever environmental philosophy becomes in its next twenty-five years.

Thus, this book represents one snapshot of a significant, lively, evolving field. As such, it cannot and does not pretend to cover every possible idea or theory. Still, by examining the strands of Callicott's theory and what he has tried to do with it, it covers a great deal. The sections of this collection fall rather naturally into place in accordance with key facets of Callicott's work. Within each section, other thinkers (philosophers, ecologists, political scientists, and scholars of religion) evaluate some aspect of that facet of Callicott's thought. In addition, most also explain their own ideas for resolving the problems they see for his position, thereby contributing new ideas to the continuing debate. So the book is about their thinking as well.

Of the seventeen essays that follow, all but six—those of Partridge, Donner, Norton, Light, Wenz, and the essay by Hester, McPherson, Booth, and Cheney—are published here for the first time; and all but one of the eleven original essays were written for this volume. The current version of the multiauthor essay was written first, and for this collection, although a later version was published before this one.

Each of our authors explains those parts of Callicott's theory that are important for her or his own analysis, but as context for what follows we need a fuller depiction of Callicott's theory. For a complete exposition, the reader should study Callicott's writings. However, here we will explain his main ideas and relate them to the essays that follow.[2]

The Land Ethic and Its Foundations

Our world faces myriad anthropogenic environmental problems. Even a partial list reminds us of their complexity and scope: global warming, the rapid elimination of tropical rain forests and with them countless species of flora and fauna, the conversion of what little wilderness remains on the planet into farmlands, and the conversion of farmlands into cities, roads, and shopping malls. One response to such problems is that we humans should change the behaviors that lead to them because, unless we do, we are harming ourselves or future generations of humans. As appealing as such a response might be, many, including Callicott, have thought it at best incomplete and at worst an invitation to continue along our present course as long as we engage in some technological tinkering that many believe will put things aright.

The missing element in this human-centered response to environmental problems, of course, is the environment itself. Although previously not a subject of direct moral concern, omitting it from our present and future ethical deliberations seems both arrogant and a blatant continuation of our past misbehavior. But the question then becomes whether and how to justify a moral concern for the environment, especially in light of the traditional Western restriction of morality to interhuman relations.

In 1948, Aldo Leopold proposed the land ethic as a response to this question, and Callicott has earned his own place in the discussion by explaining, analyzing, and defending the core ideas of that ethic. Its basic moral injunction is Leopold's famous, oft-quoted maxim: "A thing is right when it tends to preserve the stability, integrity, and beauty of the biotic community. It is wrong when it tends otherwise." [3]

But why should we accept this new moral injunction? Callicott's response is that an accurate (i.e., a scientifically informed) picture of morality shows not only that we can but also that we should accept it. The requisite scientific perspective is primarily threefold, joining evolutionary biology, ecology, and Copernican astronomy, although Callicott frequently adds his interpretation of contemporary physics. The philosophical basis for this new perspective on ethics Callicott derives from the moral theories of David Hume and Adam Smith.

The science, although not totally uncontroversial, as we shall see, is fairly straightforwardly stated: Darwinian evolution shows that we humans have become what we are, not through divine fiat, but through the same evolutionary processes that produced all the millions of other life forms on this planet. That relates us in multiple, intimate ways to the rest of nature. Ecology shows us that all those life forms are integrated into an interactive, mutual interdependence. That interdependence is part of who and what we humans are, delineating more clearly the kind of linkage we have with this world, namely, community membership. Astronomy shows us that Earth is home, that the fates of all who live here are joined inseparably on one small planet.

This bundle of scientific ideas needs a link to justify a transition from it to a moral injunction, and Callicott finds that link in the Hume-Smith tradition of moral sentiment, fortified by Darwin's account of the evolution of morality. Unlike most Western ethicists, who place reason at the center of morality, Hume and Smith instead argue that it is sentiments—emotions, feelings, both positive and negative—that provide us with our morality.[4] Importantly, those feelings, according to Hume, include an affection not only for other individuals but also for social groups or communities as a whole. Callicott convincingly argues that Darwin both knew of and used Hume's moral psychology in his account of how ethics, or altruistic behavior, could have evolved. Darwin's explanation is that those of our hominid ancestors to whom natural selection had given stronger emotional ties to their social groups developed, due to those ties, cooperative behavior. Thus their offspring were naturally selected in the evolutionary process because members of a cohesive group had a higher likelihood of surviving than individuals struggling alone.[5]

So the evolutionary description of the origins of ethics confirms the Hume-Smith theory of ethics. Recall that evolution and ecology also show us that we are part of a community that includes the rest of nature, which is not a simple collection of separate components but an integrated whole, a biotic community. Such community membership can stimulate our evolved sentiments toward perceived communal ties. Our environmental obligations arise from our emotional ties to that community, which is every bit as much our own as is our immediate family.

Callicott concludes: "Therefore, an environmental or land ethic is both possible—the biopsychological and cognitive conditions are in place—and necessary, since human beings collectively have acquired the power to destroy the integrity, diversity, and stability of the environing and supporting economy of nature."[6] In broad outline, this is Callicott's general justification, in his phrase, the foundation, of his environmental ethic. It appears throughout his work, even in recent writings where he is developing a postmodern environmental ethic.[7] More precisely, in such contexts he argues that through its use of evolutionary and ecological theory, Leopold's land ethic "opens out" on a postmodern perspective. So even there two of the main parts of his justification remain, and the others are not left far behind.

Few would dispute the general evolutionary account of our connections with the rest of nature, and the specific account of the development of ethics clearly makes sense within that Darwinian perspective. Nevertheless, Callicott's justification has problems. In the broadest terms, the metaphor of a foundation for the land ethic seems ill chosen when the same metaphor has proven problematic in other philosophic contexts and especially because foundationalism is one of the cornerstones of modernism, which Callicott rejects.[8]

More specifically, in part I, Ernest Partridge examines Hume's account of the moral sentiments and concludes that it is not an adequate basis for an environmental ethic because Hume's specifically moral sentiments originate in interpersonal relations and are attitudes toward persons. So Hume's theory would actually reinforce anthropocentrism, not a Leopoldian ecocentrism. As a counterproposal, Partridge offers as the basis for a nonanthropocentric environmental ethic "biophilic" natural sentiments, that is, positive, nonmoral emotional responses to nature, which he and others argue are part of our genetic constitution.

Smith, says John Barkdull in his essay, has a different theory of moral sentiment from that of Hume. The most relevant difference for environmental ethics is that morality, according to Smith, arises from and within close social interaction and is individually based. It thus lacks a sentiment toward society at large and so cannot fund obligations toward the community. Moreover, Barkdull argues that Smith's theory can probably not support obligations toward nonhumans because for him the general opinion gives moral principles much of their force. Because no general consensus exists on the moral standing of nonhuman nature, Smith would not see any moral obligation toward it. On the other hand, Barkdull does see some support for Callicott in Smith's view of aesthetic inspiration to improve the workings of the community. However, as was the case with Partridge's proposal, such support is decidedly nonmoral and thus diverges significantly from Callicott's account.

Robert McIntosh, an ecologist, moves the discussion to the land ethic's alleged foundation in scientific ecology. Searching a large sample of ecological literature for settled meanings of the key concepts of ecosystem, community, integrity, and stability, he finds little in that literature helpful to Callicott. All of those concepts, he claims, have diverse meanings in ecology; and that diversity raises difficulties for any philosophical appropriation of them. In addition, ecologists and philosophers of science disagree about the nature of ecology. McIntosh concludes, "The merits of ecology as the basis of an environmental ethic are unclear if its status as a science is unclear."

Although she finds much to praise in Callicott's theory, Kristin Shrader-Frechette likewise faults his use of the scientific concept of community. She too reviews some of the relevant ecological literature and claims that "there is no *scientifically/biologically* coherent notion of 'community' robust enough to ground either contemporary community ecology or environmental ethics." Her other major objection concerns Callicott's evolutionary justification of the land ethic. Callicott avoids relativism by basing ethics in natural selection: The community sentiments are not merely my subjective feelings but are possessed by all, or most of those, who survive in the social group, due to the random workings of natural selection. The trouble with that account, says Shrader-Frechette, is that the resulting ethic has no normative dimension. Altruistic feelings and the socially beneficent actions they provoke are simply natural behaviors, not free moral choices based on normative principles. The land ethic looks purely descriptive. Instead of a biologically based theory, Shrader-Frechette prefers "a metaphysical account that posits intrinsic value in nature itself. . . ."

Two additional problems for Callicott's theory come to mind. First, if our positive, community-oriented sentiments have been naturally selected for, why does the human species not exhibit more of them than it does? Our interhuman behavior exhibits at least as much aggression as it does altruism. Thus, aggression must be as basic as altruism. Whatever the social implications of that observation, it seems to show that we can never have a fully operative ecological ethical community, not simply because ideals are always impossible to achieve, but because the ideal itself runs counter to our nature, or to part of it.

Another question regarding community is this: We may and should extend our natural social sentiments to the biotic community, says Callicott, because we can see that we are part of it. Participation in the community is derived from the interdependence members of the biotic community exhibit. But *interdependence* seems too strong a word for our role in ecosystems. We are undoubtedly dependent on them, but in what way are ecosystems dependent on us? Their independence from us is not like the independence of parents from offspring who can later reciprocate love and other mutual activities that can develop into interdependency. We play no such role in any ecosystem; we seem genuinely superfluous to ecosystemic functioning. If so, however, then mutuality, a necessary constituent of community, is missing; and the call to treat the environment as community reduces to self-interest.

Intrinsic Value

Although not currently as prominent a topic in environmental philosophy as it once was, the concept of intrinsic value in nature has played a major role in the field's development. Callicott made it an important part of his position and clearly still regards it as necessary for a complete environmental philosophy. In *Earth's Insights*, one of his most recent works, he reiterates his view that "the most vexing problem of contemporary secular nonanthropocentric environmental ethics . . . is the problem of providing intrinsic value . . . for nonhuman natural entities and nature as a whole." He makes clear that a "promotion" of nature "from the instrumentally to the intrinsically valuable class" is a desideratum of a valid environmental ethic.[9] And in his introduction to the most recent collection of his essays—in which he reserves a whole section for the topic—he states: ". . . The intrinsic-value-in-nature question has been, and remains, the central and most persistent cluster of problems in theoretical environmental philosophy." He then alleges that "Nonanthropocentrists, such as practically everyone else of note in the field [besides Bryan Norton and Eugene Hargrove], agree that nature has intrinsic value. . . ."[10]

Intrinsic value is best understood in contrast with instrumental value (although an entity might have both). Instrumental value is the value something has as a means to an end. Obviously, much of nonhuman nature has instrumental value for us humans, who use it, for example, as the source of raw materials from which we build our civilizations. On the other hand, intrinsic value is the value something has in and of itself, independent of any use it might have for us or other organisms. Traditionally, philosophers have placed humans and their experiences, and not much else, in the category of the intrinsically valuable. Although such value is nonmoral, those beings that have it command special respect and moral consideration. Thus, demonstrating that nonhuman nature has intrinsic value would be a potent lever for raising the rest of nature into humanity's moral field of vision. That is Callicott's strategy.

Callicott has presented different accounts of intrinsic value, modernist and postmodernist, necessitated by his belief that most environmentalists still operate within a modernist worldview but that we are developing, and must develop, a postmodern worldview.[11] In a modernist context, science is the exemplar of knowledge. It is objective, factual, and delineates the real structure and operations of the universe. And in that delineation, it finds no values, only facts. Values exist only on the subjective side of the split between knowing subject and known object. They thus have no independent existence of their own but are created by conscious valuers.

Although no values exist outside of conscious valuing, Callicott nevertheless maintains that we can value things for what they are in themselves, that is, intrinsically. In other words, that values originate from conscious valuers does not imply that only such valuers and their experiences are valuable. We can still value things, such as the biotic community, or endangered species, for what they are in themselves. But the fact that we can thus value nonhuman nature does not show that we ought to do so. Here Callicott invokes the land ethic's foundations. We ought to value nonhuman nature for itself, he claims, because it constitutes a community to which we belong, as ecology demonstrates, and because we experience positive feelings toward our acknowledged communities, as the Humean theory of moral sentiments shows.

Callicott acknowledges that this is not full-blown intrinsic value because it allows things to be valuable only for themselves, not in themselves. Still, he thinks it sufficient for environmental ethics not only because nothing can have any greater kind of value but also because, once acknowledged, it shifts "the burden of proof from those who would protect nature to those who would exploit it only as a means." In this vision, constraints on the treatment of intrinsically valuable nonhuman nature would develop analogous to constraints on the treatment of human workers that protect them from abuses.[12]

Explicitly acknowledging the problems of modernism and the nascent postmodern transition phase we have entered, Callicott also develops another account of intrinsic value in nature that he sees as consistent with the intellectual forces driving that transition. Again, he singles out science for a central role in his theory, this time evolution, ecology, and contemporary physics. Evolution shows us that the modernist-Cartesian bifurcation of thinking subject–extended object is untenable, that we are part of nature. Ecology reinforces that change in ontological perspective and adds the crucial element that no organism is a rootless atom but is part of an interdependent system of life. Quantum theory supplies more metaphysical and epistemological force. Together with relativity theory, it "portray[s] a universe that is systematically integrated and internally related." This total integration eliminates the old modernist separation between knowing subject and known object and all its associated dichotomies, including especially the fact-value distinction. In Callicott's interpretation of the new science, all qualities are on the same ontological footing, none are objective or subjective. They are, instead, virtual, emerging on interaction between elements of the integrated universe. Thus, when we interact with the world, the qualities we "perceive" are created by that interaction. This puts values on a par with all other epistemological categories. There still is no objective intrinsic value, but "that is to concede nothing of consequence, since *no* properties in nature are strictly intrinsic. . . ."[13]

Callicott also suggests a still more radical account, conditionally interpreting the new physics as implying "that nature is one and continuous with the self." To that he adds traditional ethical theory's axiological acceptance of egoism as given. He reasons:

> If quantum theory and ecology both imply in structurally similar ways in both the physical and organic domains of nature the continuity of self and nature, and *if* the self is intrinsically valuable, then nature is intrinsically valuable. *If* it is rational for me to act in my own self interest, and I and nature are one, then it is rational for me to act in the best interest of nature.[14]

In later writings,[15] Callicott promotes the continuity of self and world and the identification of self-realization with Self-realization where the world is my self writ large; but in those later contexts he does not explicitly mention intrinsic value. However, because one such presentation is part of a book section on intrinsic value, I conclude that he would still connect Self-realization with intrinsic value.

Not surprisingly, these accounts of intrinsic value have provoked strong reactions from other thinkers. In her essay, Wendy Donner criticizes Callicott's modernist theory of intrinsic value, arguing that given its extreme subjectivism, the theory cannot "establish the conclusion that ecosystems and species are the primary bearers of value." Rather, conscious valuers seem to be the primary carriers of value. Also, Donner claims that the theory fails to give us any general guidelines for sorting or balancing our ethical duties regarding vastly different kinds of things (individual organisms, endangered species, ecosystems), all of which it counts as intrinsically valuable. Finally, she raises the specter of inhuman and inhumane decisions based on the alleged equality of intrinsic value throughout the biotic community.

Intrinsic value in nature is as equally associated with the theories of Holmes Rolston III as it is with Callicott's.[16] Rolston, in his essay, maintains his conclusion that such value is not subjective in any way, but is fully objective. Among the themes that Rolston challenges is Callicott's antidualistic naturalism. Although overcoming dualism may seem like a good idea, Rolston objects that, "Naturalizing everything naturalizes too much." Robbed of any contrasting class of the nonnatural, we no longer can sort the natural from the nonnatural, and we want to do so in guiding human behavior toward the environment. Otherwise, destructive human actions are as natural as benign ones. Rolston describes some of what he takes as clear differences between humans and nature, which we ignore at our peril.

As for intrinsic value, Rolston finds serious problems with Callicott's theory. For one thing, Callicott seems to take back his antidualism with his value theory. In saying that only we (or conscious beings) can value, he distinguishes between us and nature. In addition, Rolston analyzes Callicott's "projection" metaphor of intrinsic value and finds a serious problem. Because all the value comes from (is projected by) the conscious valuers, no value is actually located in nature. This repeats one of Donner's criticisms, but Rolston elaborates and deepens it, locating problems and confusions in Callicott's terminology and his mislocation of value. Rolston argues for his own objective account of intrinsic value because, among other things, it is simpler, discovering values already present before we humans arrive, not requiring the added process of "projection."

In the next essay, Bryan Norton disagrees with the whole project of finding intrinsic value in nature, faulting both Callicott and Rolston for assuming that the only credible response to the exploitation of nature is to assert its independent value. To Norton, the problem identified by Donner and Rolston—that Callicott's theory of value in nature actually finds none there—is due to the mistaken modernist epistemology. In its place, Norton proposes a pragmatist relational epistemology. Norton also rejects Callicott's postmodern accounts of intrinsic value, noting that Callicott himself recognizes that the version based on Self-realization still rests on the rejected Cartesian concept of self. And the account in which all features of the world are "virtual" Norton sees as a rather desperate attempt to rescue as much epistemological objectivity for intrinsic value as possible. Norton recommends instead a rejection of that pursuit in favor of a "postfoundationalist" epistemology with an ethic promoting anthropocentric but noninstrumental values.

In addition to Norton's criticisms, Clare Palmer's comments from her essay in the next section on Callicott's use of quantum physics are pertinent here. Palmer makes the important observation that Callicott never specifies on which of the several interpretations of quantum theory he bases his arguments. But each of those interpretations can have different, conflicting implications. Callicott, she points out, has simply chosen the one most compatible with his own ethics, rendering his view more ideological than philosophical.

METAPHYSICS AND METAETHICS

Palmer's criticism of Callicott's appropriation of quantum theory provides a nice entrée into a discussion of metaphysical and metaethical aspects of environmental philosophy. Such topics form an essential part of the field. One such topic came up in part I, namely, the relation between environmental ethics and scientific ecology. Callicott certainly is not the only philosopher who has seen the need to deal with metaphysical issues in connection with environmental ethics.[17] Thus, his efforts and the reactions they stimulate form a significant part of an important, wider philosophical controversy. Of course, any proposed radical revisions of our ethical traditions will provoke metaethical reflections on the nature of ethics. A central metaethical issue in the recent literature has been the ethical monism-pluralism debate. Once more, Callicott has been at the center of the debate. But first the metaphysical discussion.

Metaphysics is a continuing theme in Callicott's writings. He not only thinks that the land ethic needs a metaphysical foundation, he thinks that the new science can and will provide it. Science, he believes, has metaphysical implications that, through the elaboration of a scientific perspective into a paradigm for understanding not only the rest of nature but also human society and relations, come to permeate a culture, transforming the paradigm into a "worldview." Thus it was with modernism, a worldview that developed from classical mechanistic physics. But the new science of the late twentieth century, claims Callicott, is rapidly undermining the modernist paradigm. Specifically, he thinks that the new physics, in which the observer and observed mutually influence one another, undermines the dualism between knower and known, that it also undermines the notion that the universe is a mere collection of independent entities. Thus, in traditional metaphysical terminology, the self is not separate from the world it experiences; things are not separate entities accidentally related; rather, their relations are more important than they are. When we add, as we must, ecology to the emerging paradigm, we see that these related entities form a whole, a unity of some sort. Those are the metaphysical elements that Callicott sees at the core of the emerging postmodern worldview and that, as we have seen, undergird the land ethic.[18]

Catherine Larrère discusses Callicott's derivation of metaphysical and ethical concepts from science, comparing his thought to some French poststructuralist approaches to the emerging postmodernism. Specifically, Larrère identifies two aspects of what Callicott calls the "metaphysical implications of ecology": First, that science "enfolds" an ontology and second, that "natural philosophy is able to inform a moral philosophy." She agrees with Callicott on the first point and sees his approach as more constructive than that of some poststructuralists. But she contests Callicott's subordination of moral philosophy to natural philosophy, claiming that such a model of the relationship between the two areas is really the modernist model, that in a postmodern era we can and must grant as much authority to moral thinking (and to the humanities generally) as to science.

Going further in her criticisms than Larrère, and in addition to her comment on the interpretation of quantum theory, Clare Palmer has several questions about the metaphysical and other implications Callicott draws from science. The new, holistic, relational worldview that Callicott envisions, supported and promoted by science, might underwrite something like the land ethic. But Palmer asserts that the possibility of such a unified scientific worldview seems doubtful. Moreover, she argues that because little empirical evidence currently exists for such an emerging worldview, Callicott cannot justify his claim to a privileged place for his ethical position, which he sees as grounded in this alleged new scientific worldview. Palmer also raises serious questions about the legitimacy of moving from claims made about the quantum level to claims about the level of everyday experience. As she concludes, "[Metaphysical and ethical] positions must surely be argued in their own right, rather than relying for special support from scientific theory." Finally, she questions whether the purported new scientific worldview would, as Callicott asserts, lead directly to an environmental ethic. There seems to be no causal or logical necessity for its doing so; it might take us elsewhere.

Eugene Hargrove rejects the idea that environmental ethics needs a metaphysics in any traditional sense. Hargrove's essay is an important discussion not only of environmental philosophy, but also of the nature and function of metaphysics generally. It focuses on Callicott's "metaphysics of morals," but its cautions about metaphysics apply as well to his speculation about the new science's implications. Environmental philosophers, says Hargrove, should stick to "descriptive" metaphysics (which simply describe how people think about the world) and avoid "revisionary" metaphysics (which attempt to develop a better way to think about the world). A particularly telling and unfortunate example of the latter, claims Hargrove, is the attempted proofs of the existence of nonanthropocentric intrinsic value in nature, which he sees as contributing to the marginalization of environmental philosophy within the environmental movement. Hargrove points out some difficulties in the Hume-Darwin-Leopold-Midgley tradition for Callicott's metaphysical views: that elements of those thinkers' views do not support the land ethic or as easily do support alternative views, such as Hargrove's own version of anthropocentrism. Admitting to a metaphysical eclecticism, Hargrove says Callicott practices it as well.

That last comment raises the issue of theoretical unity, a much-debated question lately. For a variety of theoretical reasons, and especially because environmental ethics affirms obligations to several types of entities—individuals, species, ecosystems, biotic communities—many environmental philosophers have defended the idea that we need several moral principles to explain and determine our moral duties. That is moral pluralism.

Callicott's nuanced opposition to pluralism exhibits again the development of his thought. Early on, he interpreted Leopold's principle as the single overriding ethical rule.[19] So interpreted, however, it does seem to have some of the horrifying implications that Donner raises. We might, for example, be obliged to sacrifice human lives to preserve the environment. In later writings, Callicott advocates instead a version of ethics in which several principles or virtues are united in a single moral philosophy. For the latter he of course appeals to the Hume-Smith "sentimental communitarianism" that, although identifying a single basis for ethical duties (community membership), includes a "multiplicity of community-generated duties and obligations." The advantage Callicott sees in such a theoretical monism joined with a pluralism of principles is that when duties or principles conflict they can be compared and prioritized "in the commensurable terms of the common and self-consistent moral philosophy in which they are located." But Callicott remains adamantly opposed to a pluralism in which one appeals to one moral philosophy for one issue, another moral philosophy for another issue, and so on. That is because such pluralism would involve "intrapersonal inconsistency and self-contradiction."[20]

Peter Wenz, Andrew Light, and Lori Gruen critique Callicott's theoretical monism, but for different reasons. Wenz accepts Callicott's arguments against "extreme" pluralism, the view that we can jump from one moral philosophy to another to solve different types of moral quandaries. In contrast, any theory that does not provide a single formulaic solution to every moral question is "minimally pluralistic," says Wenz. He disagrees with Callicott's arguments against that variety of pluralism because no moral theory, including Callicott's, provides single, unambiguous answers to all our moral dilemmas. Moderate pluralism remains, and Wenz both defends it and claims that Callicott's theory is similarly pluralistic. Because Callicott has endorsed a plurality of principles within a single theory and Wenz says that his own moderate pluralism is a single theory, it looks as if they agree. But Wenz also claims that Callicott's "many moral principles . . . are not all derived from a single, master principle." Callicott, as we have seen, does claim that the moral principles are "unified" in communitarian sentimentalism, so the two thinkers still disagree.

Light approaches the issue from another direction. To him, the important point is not the metaethical resolution of the monism-pluralism dispute but the practical problem of gaining agreement enough among theorists to reach convergence regarding environmental practice. It is the discovery of practical solutions to environmental problems acceptable to those of different theoretical bents that is the central motivation of pluralism, he argues, not the theoretical wrangling over whether monism trumps pluralism or vice versa. So Light recommends that Callicott and others, rather than searching for a theory that combines the advantages of monism and pluralism, search instead for "compatibilism among forms of valuing" so we can find ways of cooperating on important and pressing environmental issues. Light goes on to explain how Arne Naess, the originator of deep ecology, has defended a form of pluralism that accomplishes exactly that, and does so in a manner complementary to Light's own environmental pragmatism.[21]

In her contribution, Lori Gruen explains and emphasizes the importance of context in ecofeminist theory. She contends that Callicott, in criticizing ecofeminism as rejecting the need for theory in environmental ethics, has misunderstood ecofeminism. Although she agrees with him that ecofeminists have not sought a theoretical account of intrinsic value in nature, she contends that ecofeminism does provide a theoretical perspective, but one that focuses on the analysis and critique of "the forces that contribute to the oppression of women, animals, and nature." Such forces can exist even within "supposedly emancipatory theories" in environmental philosophy, so the issue of context becomes crucial. In turn, that issue again raises the monism-pluralism debate. Gruen argues that ecofeminism emerged "to provide a critical, self-reflective and pluralistic alternative." She goes on to explain that although such an alternative seeks to honor and affirm the many voices and cultures of our world, it is not relativistic, still allowing for careful ethical assessments of others' practices.

Challenging the Implications of the Land Ethic

Given the ubiquity of environmental concerns and the wide-ranging implications of the land ethic for understanding the relation between humans and the rest of nature, that thinkers from a broad range of theoretical viewpoints have reacted to Callicott's writings is not surprising, all the more so because Callicott himself has discussed the land ethic in relation to sundry disciplines and cultural practices.

Perhaps the best known of Callicott's own take on the land ethic's implications is his polemical critique of animal-liberation from an ecocentric position. He completely rejected animal liberation because of its individualism and lack of concern for endangered species and ecosystems and because, he claimed, it absurdly implies a duty to prevent predation. Later, based on his reading of Mary Midgley's notion of a "mixed community" of humans and animals, Callicott moderated his views, proposing an alliance between environmentalism and animal liberation, connecting them via the concept of community membership.[22] But he never altered his emphasis on concern for the biotic community or his rejection of the individualism of Peter Singer's or Tom Regan's theories of ethics regarding animals.

In contrast, Angus Taylor, seeing no conflict between ecosystemic integrity and autonomy of sentient animals, presents an alternative reading of the relation between Callicott's ethic and a strong animal-liberation position. Taylor argues that both animal rights and the land ethic oblige us to leave wild animals alone, to respect their autonomous pursuit of their own natures. Callicott goes wrong, says Taylor, in at least two ways: by insisting that we can respect domesticated animals and continue "appropriating their lives and bodies for our exclusive purposes without their consent," and by ignoring the necessary connection between the rights of animals and "the flourishing of their natural environments." Taylor calls for an alliance between animal liberation and environmentalism with autonomy and ecosystemic integrity "as joint fundamental values."

Susan Power Bratton explores Callicott's views about the relation between the land ethic and Christianity. Bratton thinks Callicott, in his search for a single environmental ethic, is actually responding to the wrong question. Rather than seeking such an ethic, Bratton argues that we instead should be trying to facilitate environmental problem solving and promoting environmental sensitivity within existing moral systems. She challenges the whole project of ethical monism and of academic environmental ethics as either ignoring religious practitioners—the largest constituency of ethical study—or as insulting them, through the attempted imposition of an abstract ethic to which they are supposed to adapt. Thus, she argues, the effort to determine the effectiveness of Christian environmental ethics is not a philosophical but rather a social science question. Rejecting Callicott's criteria for an adequate environmental ethic, Bratton, based on her own empirical work, proposes seven "social benchmarks" for assessing how an ethic is expressed in a real society, which is constituted by dynamic, developing relationships. She believes such an approach will better promote beneficial environmental attitudes and behavior than will judging a religious ethic from an abstract philosophical vantage point.

Callicott has always been interested in Native American attitudes toward the environment, and in *Earth's Insights* he compares the environmental ethics of indigenous peoples throughout the world with the land ethic.[23] Consistent with his moral monism, he argues that although many indigenous environmental ethics exist, they are or can be made consonant with the land ethic, which validates them. The validation is not, he claims, an instance of Western arrogance because the land ethic is based in postmodern science, which has become a worldwide epistemological project. Lee Hester, Dennis McPherson, Annie Booth, and Jim Cheney take strong exception to Callicott's project in *Earth's Insights.* They argue that it is an attempt to subordinate indigenous people's ways of life to a distinctly Western approach to the natural world. Instead of basing ethics in metaphysics, as Callicott does, they emphasize that indigenous peoples perceive the world within an attitude of respect that concomitantly creates their worlds. So for them respect is a practical epistemology that creates an ontology. These authors' essay is simultaneously a rigorous critique of Callicott's arguments in *Earth's Insights* and a rich presentation of indigenous thought, and they suggest ways in which that thought can assist Euro-Americans to develop a similar attitude or approach separate from the domination and control characteristic of Western approaches.

Callicott vehemently opposes dualism, which he sees as separating humans from the world of which they are part. That separation, he believes, has contributed to the destruction of the nonhuman world. Based on that general view, he has argued that the concept of wilderness is dualistic in that it divides the world into the human, cultural world and the wild, natural world.[24] I also oppose dualism, but argue that Callicott goes too far in his critique of wilderness. Arguing that Callicott's rejection of dualism is itself determined by a dualist outlook, I defend the notion of wilderness as designating the nonhuman world, the world other than us. We are indeed part of nature, but we also have our own unique features that are important in working out how we should treat the rest of nature, although they do not amount to an ontological separation.

Conclusion

A volume that discussed every aspect of Callicott's environmental philosophy would be much larger than this one. It would have to include sections on environmental aesthetics, environmental education, and conservation biology,[25] to name a few topics not covered herein, and some of the coverage in this volume would need expansion. However, the major aspects of his theories are analyzed here, and in the final essay of the collection, Callicott responds to the questions, criticisms, and problems raised in the other articles. That response is ample and complex and covers all the essays just described, so I will not attempt to summarize it here.

In his introduction to *In Defense of the Land Ethic*, Callicott says that he offered that collection of essays "not only as a defense of Leopold's seminal ethic but as an invitation to critical exploration along the trail he charted."[26] The collection herein accepts that invitation both by responding to his defense and development of Leopold's ethic and by exploring what has become the trail system of environmental philosophy. The editors hope that this exploration assists readers not only in navigating Callicott's extensions of Leopold's trail and those of the other thinkers included here, but also in discovering and developing new trails that help us all learn how better to live in the land.

Notes

1. For an overview of some of the literature in environmental philosophy, see Wayne Ouderkirk, "Mindful of the Earth: A Bibliographical Essay on Environmental Philosophy, *Centennial Review* 42 (1998): 353–392. A shorter version was "Earthly Thoughts: An Essay on Environmental Philosophy," *Choice: Current Reviews for Academic Libraries*, 35 (Nov. 1997): 421–434. Interested readers can find an annually updated bibliography of the environmental philosophy literature at the Web site of the International Society for Environmental Ethics, <http://www.cep.unt.edu/ISEE.html>. The bibliography is also available on disk; purchase information is on the Web site.

2. The account that follows summarizes Callicott's explanation and justification of

the land ethic found in many of his essays and books. See for example, his "The Conceptual Foundations of the Land Ethic," originally published in J. Baird Callicott, ed., *Companion to A Sand County Almanac: Interpretive and Critical Essays* (Madison: University of Wisconsin Press, 1987), republished in Callicott's *In Defense of the Land Ethic: Essays in Environmental Philosophy* (Albany N.Y.: SUNY Press, 1989).

3. Aldo Leopold, "The Land Ethic," in *A Sand County Almanac with Essays on Conservation from Round River* (New York: Ballantine Books, 1966), 262.

4. David Hume, *An Inquiry Concerning the Principles of Morals* ([1777]; LaSalle, Ill.: Open Court, 1966), and Adam Smith, *The Theory of the Moral Sentiments* ([1790]; Oxford: Clarendon Press, 1976).

5. Charles Darwin, *The Descent of Man and Selection in Relation to Sex*, 2d ed. (New York: J. A. Hill, 1904).

6. Callicott, *In Defense of the Land Ethic*, 83.

7. Callicott, *Earth's Insights: A Multicultural Survey of Ecological Ethics from the Mediterranean Basin to the Australian Outback* (Berkeley: University of California Press, 1994), 199 ff.

8. See Bernard Williams, *Ethics and the Limits of Philosophy* (Cambridge, Mass.: Harvard University Press, 1985); Wilfrid Sellars, *Empiricism and the Philosophy of Mind* (Cambridge, Mass.: Harvard University Press, 1997); Laurence Bonjour, *The Structure of Empirical Knowledge* (Cambridge, Mass.: Harvard University Press, 1987); Richard Rorty, *Objectivity, Relativism and Truth* (Cambridge, England: Cambridge University Press, 1991).

9. Callicott, *Earth's Insights*, 21.

10. J. B. Callicott, *Beyond the Land Ethic: More Essays in Environmental Philosophy* (Albany, N.Y.: SUNY Press, 1999), 14–15.

11. This summary follows Callicott's expositions; for the modern version, see his "On the Intrinsic Value of Nonhuman Species"; for the postmodern versions see "Intrinsic Value, Quantum Theory, and Environmental Ethics." Both are in Callicott, *In Defense of the Land Ethic.*

12. Callicott, *Beyond the Land Ethic*, 18.

13. Callicott, *In Defense of the Land Ethic*, 169.

14. Callicott, *In Defense of the Land Ethic*, 173.

15. Callicott, *Beyond the Land Ethic*, 217–218; Callicott, *Earth's Insights*, 206–209.

16. See for example, his *Environmental Ethics: Duties to and Values in the Natural World* (Philadelphia, Pa.: Temple University Press, 1988).

17. For example, in defending duties toward species, Holmes Rolston first argues that they are actual entities, in *Environmental Ethics.* Metaphysical theses also form part of the theories developed in: Val Plumwood, *Feminism and the Mastery of Nature* (London: Routledge, 1993); Bill Devall and George Sessions, *Deep Ecology: Living as if Nature Mattered* (Salt Lake City, Utah: Gibbs Smith, 1988); Eric Katz, *Nature as Subject: Human Obligation and Natural Community* (Lanham, Md.: Rowman and Littlefield, 1997); Kate Soper, *What Is Nature? Culture, Politics, and the Non-Human* (Oxford: Blackwell, 1995). Keekok Lee has argued that environmental philosophy must begin with ontology rather than axiology, in *The Natural and the Artifactual: The Implications of Deep Science and Deep Technology for Environmental Philosophy* (Lanham, Md.: Lexington, 1999).

18. See chap. 9 of Callicott, *Earth's Insights*, "A Postmodern Evolutionary-Ecological Environmental Ethic"; also "The Metaphysical Implications of Ecology," "Intrinsic Value, Quantum Theory, and Environmental Ethics," both in Callicott, *In Defense of the Land Ethic*; and "After the Industrial Paradigm, What?" in Callicott, *Beyond the Land Ethic.*
19. For example, in "Animal Liberation: A Triangular Affair," in Callicott, *In Defense of the Land Ethic.*
20. Callicott, *Beyond the Land Ethic*, 13, 11, 10.
21. See Light's contributions to *Environmental Pragmatism*, Andrew Light and Eric Katz, eds. (London: Routledge, 1997). Others' contributions to the same anthology are of course relevant here.
22. "Animal Liberation: A Triangular Affair," and "Animal Liberation and Environmental Ethics: Back Together Again," in Callicott, *In Defense of the Land Ethic.*
23. See also his essays, "Traditional American Indian and Western European Attitudes Toward Nature: An Overview," and "American Indian Land Wisdom? Sorting Out the Issues," in Callicott, *In Defense.* See also Thomas W. Overholt and J. Baird Callicott, *Clothed-in-Fur and Other Tales: An Introduction to An Ojibwa World View* (Washington, D.C.: University Press of America, 1982).
24. J. Baird Callicott, "The Wilderness Idea Revisited: The Sustainable Development Alternative," *The Environmental Professional* 13 (1991): 235–247. Callicott also discusses other problems with the concept of wilderness in this essay.
25. On aesthetics, see his "The Land Aesthetic" in *Companion to "A Sand County Almanac": Interpretive and Critical Essays*, ed. J. B. Callicott (Madison: University of Wisconsin Press, 1987). On education, see J. B. Callicott and Fernando J. R. da Rocha, eds., *Earth Summit Ethics: Toward a Reconstructive Postmodern Philosophy of Environmental Education* (Albany, N.Y.: SUNY Press, 1996). On conservation biology, see his essay "Whither Conservation Ethics?" in *Conservation Biology* 4 (1990): 15–20.
26. Callicott, *In Defense of the Land Ethic*, 12.

Part I

The Conceptual Foundations of the Land Ethic

1

Ecological Morality and Nonmoral Sentiments

ERNEST PARTRIDGE

Arcturian zoologists visiting this planet could make no sense of our morality and art until they reconstructed our genetic history—nor can we.

—Edward O. Wilson, *Biophilia*

MORAL SENTIMENTS

Radical environmentalists have often characterized Homo sapiens as a cancerous mutation, heedlessly devouring the planetary body that nurtured and sustains it, and thus veering toward its own destruction and that of its ecosystemic host.

If this bleak scenario is to be reversed, a key ingredient of our collective rescue must be a mix of scientifically informed insight into the consequences of our assaults upon the planet; a clear view of our duties to our species, the ecosystem, and the future; and finally the motivation to do what that duty demands of us. Of these, the third, *motivation*, and the sentiments that support it, has arguably received the least attention.

In several of his essays, J. Baird Callicott has enriched Aldo Leopold's visionary land ethic with the insights of critical and normative ethics, thus bringing Leopold's vision into the arena of philosophical debate and scholarship. To his credit, Callicott has recognized the essential role of moral psychology to a cogent environmental ethic.

Although I share Callicott's conviction that an environmental ethic cannot stand without a theory of sentiments, I dispute his suggestion that David Hume's theory of moral sentiments adequately functions in this role.[1] To the contrary, I contend that Humean *moral* sentiments are more likely to reinforce anthropocentrism and alienate humans from nature. If moral sentiments are to aid the ecological moralist, they must do so in a secondary way by binding human communities and motivating them to appropriate action in the defense of their natural contexts and heritage. However, for a primary motivational support of environmental ethics, we must look to the *nonmoral* sentiments. In this essay, I close with a suggestion as to where we might find those requisite nonmoral sentiments.

In several publications Callicott has attempted to show that Leopold's land ethic "actually has a legitimate ancestry in the Western philosophical canon . . . traceable through [Charles] Darwin [in the *Descent of Man*], to the Scottish Enlightenment in the eighteenth century," notably the moral philosophy of Adam Smith and David Hume.[2] He thus outlines "The Conceptual Foundations of the Land Ethic," in his essay of that title:

> Its conceptual elements are a Copernican cosmology, a Darwinian protosociobiological natural history of ethics, Darwinian ties of kinship among all forms of life on earth, and an Eltonian model of the structure of biocenoses all overlaid on a Humean-Smithian moral psychology. Its logic is that natural selection has endowed human beings with an affective moral response to perceived bonds of kinship and community membership and identity; that today the natural environment, the land, is represented as a community, the biotic community. . . .[3]

If we are to assess this claim, a review of some elements of Hume's moral philosophy is in order. First, Hume posits that moral judgment is based, not on reason, but on "some internal sense or feeling which nature has made universal in the whole species." In this crucial assertion, we find that to Hume, morality is strangely both subjective ("internal") and "universal" because these "moral sentiments" issue from "the original fabric and formation of the human mind, which is naturally adapted to receive them."[4] Note the explicit reference to the "natural" foundations and adaptations of the human mind and morality. I have much to say about this point later.

"Morality," writes Hume, "is determined by sentiment. It defines virtue to be *whatever mental action or quality gives to a spectator the pleasing sentiment of approbation*, and vice the contrary."[5] Prominent among the moral sentiments mentioned by Hume are generosity, love, friendship, esteem, compassion, gratitude, guilt, shame, contempt, and hatred. Primary among these are the sentiments of *benevolence* and *sympathy*—in fact, the latter might better be regarded as the capacity necessary for the generation of the sentiments. Note how all these sentiments are personal, that is, either reflecting or referring to qualities of persons.

Thus, if I understand him correctly, Callicott is attempting to demonstrate that Humean moral sentiments emerged from "the original fabric and formation of the human mind," as Hume himself put it in words that Charles Darwin could and apparently did embrace. Such sentiments, argues Callicott, can extend out from the individual to attach to his immediate family and friends, then to the society beyond, and finally may affirm the life community itself and thus support a normative environmental ethic.

I believe this view to be unworkable because (1) the application of "moral sentiments" ends with our "moral community," which (2) can be no more than a community of persons, or at most, of sentient beings, due to the profound disanalogies between such "moral communities" and Leopold's "natural community" of ecosystems. These points require argument. I begin with moral sentiments.

Just what are *moral sentiments*? Let's take the phrase one word at a time. First, I interpret the crucial term *moral* in a manner I believe to be fairly standard among contemporary moral philosophers. The adjective *moral* must have, lurking at least somewhere in its context of application, some deliberative agent or community of agents, which is to say, a "person" (although not necessarily a human). *Moral* implies responsibility, accountability, praise, and blame. In essence, a moral judgment is a judgment that reflects upon the worth of a person. Persons of moral worth are called "virtuous," and persons of little worth are called "wicked." Acts that reflect well on persons are "right" and their opposites are "wrong." On a planet without persons, however teeming with sentient but nondeliberative and nonreflective life, there will be "goods" and "bads," but no morality—no right and wrong, no justice, no duties, no rights. Put bluntly, if the latest data of human evolution are to be believed, morality emerged upon the Earth within the past million years—possibly within the past few tens of thousand years.

Moral sentiment, then, is simply an emotional and evaluative attitude toward a person, persons, or their institutions. Positive and negative moral sentiments toward oneself include, respectively, *self-esteem* and *guilt* or *shame*. Toward others, these sentiments are respectively called *admiration* and *indignation* or *contempt*. Of particular interest to Hume, and thus to Callicott, are the moral sentiments of *sympathy* and *benevolence*.

We morally praise and blame people with regard to their treatment of other persons. The traditional virtues (i.e., *courage*, *charity*, *benevolence*, *trust*, and *fidelity*) testify to the command of our will and signify our recognition of the worth of other persons. The deadly sins (i.e., *pride, lust, anger, gluttony, sloth, envy,* and *greed*) issue from our depersonalization of our brethren and stigmatize the willful crippling of our moral potential.

The worth of persons—of oneself and of those with whom we deal—is the paradigm context of moral evaluation. The invasion of personal interest and the destruction of personal property have traditionally been regarded as paradigms of immorality. By extension, the infliction of pain upon defenseless, sentient nonpersonal beings has been seen as a penumbral immorality.

With this elucidation, I submit that the problem of basing a normative environmental ethic on *moral* sentiments becomes clear. *Moral* sentiments seem to require *persons* in the equation. But what if persons are not apparent among the objects of our concern? We can ask: "Why does the clear-cutting of a primeval forest, the damming of a wild river, or the extinction of a species, violate a normative environmental ethic?" If these are moral wrongs, then one must presumably show that the agents responsible have done something that reflects poorly upon them as persons, due perhaps to their wrongful treatment *of* persons. Yet all this environmental destruction might be done *on behalf of* persons: the rain forest cut on behalf of the poor farmers, dams built to provide cheap and abundant power, and so on.

To state that the willful destruction of nature is morally wrong, presupposes an underlying theory of value that supports principles, the violation of which reflects unfavorably upon the worth of the agent responsible for this destruction. As the precondition for moral evaluation, such a theory must be a theory of *nonmoral* value, otherwise the theory will be circular.[6] Thus, if this theory is based upon sentiments, then these must be *nonmoral* sentiments.

At this point, two theoretical roads diverge: along one, we return to a familiar anthropocentrism by identifying *nonmoral value* as pleasure/pain, or human potential and welfare, or some other "good for people"—choose your theory. Along the other road, we might seek intrinsic values in nature, a vast and fascinating realm of inquiry. The second road, I believe, is far more promising for environmental ethics, and Callicott has often explored it in promising and suggestive ways.

Unfortunately for the argument offered by Callicott, Hume appears to have had the first road in mind. As Callicott correctly points out, Hume's moral sentiments have their origin in interpersonal relationships. These sentiments are evoked by our recognition of the personhood or sentience in others. *Personhood* is not only the source of these moral sentiments but also its limit. Accordingly, the Humean sentiment of *benevolence* is not directed toward insentient nature, much less toward abstractions such as species or ecosystems.[7] Nor can Humean *sympathy* connect with objects in or conditions of impersonal nature. Hume could not have been more explicit concerning this point than when he wrote: "Inanimate objects . . . can never be the object of love or hatred, nor are [they] consequently susceptible of merit or iniquity."[8] Thus, the Humean moralist will favor the logger and his dependent family over the old-growth forest, the abalone fishermen over the sea otters, the Lake Powell water skiers over the Glen Canyon wrens. I submit that the uses to which Callicott is putting moral sentiments would astonish David Hume.

Humean moral sentiment is a poor theoretical stream in which to fish for a land ethic.

Human and Natural Communities

Like Callicott and many other ecophilosophers, I find Leopold's "natural community" metaphor to be attractive and have often used it. Occasionally, some of my colleagues have warned me not to be beguiled by this metaphor. Reading Callicott, I begin to see what they had in mind.

No one can read Leopold without recognizing immediately and vividly the aptness of the community metaphor. As in human societies, the individual "members" survive and flourish only as they interact and respond, share and cooperate (even in the "cooperative" act of predation), and thus sustain the "community"—a whole that is more than the sum of its individual parts, in fact that is best conceived, not in terms of its component parts, but in terms of its internal relations and processes.

So much for the compelling analogies. One fundamental disanalogy remains: the human community alone is characterized by reciprocity among moral agents. Thus, *rights, duties, justice*, and *responsibility* belong exclusively to the axiological vocabulary of human communities. These terms are meaningless in the natural community unless that community is touched by the human (or better, the *personal*).[9]

If the reach of moral sentiment stops at the barrier of personhood or, at most, of sentience, does not the extent of the moral community likewise end with those beings who can reciprocate the bonds of moral consideration, or at least have the bare neural equipment to care how they are treated? Callicott thinks not and for reasons now familiar to us. In "Intrinsic Value, Quantum Theory, and Environmental Ethics," he writes:

> Hume suggests that the values you project onto objects are not arbitrary, but arise spontaneously in you because of the "constitution of your nature.". . . Leopold masterfully played upon our open social and moral sentiments by representing plants and animals, soils and waters as "fellow members" of our maximally expanded "biotic community." Hence, to those who are ecologically well-informed, non-

> human natural entities are inherently valuable—as putative members of one extended family or society. And nature as a whole is inherently valuable—as the one great family or society to which we belong as members or citizens.[10]

Here Callicott boldly goes where few moral philosophers have gone before, carrying his community metaphor to the far end of the field.

A critic of Callicott may reply:

> It is just the differences between human and natural "communities" that cause me to reject this extension. Extend out from human communities, and you leave the domain of cognition and reciprocation among equals, to that of mere sentience, and then, into the domain of insentience and nonlife. As you do, you shed the stringency of your moral imperatives. Thus, as my neighbor cares how I treat him and his property, so then must I respect his concerns, as I demand that he respect mine. To assure this mutual respect and restraint, we form communities regulated by laws. But that redwood and that river don't care in the least how I treat them—so why should I? Granted, if I despoil the tree and the river, and thus violate the "integrity, stability and beauty" of the so-called ecological "community" of which they, and I, are a part, I will also impoverish my world and that of my neighbors and posterity. So I'll keep on paying my Sierra Club dues, and I'll agree to march on Washington. But I'll do all this for my sake, and that of my neighbors and posterity—not for the "sake" of the tree and river which, strictly speaking, have no "sake."[11]

Callicott correctly points out that it is scientific knowledge that makes us "ecologically well-informed" by teaching us that the ecosystem is a figurative "community" in the sense of a cooperative scheme of interacting parts, and of information, energy, and nutrient distribution. But the social sciences also point out significant dissimilarities between ecosystems and human communities of persons, with their complex systems of moral controls (e.g., reciprocating rights and duties, procedural and distributive justice, sanctions, moral sentiments.) To be fully "well-informed" is to be aware both of the similarities and the differences of these two "communities."

Nevertheless, the attempt to extend, by analogy, our loyalty to our human community over to the natural, is based on the presupposition that our *human* community deserves our prima facie loyalty (surely one of the most fundamental assumptions of political philosophy). Notice how Callicott uses this presumptive "community loyalty" to derive, by extension, a (deontological) *ought* from the ecological *is*, as he asks why we should, in Leopold's indelible words, "preserve the integrity, stability, and beauty of the biotic community." Callicott replies: "Because (1) we all generally have a positive attitude toward the community or society to which we belong; and (2) science has now discovered that the natural environment is a community or society to which we belong, no less than to the human global village."[12]

But *should* we "have a positive attitude"? Unfortunately for Callicott's argument, a "positive attitude" is a notoriously poor "is" from which to imply an "ought." For instance, saying that "P has a positive attitude toward his or her society," and then saying that "P's society is unjust" (or otherwise "bad") makes perfectly good sense. Still worse, doing so is commonplace. Consider, for example, Eichmann's attitude toward the Nazi society. Fundamental to our political traditions is the conviction that our political institutions must measure up morally. If they do not, we strive to reform them, and failing that, we are entitled to abolish them. "Community" is not a self-authenticating good.

How, then, do we reply to those who say that "Mother Nature is a monster," and her so-called "community" deserves to be wiped out and supplanted by the sort of artificial environments beloved by James Watt (and others . . .)? One might reply that if we attempt to obliterate nature, nature will strike back and obliterate us instead. But even if one accepts this retort (as I do), if that is all one has to say on behalf of the land ethic, that ethic reduces to "enlightened anthropocentrism." Surely Callicott and Leopold want more from their land ethic than it. I know that I do.

Clearly, what we need is some indication that each of us is, in Leopold's eloquent words, a "plain member and citizen" of the land community, notwithstanding the fact that our "fellow citizens" in this community are unreflective, inarticulate, and in most cases, insentient. The scientific evidence that we stand in fundamental interdependence with "the life community" of nature is, I submit, conclusive. But this conclusion merely bids us that, in our dealings with nature, we should be prudent at best—that we should "obey" the "laws of ecology" for our own good: "enlightened anthropocentrism" again.

This is not an environmental ethic that Leopold or Callicott can accept; nor can I. *Fear* and *apprehension* of nature, and of its retaliation upon us for our poor management, are precisely the opposite of sentiments sought by the ecomoralists. What they celebrate is an ethic founded on the gentler sentiments of affirmation, wonder, and love. Are such sentiments toward nature appropriate or even possible? I believe that they are not only possible, but also that they may even be essential to a viable environmental ethic, which is to say, to our continued membership in the natural "community."

Interestingly, I suspect that Hume, Darwin, and Callicott have each made significant contributions toward the articulation of an empirical and cognitivistic theory of ecomorality, based upon natural (but *not* "moral") sentiments. Briefly, the theory is as follows: We are genetically "programmed" to respond to nature with the sentiments of affirmation, wonder, and love because nature supplies the environment that selected our genes and thus shaped our neurological and cognitive equipment. However intuitively attractive the theory may appear, it rests on some poorly validated conjectures about the origin and status of certain fundamental responses to nature. Yet, if supported by subsequent empirical investigation, it just might be the "theory of sentiments" sought by Callicott to "support a normative environmental ethic." Concerned that I just might have been all too successful in my critique of Callicott's worthy search for "moral sentiments" in defense of the land ethic, I turn now to the task of suggesting an alternative theory of sentiments.

Natural Nonmoral Sentiments

In a celebrated and oft-quoted letter, Wallace Stegner writes:

> Something will have gone out of us as a people if we ever let the remaining wilderness be destroyed; if we permit the last virgin forests to be turned into comic books and plastic cigarette cases; if we drive the few remaining members of the wild species into zoos or to extinction; if we pollute the last clear air and dirty the last clean streams. . . . so that never again can we have the chance to see ourselves single, separate, vertical and individual in the world, part of the environment of trees and rocks and soil, brother to the other animals, part of the natural world and competent to belong to it.[13]

Just what will we have lost? *Nothing,* replies Martin Krieger in his notorious paper, "What's Wrong with Plastic Trees?" After all, we are "plastic people," that is, infinitely malleable. We can adapt to anything, and like it. . . .[14] On the contrary, writes botanist Hugh Iltis, "like the need for love, the need for nature, the need for its diversity and beauty, has a genetic basis. . . ."[15] E. O. Wilson elaborates:

> The brain evolved into its present form over a period of about two million years, from the time of *Homo habilis* to the late stone age of *Homo sapiens,* during which people existed in hunter-gatherer bands in intimate contact with the natural environment. Snakes mattered. The smell of water, the hum of a bee, the directional bend of a plant

> stalk mattered. The naturalist's trance was adaptive: the glimpse of one small animal hidden in the grass could make the difference between eating and going hungry in the evening. And a sweet sense of horror, the shivery fascination with monsters and creeping forms that so delights us today even in the sterile hearts of the cities, could see you through to the next morning. . . . Although the evidence is far from all in, the brain appears to have kept its old capacities, its channeled quickness. We stay alert and alive in the vanished forests of the world."[16]

And so the issue is joined.

Almost two decades ago I gave this hypothesis a name, ***bio-humanism,*** and today that term is in use in a language community of approximately one. With much more success, Wilson gave the theory the name *biophilia,* which he used as the title of his book, published in 1984. The debate that has raged over the issue is a fascinating story in itself, but one that I must bypass.[17] Suffice to say that little progress has been made toward a resolution twenty-three years after Krieger threw down his plastic gauntlet and Iltis led the countercharge on behalf of our genes and their allegedly favored habitats. Wilson, a strong advocate of the theory, admits that "the subject has not been studied enough in the scientific manner . . . to let us be certain about it one way or the other,"[18] and Paul Ehrlich adds that such a demonstration "would be a task beyond the scope of today's biology."[19]

If, in fact, our genes beckon us home to our natural origins, throngs of noteworthy individuals seem able to ignore these siren songs, not only with little apparent harm, but even with some enthusiasm. The story is told that James Watt, then U.S. Secretary of the Interior and thus the overseer of the national parks, pleaded after three days of a two-week trip through the Grand Canyon, to be rescued from that dreadful wilderness. A park service helicopter was dispatched to pull him out.

Notwithstanding such puzzling counterexamples, I assume that there is at least something to the biophilia hypothesis—that, to use Paul Shepard's vivid image, the destruction of nature is an "amputation of man."[20] How we can live in a totally artificial environment, detached from the environment that selected our genes and shaped our genome, without going bonkers, remains to be determined. I only suggest that among those genes that hardwire our nervous system, are a few that dispose us toward having positive "natural sentiments" toward undisturbed nature, and conversely, to suffer when deprived of our primeval landscapes. From this "biophilic" nervous system has issued the great works of art, literature, and science that celebrate nature. *The Pastoral Symphony*, *La Mer*, *The Starry Night*, *Walden, A Sand County Alma-*

nac, and *Desert Solitaire* are all voices of nature speaking back to us and through us.

We are back to the "natural sentiments," this time the sentiments of *wonder, delight, serenity,* and throughout of *affirmation.* But these sentiments are not directed to persons; nor do they reflect on our worth as persons, although they may indicate our state of neurological health. In short, "biophilia" may provide us with the sort of nonmoral psychological equipment that we are looking for. If so, then what follows?

First of all, it would seem to follow that being in tune with nature is a sign of *health* that can offer a hard objective reference into the contentious arena of ethical debate.

Granted, the analysis of the concept of *health* is open to considerable debate among medical ethicists. However, little debate occurs among doctors and their patients. To paraphrase Justice Powell, we may not know how to define *health,* but we all know what it is when we have it, and even more acutely, when we do not. Moreover, the further we get "down" Maslow's pyramid from "social health" to "mental health" to "emotional health" and then to "physiological health," the less controversial is the concept of health. To be "unhealthy" is to be "diseased," which means to suffer pain (more than normally), to lack vitality, to have a diminished life expectancy (due to physiological conditions), and all other factors equal, to be less happy. If someone were to say that by this analysis, a drug addict or an alcoholic is more "healthy" because he or she is euphoric, then one need only add the qualifier "in the long term" to dismiss that sophistry.

Thus, the clear implication of biophilia is as follows: an *artificial world* is a world in which one is less "healthy" than in a world with nature abundant, conspicuous, and itself "healthy." Thus, the destruction of nature deprives us of our "health." Perhaps we can live without it—but we cannot live as well. It follows that if the existence of natural environments and landscapes is, like an essential "nutrient" in our diet, necessary for our health, then we might have within our reach a naturalistic theory of sentiments and, hence, of evaluation. But not yet of *moral* evaluation. However, this step is not far behind. To return to our medical analogy, although health is a nonmoral value, it is a basis for evaluating moral virtue and wickedness: virtue in the endeavor to maintain and restore the health of others for their sake and wickedness in the careless disregard of the health of others. Thus, if the preservation of nature is essential for the health of human individuals, societies, and posterity, then "environmentalists," as they strive to protect and restore nature, are engaged in a moral enterprise.

Upon the foundation of these nonmoral natural sentiments—call it *psycho-eco-health*—some familiar moral issues arise. For example, wilderness and natural environments are, as we well know, becoming increasingly scarce. This situation raises moral questions of distributive justice, not unlike those questions raised with regard to scarce economic resources.

Another moral issue concerns the social institutions best suited to optimize psycho-eco-health. Herein, the Humean moral sentiments such as *benevolence* and *sympathy* can be enlisted to preserve nature, the font of nonmoral biophilic sentiment, as we address the question, "How shall we, as a human community, regard and treat the natural environment?" It will not do for each of us, in response to our private biophilic sentiments, to go at it alone. Doing so would only introduce the well-known "tragedy of the commons," for if we fail to adopt a robust social contract regarding our treatment of the environment—if, that is, we attempt to protect nature through the uncoordinated and individual volitions of "6 billion points of light"—then clearly the natural environment has had it. Accordingly, if nature is to be protected and preserved, it must be done through a social compact to protect it ("mutual coercion, mutually agreed upon," as Garrett Hardin puts it), at the cost of foregoing many human freedoms and benefits.

By empathetically recognizing in others the biophilic "need for nature" that we feel in ourselves, we may be moved, through Humean benevolence, to preserve nature for all, and for posterity. Thus, moral sentiments, which are incapable of attachment *to* nature (for reasons argued earlier), can be enlisted *on behalf of* nature—given, of course, a foundation of nonmoral "natural sentiment" that affirms nature.

But are we not thus becoming anthropocentrists, albeit "enlightened anthropocentrists," in spite of ourselves? Is this not an argument to preserve nature for humanity's sake, notwithstanding the acknowledgment that humanity has a stake in preserving the origins of its own genome? That criticism stands on an implicit human-nature dichotomy that is severely undermined by the biophilic hypothesis. That hypothesis, after all, posits a "naturalness" to "human nature" that obscures the boundary. As Callicott eloquently argues in his "Quantum Theory" essay,[21] the very notion of the subject-object dichotomy may now be outmoded. After all, Shepard reminds us, "ecological thinking requires a kind of vision across boundaries."[22] Thus, the new physics, the ecological vision, and now biophilia all affirm the Zen maxim that "the World is my Body." *Enlightened anthropocentrism?* The more the "enlightenment," the less the anthropocentrism.

To repeat a theme that I have often urged, just as a "moral paradox" exists in our personal relations, a "paradox of ecomorality" also exists. In the first case, the lover's life is enhanced to the degree that he or she cares less for himself or herself and genuinely focuses concern on the well-being of the loved one. The "game is lost" when he or she starts to ask, "What's in this for me?"

Similarly, only if we genuinely treat the natural environment with respect and restraint, finding and cherishing values in it that we regard as *intrinsic*, can we flourish in that environment and deserve to do so.[23] As Leopold so wisely observed, "We abuse land because we regard it as a commodity belonging to us. When we see land as a community to which we belong, we may begin to use it with love and respect."[24] I think he might have said, just as well, that "if we treat land with love and respect, we might be less likely to see it as a commodity and be more inclined to value it for itself." Crass anthropocentrism is little more than asking of the land, "What's in it for me?" Biophilia suggests that a fundamental genetic basis may exist for this sentiment of love and respect for the land, a sentiment that may be essential if we are to act on an ecological "moral paradox."[25]

These biophilic considerations suggest an environmental ethic that may be more Aristotelian than Humean in that the "goodness" of being in tune with nature (i.e., of living in a surrounding that we evolved from and preserving the conditions of our evolution) is a goodness interpreted as a consistency with human nature and fulfillment. Furthermore, it is an ethic that endorses the actualization of human potential because it suggests that we are most likely to flourish in a natural environment, just as an acorn can best manifest its potential "oakness" in a biome that is conducive to the flourishing of oak trees. Thus, we can accomplish our fullest potential in the Aristotelian sense—have the best kinds of lives—if those lives can develop in an environment that is genetically natural to us, which is to say an environment to which we are "attuned." This, I think, is an essential claim of biophilia that Aristotle might recognize and, apprised of the facts, even endorse.

But what if biophilia is false? What if we are, as Krieger claims, "infinitely malleable"? If so, then I believe that the prospects for a robust environmental ethic will be severely diminished. Yet, even then, all is not lost. We can still choose between an artificial and a natural world (or various mixes between). Also, we can choose between learning to like plastic trees or live trees. We can be all of whatever we can be. We can, that is, *design* affirmative nonmoral sentiments toward nature or abolish them, and in the short term at least, be none the worse for either choice. I strongly suggest the former: to design ourselves as if we were designed by nature to affirm nature. Why? Because, if left to its own processes, a natural world is stable, self-regulating, and permanent and thus a safer place for us. Unlike a totally artificial world, a natural world need not be constantly managed and will not destroy us by falling apart from neglect and disrepair. If nature seems fragile now and thus threatening to us, it is not so because of its inherent weakness. It is fragile and threatening because of the assault of our artifice upon it.

The truth of the biophilia hypothesis is still at issue, and serious investigation is long overdue. In the meantime, nature is retreating at an alarming and accelerating rate—tropical forests, species, coastal wetlands, migratory routes, the common atmosphere. Wilderness areas and species once lost, cannot be reclaimed. What we have done and are doing to our planet cannot easily be undone. Thus, unless and until we can be confident that there is no need for what we are casually destroying (i.e., no need coded in our common genes and designed into our nervous system) we would best be very cautious toward our natural estate. Far more cautious, I submit, than we are today.

Summary

If the foregoing arguments are successful, we must reluctantly reject Callicott's suggestion that a normative environmental ethic can be supported by a theory of moral sentiments because moral sentiments arise from interpersonal relations and thus are fated to be confined to communities of persons. Nonetheless, moral sentiments can motivate, coordinate, and implement appropriate ethical behavior and policy toward the environment. This second order application of moral sentiment to an environmental ethic is of little use, however, without an underlying *nonmoral* theory of environmental value and responsibility. Interestingly, Hume, Darwin, Leopold, and Callicott, despite the false start noted earlier, may nonetheless be close to the mark. They are right to suggest (either directly or by implication) that we must ground our environmental ethic in appropriate attitudes and sentiments. But, I have argued, these must be sentiments of a *nonmoral* kind. I have offered the hopeful suggestion that the sentiments we need to validate an environmental ethic and motivate ethical environmental policy in fact may be fundamental to our physical and neural constitution, having evolved through natural selection amidst the very natural environment that we must now preserve. If so, then this theory of *nonmoral sentiments* might support a normative environmental ethic. Finally, to end on a hopeful note, these fundamental natural sentiments and affections just might give us the motivational substance that we manifestly need in the face of the enormous environmental responsibilities before us, brought on by our folly and greed.

Notes

1. J. Baird Callicott, "Can a Theory of Moral Sentiments Support a Genuinely Normative Environmental Ethic?" *Inquiry* 35 (1992): 183–198.

2. Callicott, "Can a Theory of Moral Sentiments," 184.
3. J. Baird Callicott, "The Conceptual Foundations of the Land Ethic," in *In Defense of the Land Ethic*, ed. J. Baird Callicott (Albany, N.Y.: SUNY Press, 1989), 93.
4. David Hume, *An Inquiry Concerning the Principles of Morals* (New York: Bobbs-Merrill, 1957), 5.
5. Hume, *An Inquiry Concerning the Principles of Morals*, 107.
6. Clearly, simply arguing that "X is immoral because it violates a moral rule Y, which is immoral" will not do. Eventually, an account of "morality and immorality" must be given that does not repeat the *definiendum* in the *definiens.*
7. In fact, Hume had no inclination to apply moral sentiments to abstractions and collectives, such as "the human community." In *A Treatise of Human Nature* (New York: Doubleday-Dolphin, 1961), 433–434, bk. 3, pt. 2, sec. 1, he wrote: "In general, it may be affirmed that there is no such passion in human minds, as the love of mankind, merely as such, independent of personal qualities, of services, or of relation to ourself."
8. Hume continues, "A young tree which overtops and destroys its parent stands in all the same relations with Nero when he murdered Agrippina, and if morality consisted merely in relations would, no doubt, be equally criminal." *An Inquiry Concerning the Principles of Morals*, 111.
9. These bold pronouncements have the tincture of dogmatism. Yet a meticulous demonstration of it would lead us hopelessly astray from the primary theme of this essay. Fortunately, I have elsewhere defended the connection between moral agency, language, and community and am willing to allow these other works to stand in support of this paragraph. See in particular, my "Posthumous Interest and Posthumous Respect," *Ethics* 91, no. 2 (1981): 243–264; "Nature as a Moral Resource," *Environmental Ethics* 4 (1982): 175–190; "Three Wrong Leads in a Search for an Environmental Ethic," *Ethics and Animals* 3 (September 1994): 61–74; "Why Care about the Future?" in *Responsibilities to Future Generations*, ed. Ernest Partridge (Buffalo, N.Y.: Prometheus Books, 1981).
10. J. Baird Callicott, "Intrinsic Value, Quantum Theory, and Environmental Ethics," in *In Defense of the Land Ethic*, 162–163.
11. Here I adopt and apply (as I have before) Joel Feinberg's "interest theory of rights," as defended in his influential essay, "The Rights of Animals and Unborn Generations," in *Philosophy and Environmental Crisis*, ed. William T. Blackstone (Athens: University of Georgia Press, 1974).
12. J. Baird Callicott, "Hume's *Is/Ought* Dichotomy and the Relation of Ecology to Leopold's Land Ethic," in *In Defense of the Land Ethic*, 127.
13. Wallace Stegner, *The Sound of Mountain Water* (Garden City, N.Y.: Doubleday, 1969), 146–147.
14. Martin Krieger, "What's Wrong with Plastic Trees?" *Science* 2 (February 1973): 453.
15. Hugh Iltis, "To the Taxonomist and the Ecologist, Whose Fight Is the Preservation of Nature," *BioScience* 16 (1967): 887. In numerous articles and lectures in the early 1970s, Iltis was among the first, perhaps the first, to articulate and defend this theory. It is a great pity that his energy and devotion did not prompt an appropriate response among scientific researchers and environmental philosophers.
16. Wilson, *Biophilia* (Cambridge, Mass.: Harvard University Press, 1984), 101.

17. For a sample see Stephen R. Kellett and Edward O. Wilson, ed., *The Biophilia Hypothesis* (Washington, D.C.: Island Press, 1993).
18. Wilson, *Biophilia*, 84.
19. Paul R. Ehrlich, "Shared Sensibilities," *Natural History* (November 1984): 92.
20. Paul Shepard, "Ecology and Man," in *The Subversive Science*, ed. Paul Shepard and Daniel McKinley (Boston: Houghton Mifflin, 1969), 4.
21. Callicott, "Intrinsic Value, Quantum Theory, and Environmental Ethics."
22. Shepard, "Ecology and Man," 4.
23. Iltis states the full force of the paradox: "Not until man places man second, or, to be more precise, not until man accepts his dependency on nature and puts himself in place as part of it, not until then does man put man first! This is the greatest paradox of human ecology." H. H. Iltis, "Man First, Man Last: The Paradox of Human Ecology," *BioScience* 20 (1970): 820. The issue of the status of values in nature, here reduced to the confines of a sentence, is treated with more care and elaboration in my "Values in Nature," *Philosophical Inquiry* 8, no. 1–2 (Winter–Spring 1986): 96–110.
24. Aldo Leopold, *A Sand County Almanac* (New York: Sierra Club/Ballantine, 1970), xviii–xix.
25. Time permits only a statement of this paradox. The elaboration and justifications lie elsewhere. My own attempts along this line include "Why Care about the Future?" in *Responsibilities to Future Generations*, ed. Ernest Partridge (Buffalo N.Y.: Prometheus Books, 1981), 203–219; "Environmental Ethics: Obstacles and Opportunities," in *Environmental Consciousness*, ed. Robert C. Schultz and J. Donald Hughes (Washington D.C.: University Press of America, 1981), 325–350; and "Nature as a Moral Resource," *Environmental Ethics*, 6 (1984): 101–130.

2

How Green Is the Theory of Moral Sentiments?

John Barkdull

A major debate in environmental ethics is whether Western philosophical, moral, political, and religious traditions inherently oppose ecological values. Lynn White's widely influential essay "The Historical Roots of Our Ecologic Crisis" (1967) initiated an intense debate (Attfield, 1983; Green, 1995; Passmore, 1974) over White's assertion that Western moral and intellectual thought virtually guarantees disregard for the environment. Modern Christianity lacks a concept of sacred place and thus sanctions heedless land and resource use. Animals occupy a subordinate position in Western religion and philosophy; lacking souls, they deserve no moral consideration. Cartesian dualism separates humans from nature, further encouraging an exploitive attitude toward other creatures and the ecosystem. Barring a global turn toward Zen Buddhism, White recommends that the West return to an earlier minority tradition in Christianity, specifically to "model ourselves upon the heretical beliefs of St. Francis of Assisi, on which all creatures, whether animate or inanimate, have souls and are to be respected as such." Similarly, deep ecologists, social ecologists, ecofeminists, and postmodern theorists locate in Western ideas the origins of the global environmental crisis (Zimmerman, Callicott, Sessions, Warren, & Clark, 1993). If these analysts are correct, averting catastrophe will require abandoning the moral and political philosophies shaping the modern polity and market economy. Thus, environmental education faces a daunting task and the prospects for Western civilization to avert an environmental apocalypse are grim.

Against this challenge, J. Baird Callicott, one of the leading figures in environmental ethics, has attempted to show that Aldo Leopold's land ethic has deep roots in Western thought, and extending moral ideas to other creatures and to ecological wholes is part of the natural progression in the West's moral development. Specifically, Callicott argues that the land ethic connects (by way of Charles Darwin) to the moral theories of David Hume and Adam Smith. Therefore, the land ethic, although a major advance toward ecological consciousness and ethical concern, does not entail abandoning Western ideas.

Yet Callicott's attempt to link the land ethic to Adam Smith's moral philosophy does not entirely succeed. Although principles immanent in Smith's *Theory of the Moral Sentiments* (1790/1976, hereafter *TMS*) do suggest greater concern for the environment than one would expect from capitalism's preeminent philosopher, these principles are less moral than aesthetic. Thus, the pathway to Callicott's general conclusion—Adam Smith's moral philosophy contains the seeds of a philosophy of environmental concern—is more roundabout than Callicott's reading of Hume would indicate because Callicott takes Smith and Hume, who both based their moral theories on sentiment and sympathy, as saying more or less the same thing, ignoring significant differences between the two. Smith's moral theory points toward environmental concern arising from appreciation of beauty and order rather than a moral sentiment directed toward biotic communities. In short, Callicott's proposal to trace the land ethic to venerable traditions in Western moral theory has promise, but it requires rethinking to be a convincing response to those who argue for radical transformations in Western thought to achieve an ecologically sound society.

With these considerations in mind, then, would Adam Smith be an environmentalist? If so, how green would he be? Would he advocate biocentric ethics or would he see the environment as a storehouse of resources for human use? Would Smith assert that environmental values ought to override the "system of natural liberty" that was at the heart of his economic theory? To answer that Smith would advocate environmental protection over commercial values runs against popular stereotypes of Smith. Nonetheless, if Callicott is at all correct, Adam Smith, were he alive today, would be green, if not deep green, and in turn might call for significant limitations on markets for environmental protection. Contrary to rhetorical deployments of Smith's name in support of laissez-faire, he might even support considerable reform of a world political economy now ostensibly based on ideas he advocated in *Wealth of Nations* (Smith, 1776/1937). Indeed, Earth First! cofounder Dave Foreman, advocate of monkey wrenching and other forms of resistance to corporate exploitation of nature, also claims Leopold as inspiration (Foreman, 1991, pp. 139–140). Thus, the green reading of moral thought that places Adam Smith in line with Leopold implies a startling reinterpretation of Smith's position in Western political and moral philosophy.

Conflicting readings of Adam Smith are nothing new, but exploring the implications of his work for environmental ethics and policy has received little notice. Scholarship on Smith has focused largely on political economy and the role of government in regulating the market (see e.g., Harpham, 1984; Hont & Ignatieff, 1983; McNally, 1993; Minowitz, 1993; Muller, 1993; Wilson, 1989; Winch, 1978). Various studies attempt to recover the "real" Adam Smith against his caricature as unrelenting advocate of self-interest and the invisible hand (Meyerson, 1989; Pack, 1991). Hence, to establish that Smith is no libertarian—that he was highly skeptical of the merchant class's motives and endorsed an activist government to protect the common welfare—would be to plow old ground. Yet none of this literature suggests that Smith's critique of commercial society in combination with his moral theory would endorse a biocentric land ethic.

Although Smith's moral theory does have such potential, Callicott goes wrong in ignoring important differences between Hume's and Smith's views of moral psychology (Hope, 1989, p. 83). As Raphael and Macfie relate, Smith assumed that Hume and Hutcheson had established sentiment as the basis for morality. Smith's contribution was to provide an analysis with more nuance and detail in that he distinguished more types of moral feeling. Different forms of the sense of propriety, sympathy distinguished from propriety and merit, and all these from the sense of duty imply "several kinds of moral approbation, a variety of moral feelings or sentiments" (Raphael & Macfie, 1976, p. 14). Hence, to trace the land ethic to Adam Smith requires direct attention to Smith's writings, rather than to assume that Hume and Smith would agree. Callicott establishes his case almost entirely by reference to Hume, not Smith; so the question arises, is placing capitalism's most venerated philosopher in the same line of subversive thought leading to the land ethic sensible? Would Smith place community on par with or even before the individual, and the biotic community on par with or before human interests?

Adam Smith and the Land Ethic

How can Smith's ethical theory, one which presumably complements Smith's endorsement of a reformed commercial society (Raphael & Macfie, 1976, pp. 20–25), stand as a precursor to the land ethic? At first glance, nothing in *TMS* challenges anthropocentrism or individualism. Yet, Callicott links Leopold to the Scottish Enlightenment by way of Darwin's evolutionary account of the prevalence of altruism in human society. Looking beyond Leopold, he sees a connection between the Scottish Enlightenment and sociobiology. For Callicott, evolutionary theory explains the development of moral sentiments and the possibility of altruism; without such potential in human beings, society would be impossible. Because morality makes society possible, humans gain an advantage in the struggle for survival. Whereas Smith attributes the fortuitous arrangement of the moral sentiments to the "great Director of nature," Callicott gives credit to nature alone, but this only reflects the fact that Darwin lived after Smith. In either case, we are endowed with natural moral sentiments, including the capacity to sympathize with others, to take their part and thus to feel obligations toward them regardless of personal gain, and the consequence is human success as a species. For Smith, our moral feelings are carefully calibrated so that, without intending it, we behave such that the social order persists and happiness is enhanced (see also Wilson, 1993a, 1993b). This capacity for moral sentiment makes possible, Callicott claims, the extension of conscience from the human community to the larger biotic community once we become aware of our place in the ecosystem. Reason's role is informative, not regulative; that is, reason—although it cannot, by itself, provide compelling moral rules (as Kant imagined)—can show the connections between things and thus can change the objects that are properly given moral consideration. In short, the facts of ecological science yield the oughts of the land ethic.

Specifically, the land ethic rests, according to Callicott, on "three scientific cornerstones" that reveal these connections: ". . . Copernican cosmology, a Darwinian protosociobiological natural history of ethics, Darwinian ties of kinship among all forms of life on earth, and an Eltonian model of the structure of biocenoses all overlaid on a Humean-Smithian moral psychology" (Callicott, 1989, pp. 82–83). The result is, or ought to be, holistic moral considerability for the biotic community as such, not merely as instrument of human welfare.

The public policy implications are significant: Valuing the land for its own sake rather than as means to further human interests could mean, for instance, that cost-benefit analysis would not guide decisions affecting the environment in that cost-benefit analysis "requires that *everything* be reduced to cash" (Adams, 1996, p. 2, emphasis in original). Moreover, plowing the ground, cutting down forests, diverting streams and rivers, and leveling mountains could be moral wrongs, regardless of the short-term benefit for humans. The land ethic instead calls on us to place the preservation of evolutionary processes ahead of human desires, including those registered in the market. This seems to contrast sharply with Adam Smith's hope that his work would contribute to the progress of industrial production, cultivation of the soil, and even to a larger human population.

Equally significant, Callicott notes that Hume posited a moral sentiment in regard to the "publick interest" that could be encouraged by proper education. Thus, by analogy, awareness of our place in an even larger community than human society, a community that includes other species and inanimate objects, can arouse the same sentiment as that directed toward the public interest. Recent scientific discoveries would tell us of a new "publick" appropriate to our moral passions: the biotic community (Callicott, 1989, p. 126). In sum, Callicott says that the land ethic, drawing on Hume and Smith, points toward consequences, duties, and innate moral sentiments all tending to support the principle that whatever enhances the integrity, stability, and beauty of the land is right.

Yet, if the land ethic resonates with Adam Smith's work, it is on very different grounds than these. Callicott ignores the kind of close analysis Smith undertook in creating his science of morals and thus the differences with Hume. The question here is not whether Callicott was faithful to Smith's philosophy, but whether Callicott has overlooked important themes in Smith's work that would lend new insight to the problem of Western philosophy and environmental ethics. One cannot infer from the general proposition that morals arise from feeling that Smith would posit moral sentiments toward ecological wholes. Rather, one must demonstrate how, if at all, such moral sentiments would arise in a way that is faithful to Smith's own method in order to tie his moral theory to the contemporary land ethic. All the more so is detailed demonstration needed in that, at least from the standpoint of Smith's ethical theory, Callicott's argument turns on a rather weak point—the presumed analogy between moral sentiment directed toward society and the possibility of a similar attitude toward the ecosystem. If, as a close reading of *TMS* suggests, Smith does not posit a more or less direct moral sentiment toward society, then the analogy to moral sentiment for the biotic community fails. What, then, did Smith say about how morals develop in society?

THE VIRTUES, PASSIONS, SENTIMENTS, AND SYMPATHY

Smith, by contrast to Hume (and Hutcheson), emphasized conscience over either benevolence or self-interested concern for social order. Although following Hume in basing moral approval on sentiment, he offered a rival theory to Hume's, not a mere restatement. Smith presents his general thesis regarding the moral sentiments in the first line of *TMS*: "How selfish soever man may be supposed, there are evidently some principles in his nature, which interest him in the fortune of others, and render their happiness necessary to him, though he derives nothing from it except the pleasure of seeing it." Clearly, Smith, like other classical liberals, did not believe that the self-interested pursuit of gain was the only or even the predominant motive for action (Holmes, 1995, pp. 42–68). Rather, self-interest (properly understood) can be a moderating influence, reducing the effect of passions and "enthusiasm" (Clark, 1992; Farr, 1988) on politics and society generally. More generally, the opening line of *TMS* informs us that Smith assumes altruism is part of human nature.

The major elements of Smith's theory are the virtues, the passions, the moral sentiments, and sympathy. Establishing the relation among these concepts constitutes the subject matter of *TMS*. The virtues include prudence, beneficence, justice, and self-command. The passions, which are the basic springs of action, include those arising from the body (e.g., hunger, desire for sex, need for comfort), those arising from a habit of mind or imagination (e.g., love), the unsocial passions (e.g., resentment, anger, hatred), the social passions (e.g., generosity, humanity, kindness), and the selfish passions (e.g., grief and joy resulting from our own condition). The moral sentiments are the senses of propriety and of merit. Finally, sympathy—the capacity to take the part of another and to see either his actions or your own through his eyes—and the prompting of the "Impartial Spectator" recommend virtuous behavior to us (except for self-command, which is recommended primarily by the sense of propriety). Two important purposes of the theory Smith forwards are to provide reason for acting virtuously and to oppose "licentious systems" that advocate acting hedonistically or selfishly.

The "great Director of nature" has constituted us to promote happiness and social harmony, but only indirectly. Immediate sense and feeling are the ultimate grounds of moral evaluation, not any calculation that this or that action or response will serve a larger aim, such as human happiness or social order. Humans lack the capacity to reason out moral and ethical rules adequate to sustaining social order. Rather, for Smith, the task of moral philosophy is to discern those rules that do exist and that seem to contribute the most to order and personal happiness; given this, recommending certain rules over others is possible, but only because observation shows them to work in practice. Although morality is an aspect of human nature, specific rules can vary at different times and in different places. Furthermore, moral rules are learned inductively from observation of numerous instances in which certain emotional responses (sentiments) attend a situation of a certain type. Like language, morality is a social product, and individual performances constitute the substantive expression of the collective phenomenon. Thus, Smith can argue that moral rules are discovered inductively and developed in society, and yet that they do not depend on caprice, custom, fashion—or the self-regarding desires of autonomous individuals—any more than language could be arbitrarily constructed anew with each generation. In all times, however, morals arise from emotion and feeling, not primarily from attempts to construct appropriate rules through the application of reason. Smith writes:

> But though reason is undoubtedly the source of the general rules of morality, and of all the moral judgments which we form by means of them, it is altogether absurd and unintelligible to suppose that the first perceptions of right and wrong can be derived from reason, even in those particular cases upon the experience of which the general rules are formed. These first perceptions, as well as all other experiments upon which any general rules are founded, cannot be the object of reason, but of immediate sense and feeling. It is by finding in a vast variety of instances that one tenor of conduct constantly pleases in a certain manner, and that another as constantly displeases the mind, that we form the general rules of morality. But reason cannot render any particular object either agreeable or disagreeable for its own sake. Reason may show that this object is the means of obtaining some other which is naturally either pleasing or displeasing, and in this manner may render it either agreeable or disagreeable for the sake of something else. But nothing can be agreeable or disagreeable for its own sake, which is not rendered such by immediate sense and feeling. If virtue, therefore, in every particular instance, necessarily pleases for its own sake, and if vice as certainly displeases the mind, it cannot be reason, but immediate sense and feeling, which, in this manner, reconciles us to the one and alienates us from the other. (*TMS*, p. 320)

That is, induction informs us of the general rules underlying the "vast variety of instances," and reason may tell us which object leads to pleasant and which to disagreeable reactions, but reason cannot be the ultimate basis for morality. Humans are too limited to hope to design anything so complex as a society, much less to expect rules derived from rational inquiry to have sufficient force to affect behavior. On this, Hume, Smith, and Callicott agree. The question is whether these "first perceptions of right and wrong" include aversion to damaging the integrity, stability, and beauty of the land, even to serve human ends for, say, economic development.

Callicott and Smith diverge in that Smith's theory is highly personal. For Smith, moral rules are learned and applied in very concrete, specific ways in close personal interactions with other human beings. We observe both the cause and the effect of an action on individuals, and we infer their responses. (The action itself lacks meaning without regard to these elements.) We give either approbation or disapprobation to the motives of the actors involved. We give merit or demerit to the consequences. As Hope observes, this differentiates Smith from Hume, in that for Hume only consequences matter, not motives (Hope, 1989, p. 88). The same act can be judged differently depending on the motive, so that one can disapprove foolish generosity, give merit to inadvertent good done to another, or ameliorate demerit for an attempted good deed gone sour. Yet, good intentions alone are not enough to gain high praise; the effect must be seen as well. An act is most virtuous, then, when motivated well and executed well. Our responses then arise from a "double sympathy" with the agent and with the acted upon, as well as approval of the observed consequences of the act. Evaluations of approval and merit stem from placing oneself, by an act of imagination, in the positions of those involved, which in turn depends upon the presupposition that one's own perceptions and feelings are much like those of other human beings. Moral judgment, then, arises from spontaneously occurring feelings felt on observing or being a part of a particular social transaction.

From this arises Smith's main theoretical concept, a complex understanding of sympathy. Smith extends sympathy beyond offering approbation to benevolent motives to entail the general capacity to adopt another's point of view. Thus a person can sympathize—share a correspondence of sentiments—with evil intentions, anger, fear, and so on, not only with good will. Furthermore, the mature person should be able to adopt the point of view of others when judging his own behavior, and this capacity produces conscience, a sense of shame, mortification, and the desire to be praiseworthy as well as praised. This gives rise to the notion of the "Impartial Spectator," the informed and sympathetic observer who would be best positioned to evaluate the degree of approbation and merit one's own action deserves. Our observations of others' reactions to our behavior, and our observation of our own reactions to others' behavior, inform us of what is virtuous and what is wrong in our own case. Again, this development of conscience depends on interactions with other human beings over a considerable period of time.

Because of our natural desire for approval and to be worthy of approval, the result is moderation of impulses and passions, proportionality in our estimation of our own place in the world, and a capacity to adjust passion and behavior to the requisites of social order and well-being. Consequently, self-sacrifice becomes thinkable and even obligatory, whereas morality based on a strict notion of self-interest would find the idea absurd. The pursuit of self-interest finds moderation in due concern for the welfare of others as seen from the impartial spectator's perspective; that is, because of our desire to be worthy of approval, our sense of self-approbation, we develop a sense of duty that often overrides self-interest. Again, this rests on imagining what other people would think of one's action. Although Smith does not reduce morality to a matter of majority opinion, nonetheless, even in evaluating one's own actions, the standpoint of other observers (even if imagined) is what recommends virtuous behavior.

Yet, if moral judgment and achieving moral behavior depends upon a "correspondence of sentiments" as Smith claims, then it is difficult to see how a moral relationship can develop between dissimilar creatures, much less abstract entities such as ecosystems. For Smith morality arises from close social interaction and from the immediate responses we feel in such interaction. Moral rules are good in that they lead indirectly to society's survival and, to some degree, to happiness. In practice, we aim at neither of these; we experience approval and disapproval of what we observe (including our own behavior) in regard to specific individuals engaged in specific acts. No rules derived from rational procedures or calculations of effect are needed for moral responses to be elicited. Smith's task, then, was to detail the systematic nature of these responses, to categorize the passions, sentiments, and virtues and to show by close analysis how they work in practice. Although these complex, highly subjective, and personal psychological interactions can extend to pets and other animals, their primary focus is on how humans relate to one another. Smith discusses how we perceive and evaluate motives such as pride, vanity, love, desire, and more. Given that the moral theory just described is evidently human-centered and depends on close personal relationships, on immediate sense and feeling, how can it imply a land ethic?

The Publick Interest

Callicott asserts that, for Hume, moral sentiments can extend to concern for the "publick interest," and this can be generalized to include the larger "publick" of the biotic community. If so, then recognizing that animals, plants, and even mountains and streams are part of an integrated community can create a sort of sentiment that grounds moral obligation. Ernest Partridge challenges this proposition with Hume's own words: "Inanimate objects . . . can never be the object of love or hatred, nor are [they] consequently susceptible of merit or iniquity" (Hume, 1957, quoted in Partridge, 1996, p. 153). Yet Callicott can retort that, regardless of Hume's views so many years ago, the objects of moral sentiment have now changed due to the findings of ecological science. More to the point is how Smith's theory of moral sentiments derives concern for the public interest. In Smith's analysis, the analogy between care for society and the biotic community does not hold; the most likely moral basis for the land ethic would be the promptings of the impartial spectator, but this shows the limits of sentiment based moral philosophy; and the aesthetic argument for the land ethic is more convincing than the attribution of a moral sentiment toward the community.

The central question for the land ethic is how the biotic community comes to have intrinsic value, so that the central principle of the land ethic is binding. For Callicott the answer lies in a "Humean/Darwinian bioempathetic moral metaphysic, based upon naturally selected moral sentiments" (1989, p. 152). Drawing an analogy between empathy for the human community and for the biotic community, Callicott concludes:

> Now, as Hume has observed, not only have we sympathy for our fellows, we also are naturally endowed with a sentiment the proper object of which is society itself. Ecology and the environmental sciences thus inform us of the existence of something which is a proper object of one of our most fundamental moral passions. The biotic community is a proper object of that passion which is actuated by the contemplation of the complexity, diversity, integrity, and stability of the community to which we belong. (1989, 126)

This raises the question, which passion is it that contemplation of complexity, diversity, integrity, and stability excites? Is it a moral passion? For Smith, if not for Hume, the answer would appear to be no. Rather, for Smith, concern for the public good derives largely from love of beauty and order, not from moral obligation.

Smith does not assert a natural sentiment toward society as such, but instead that more immediate responses toward individuals lend themselves to the maintenance of social order in line with the description just given of his theory. Indeed, Smith argues against Hume's claim that man has a "natural love for society, and desires that the union of mankind should be preserved for its own sake, and though he himself was to derive no benefit from it" (*TMS*, p. 88). To focus on one especially important issue—justice—Smith does not believe that concern for society's well-being animates moral approbation of justice to any great degree. Instead, Smith reiterates his general principle that immediate sense and feeling determine moral judgment, including a desire for justice, and any good social consequences are unintended. Thus, our eagerness to punish injustice rests on our capacity to sympathize with the indignation and resentment of the individual suffering harm, not on a general love of society. Few even reflect on the role of justice in maintaining social order, and concern for justice "does not, in common cases, arise from that which we take in the fortune and happiness of society" (*TMS*, p. 89). To the contrary, our capacity to sympathize with the misery of the condemned often leads to a desire for mercy. We must override natural sentiments of pity to approve of carrying out extreme punishment—even though reason tells us that such action can help to preserve society. Whereas Smith speaks of "a more enlarged compassion" for mankind, he quite clearly indicates that we must consciously remind ourselves of this larger compassion to overcome more natural, immediate sentiments directed toward individuals. Indeed, Smith concludes that our concern for social order is so weak by comparison to sympathy with individuals that we must rely on God to carry out punishments we are unwilling to inflict: "For it well deserves to be taken notice of, that we are so far from imagining that injustice ought to be punished in this life, merely on account of the order of society, which cannot be otherwise maintained, that Nature teaches us to hope, and religion, we suppose, authorises us to expect, that it will be punished in a life to come" (*TMS*, p. 91). For the most part, nature has arranged it so that our immediate concern for the fate of individuals serves to maintain the good order of society, but Smith denies, in this context, that we have a strong concern for society itself. Accordingly, if the land ethic depends on the analogy between moral sentiment for society to moral sentiment for the biotic community, we see that for Smith society is barely a "proper object" of our moral passions. Thus, the derivative moral sentiment toward the biotic community would seem to be exceedingly weak for the average person.

Moreover, Smith doubts that our moral compass should extend far beyond our own friends and family. He criticizes those "whining and melancholy moralists" who insist that we should value others' happiness as highly as our own and who reproach us for being happy while others are in misery. This only leads to insistence that our own spirits should be habitually dejected, something "altogether absurd and unreasonable." Concerning ourselves with the fortunes of "those with whom we have no acquaintance or connexion, [*sic*] and who are placed altogether out of the sphere of our activity" produces nothing useful. Smith concludes, "That we should be but little interested, therefore, in the fortune of those whom we can neither serve nor hurt, and who are in every respect so very remote from us, seems wisely ordered by Nature; and if it were possible to alter in this respect the original constitution of our frame, we could yet gain nothing by the change" (*TMS*, p. 140). The question at stake here is how much modern communications, ease of travel, and interdependence have shortened the distance between us and others. In today's world, one can hardly posit a significant environmental policy choice that would fail to serve or hurt individuals on distant parts of the planet—or, by the same token, future generations. Yet, one might take Smith to mean that anxiety on the account of tropical rain forests of which one has had no experience or acquaintance "would be perfectly useless, and could serve no other purpose than to render miserable the person who possessed it."

In truth, Smith is ambivalent on this point. To show that natural moral sentiments frequently override self-interest, that we must maintain proportion between our own needs and those of others, even those remotely connected to ourselves, he provides the following example. Let us suppose, he offers, that all of China were to be destroyed suddenly. When all the fine philosophy of the most sensitive man of humanity in Europe was over, "the most frivolous disaster which could befal[l] himself would occasion a more real disturbance. . . . If he was to lose his little finger tomorrow, he would not sleep to-night; but, provided he never saw them, he will snore with the most profound security over the ruin of a hundred millions of his bretheren, [*sic*] and the destruction of that immense multitude seems plainly an object less interesting to him than this paltry misfortune of his own" (*TMS*, p. 136). Plainly our inherent moral parochialism lends itself to passing concern with the fortunes and tragedies of millions, as long as they are out of sight. Nonetheless, not only the rest of mankind, but the individual concerned as well, would find a man a moral monster to suggest that to prevent this misfortune to himself, the fortunes of the hundred million should be sacrificed. The sacrifice of the finger—and even more—is to be expected, a duty of benevolence. Yet, the unlikeliness of the situation suggests that only in extreme cases might Smith expect that moral sentiment would move us to act on behalf of unknown strangers.

Elsewhere, Smith does assert love of one's own country, but even here the extent of moral sentiment is quite limited. Because our fortune, that of our family, and that of our friends depends on the prosperity and safety of the state in which we live, it is "endeared to us, not only by all our selfish, but by our private benevolent affections" (*TMS*, p. 227). Moreover, its glory and honor seem to reflect on us, and we insist on our own country's superiority over all others. Yet, this sentiment does not extend readily beyond our own nation. "The love of country," Smith observes, "seems not to be derived from the love of mankind. The former sentiment is altogether independent of the latter, and seems sometimes even to dispose us to act inconsistently with it" (*TMS*, p. 229). If extending moral sympathy across borders is so difficult, how much more difficult to do so across species! Moral parochialism, which Smith takes as natural, would seem to speak against the land ethic's universal—or even bioregional—care. Advocates of the land ethic might respond that their entire point is that love of country is but a step along the way to love of biotic community, due to increasing knowledge. Yet the concept of a universal community of mankind was not lacking in Smith's day. It is not obvious why, if the extension of moral considerability is so readily done in regard to ecosystems, Smith's contemporaries could not do so in regard to other human beings.

These observations regarding love of nation preface Smith's larger argument against "universal benevolence." Our good will, Smith writes, knows no boundary, and "we cannot form the idea of any innocent and sensible being, whose happiness we should not desire, or to whose misery, when distinctly brought home to the imagination, we should not have some degree of aversion" (*TMS*, p. 235); clearly, Smith is extending this sentiment to all sentient beings, not only humans. Yet, this natural impulse does not serve as the basis for ordinary morality. Even the most patriotic and self-sacrificing citizen is "allotted a much humbler department, but one much more suitable to the weakness of his powers, and to the narrowness of his comprehension; the care of his own happiness, of that of his family, his friends, his country." Against this, "the administration of the great system of the universe, however, the care of the universal happiness of all rational and sensible beings, is the business of God and not of man" (*TMS*, p. 237). For Smith, then, our natural inclination to see to our own immediate affairs is for the best. In fact, extending care beyond one's friends, family, and country only causes the individual misery, as he contemplates the vast amount of suffering in the world.

To be sure, Callicott, like Smith, also says that moral commitment varies according to the nearness of the community. Callicott writes of "nested communities" entailing different degrees of obligation. The land ethic "neither replaces nor overrides previous accretions" of moral obligation. "Family obligations in general come before nationalistic duties and humanitarian obligations in general come before environmental duties" (1989, pp. 93–94). But what remains of the land ethic, given this qualification? The entire import of Callicott's defense of the land ethic would lie in the phrase *in general*: In general, the land ethic occupies last place in the moral hierarchy. So when, specifically, do environmental duties override family duties? The general tenor of this passage places maintaining the integrity, stability, and beauty of the land far down the list of priorities, which would reduce the land ethic to irrelevance in regard to political and economic policy. Yet, if Callicott wishes to maintain the stronger implications of the land ethic—that, for instance, doing well for one's family is not moral if it also harms the integrity of the land—then one must infer that the land ethic overrides other obligations quite often and in quite profound ways. In light of Callicott's voluminous writings defending the land ethic's stronger demands, I take the qualification as tentative and exploratory rather than indicative of Callicott's larger views, so that the difficulties in reconciling his and Smith's moral theories remain.

VIRTUE BY CONSENSUS?

Another important issue in reconciling the land ethic with the theory of moral sentiments is the relation between general opinion and particular moral precepts. Smith would appear to offer a moral theory based largely on the widely felt judgments of mankind. Moral judgments gain force because most people feel similar degrees of approbation or disapprobation, merit or demerit, when contemplating a given situation. Smith's argument hinges on the proposition that social order results from a fine, if unintended, adjustment of widely shared natural moral sentiments. Similarly, Callicott posits that the evolutionary process has given humans the shared capacity for moral sentiment, and this provides a kind of functional equivalent to objective standards. Yet, Callicott does not claim that the shared capacity for moral sentiment implies any specific a priori content. This implies that the moral sentiments can go anywhere, which would reduce the claim that the land ethic is consistent with a theory of moral sentiments to a triviality. Against this, Callicott also claims that consensus would rest on "matters of fact and scientific revelation" (1992, p. 196), such that any who share the same set of facts and understanding of causal connections will arrive at similar moral judgments. Consequently, if the land ethic is not prevalent, then the reason is that not enough people possess sufficient ecological knowledge, and he calls for environmental education to bring about the necessary transformation: "When this basic concept of ecology is taught at all levels of education, from story and song in early childhood education to abstract, theoretical mathematics, science, and philosophy in higher education, the land ethic may be transformed from one man's dream to all mankind's reality" (1989, pp. 236–237).

Yet, this is entirely speculative. Smith could observe normal human reactions to the eternal events in human society—murder, theft, love, generosity, and the like. Thus, Smith's moral theory is empirical in a way Callicott's is not. Accordingly, if the theory of moral sentiments is indeed a brand of virtue by consensus, then because the land ethic is far from general acceptance it does not meet an important criterion for having normative force. To be sure, reason informs us of the proper objects of our moral passions. Consequently, the intense deep green experience, such as Callicott had on viewing the polluted Mississippi (1989, p. 114), could become general with the extension of ecological education to an ever wider human audience. Presumably, the universality of moral sentiments opens the possibility of most humans feeling the same kinship with the biotic community as Callicott. Yet, the question is whether the land ethic has normative force *now*. Whereas calling for creating a new consensus of feeling may be sensible, one cannot refer to its hypothetical existence to establish the objective existence of its norms. Clearly, the contemporary consensus of feeling does not grant moral standing to inanimate objects, such as rivers and streams, or to the ecosystem as such. That view seems to be confined to a handful of committed environmentalists who have experienced such feelings as Callicott's firsthand. Although Callicott may hope that his own moral sentiments may one day become those of the majority, currently the "normal" moral feelings in society seem to be anthropocentric, and we have no way to know if that will change. Uniquely intense personal experiences do not add up to a moral consensus, and, rather than encourage them, Smith might discount them as examples of misplaced "enthusiasm" (Farr, 1988) for which the actually existing consensus of feeling would be the remedy.

Nor is it at all certain that anyone equipped with the same knowledge of ecological science as Callicott would have similar moral sentiments. Some critics of environmentalism simply reject the notion that morality should serve any but human ends. Furthermore, one must doubt whether the kind of moral education Callicott calls for will ever come to predominate in Western society. Although ecological science will remain a mainstay of college education, it will most likely continue to occupy its own niche in the academy alongside the other disciplines, including economics, dairy science, and industrial engineering. Childhood stories and songs will in all likelihood continue to focus on handsome princes and maidens in distress, ugly ducklings and gingerbread houses. In short, the institutional and cultural barriers to the land ethic tend to indicate that the human consensus of feeling will remain anthropocentric.

Nonetheless, all this said, Callicott's view is not incompatible with Smith's. Indeed, Callicott's land ethic gains from reliance on Smith's moral theory rather than Hume's, but to make room for minority opinions requires some retreat from sentiment-based theory. Because Smith does not share Hume's reliance on the consensus of human feeling, Smith's theory offers a way beyond the apparent contradiction between "virtue by consensus" and individual conscience. Specifically, Smith's concept of the impartial spectator can be deployed to support moral standards that run contrary to the normal, general consensus of feeling, including an environmental ethic (Taliaferro, 1988). For Smith, we act virtuously both because doing so in fact generates social approbation and because we desire to be worthy of such approbation, regardless of whether we receive such praise (*TMS*, pp. 113–134). Because public opinion can be fickle and wrongheaded, we seek the approbation of an impartial, fair-minded observer. If no one agrees with one's actions although one knows them to be correct, then the opinion of an imagined spectator must suffice.

Accordingly, greens who base their ethics on intense personal experiences can appeal to their own impartial spectator and insist that the green perspective would become general were everyone to exercise impartial judgment, knowing the facts of ecological science. Thus, Smith, more than Hume, would support minority views such as the land ethic.

Yet, while beginning to reconcile Smith's theory of moral sentiments with the land ethic, the result of resort to the impartial spectator is to reveal the limits of sentiment-based moral theory. The land ethic rejects deriving norms from the precepts of reason in favor of moral sentiment; the role of reason is informative. Minority opinions, of which the land ethic is one, can call on the impartial spectator to overcome the subjectivism inherent in a majoritarian ethic and to support their moral claim. Yet, because the land ethic is a distinctly minority position, the impartial spectator must be appealing to something other than the consensus of feeling among normal people for claims based on it to have normative force. For Smith, "It is *reason*, *principle*, conscience, the inhabitant of the breast, the man within, the great judge and arbiter of our conduct" that tell us that self-love carried to extremes (among other moral failings) is wrong (*TMS*, p. 137, emphasis added). In this way, Smith avoids the cultural relativist implications of his moral theory, but if reason and principle, then which reasons and which principles; that is, on what basis does the impartial spectator oppose the general opinion of society? Categorical imperatives? Striving for the best consequences? Justice? Which standards would apply is not clear, but it is clear that specifying any would return Smith—and thus the land ethic, to the extent it must depend on the impartial spectator to avoid majoritarian relativism—to debates between and among mainstream moral philosophers. For here reason is not only informative, it also provides principles for action.

Why is this a problem for the land ethic? Because Callicott based the land ethic on the theory of moral sentiments to avoid the egocentric, individualist commitments of deontology and utilitarianism. If, now, dispassionate universal principles are to be invoked, advocates of the land ethic must produce the specific principles for evaluation and critique. It would be most surprising if these principles turned out to be nothing more than matters of scientific fact.

In sum, although the analogy to the public interest does not hold, the land ethic is not incompatible with Smith's version of the theory of moral sentiments. Nonetheless, close consideration reveals that to avoid cultural relativism, Smith and the land ethic approach the limits of sentiment-based moral theory. Whereas both the land ethic and Smith's theory appear on first inspection to rely on moral consensus, thus inviting dismissal of the land ethic for having no substantial following on the one hand and charges of relativism on the other, the impartial spectator provides room for minority opinions.

Yet, as important as this is for both Smith's aim to introduce a new economic philosophy and for Callicott's aim to introduce a new ecological philosophy, resort to the "man within" seems to show that the theory of moral sentiments cannot meet the charge of relativism without unstated general principles that exist apart from the normal consensus of feeling. Otherwise, the normative force of the land ethic must rest on the hypothetical existence of a future consensus of feeling that biocentric communities deserve moral consideration, a rather weak reed. Thus, although the impartial spectator links the land ethic to Smith's theory of moral sentiments, as Callicott suspected, the moral argument becomes rather unconvincing. Fortunately, a more promising avenue lies open.

The Aesthetic Dimension

Smith asked why some individuals feel driven to perform great works of public service, even though personal material gain might be small. His general answer is to observe "the beauty that action derives from its utility" (Minowitz, 1993, p. 47). More specifically Smith notes an odd characteristic of human psychology regarding our evaluation of objects, which is that the "exact adjustment of the means for attaining any conveniency or pleasure, should frequently be more regarded, than that very conveniency or pleasure" (*TMS*, 179–180); the utility lies as often in the means as in the consequences. Thus, from attention to the most frivolous items—a timepiece, an arrangement of furniture—to the most ardent striving after a life of opulence, the driving force is the imagined convenience they provide; the thought of such fitness and usefulness is more pleasing to the mind than the actual experience. Therefore, a poor man will exert great effort to attain the imagined ease of the wealthy while disregarding comfort that was ready to hand all along: good friends, restful sleep, and decent food, generally gaining the desired state long after he is of an age to enjoy it and despite innumerable sacrifices and privations. Why, other than a peculiarity of the imagination, would anyone make such a bad bargain? It is certainly not from a rational calculation of self-interest. Yet, for Smith, "it is well that nature imposes on us in this manner" (*TMS*, p. 183) because without this peculiarity, nothing would have spurred us:

> to cultivate the ground, to build houses, to found cities and commonwealths, and to invent and improve all the sciences and arts, which ennoble and embellish human life; which have entirely changed the whole face of the globe. . . . The earth by these labours of mankind has been obliged to redouble her natural fertility, and to maintain a greater multitude of inhabitants. (*TMS*, pp. 183–184)

The connection to public policy is this: Smith asserts that the same love of "conveniency" leads to public spirit. Those who exert themselves for public improvement rarely do so out of sympathy and concern for other citizens. Rather:

> The perfection of polic[y], the extension of trade and manufactures, are noble and magnificent projects. The contemplation of them pleases us, and we are interested in whatever can tend to advance them. They make part of the great system of government, and the wheels of the political machine seem to move with more harmony and ease by means of them. We take pleasure in beholding the perfection of so beautiful and grand a system, and we are uneasy till we remove any obstruction that can in the least disturb or encumber the regularity of its motions. (TMS, p. 185)

Accordingly, to ignite the passion for social improvement, one must describe and explain "the connexions and dependencies" of the many parts of the system, show how a grand new system may be introduced and set to work, so that even the least public-spirited person will see how "all the several wheels of the machine of government may be made to move with more harmony and smoothness, without grating upon one another, or mutually retarding one another's motions" (*TMS*, p. 186). In other words, one must appeal to an aesthetic sense to inflame public spirit. Informing others of the material improvements to be had from a new system or a new policy—the good one can do for other members of the community—will not make much impression; for example, no one had less humanity and more public spirit than "the celebrated legislator of Muscovy," Peter the Great. Yet, again due to unintended consequences, and although no moral end is sought, although the aim is an imagined convenience and a smooth functioning of all the parts of the great machinery of society, so is the greatest happiness of society served.

Thus, whereas Smith's views on our peculiar attitude toward utility—that the utility is in the means rather than the end—can indeed imply considerable concern for society, and by analogy the biotic community, it remains that this is not a moral sentiment. Nowadays, we might conclude that the science of ecology demonstrates a fundamental failing in the workings of the great system of nature. Human society, now having transformed the "rude forests of nature into agreeable and fertile plains," reveals itself to be a dysfunctional part of a much larger system. Ecological science, then, can excite in us care and concern that this obstruction to the regularity of nature's motions be remedied. Yet, this need not proceed from any particular sympathy for other humans or for the members of other species, nor need it imply extension of conscience to the biotic community. Instead, it depends on aesthetic sentiments, not moral. To be sure, this is not entirely contrary to the land ethic's central principle: "A thing is right when it tends to preserve the integrity, stability, and *beauty* of the biotic community," and again: "Examine each question in terms of what is ethically and *esthetically* right, as well as what is economically expedient" (Leopold, 1949/1987, p. 224, emphasis added). The land ethic and Smith's theory of sentiments do share important common ground. The impartial spectator allows, on occasion, individual conscience to predominate over general opinion, which gives some support to contemporary advocates of the minority biocentric view. More important, they share a common regard for the importance of beauty and harmony in our evaluations and as motive to action.

The conclusion that *TMS* points toward an aesthetic basis for environmental ethics resonates with Eugene C. Hargrove's work (1989), which argues that destroying nature is equivalent to destroying a work of art. By the same token, preserving nature would have similar moral import as protecting a work of art. Beauty is an irreducible good, and if humans have a duty to increase the good in the world, then they have a duty to protect and preserve the beautiful. The duty to protect the physical processes that allow the production of natural beauty can even take precedence over the duty to preserve works of art. (Hargrove, 1989, p. 198).

Against those who doubt that environmental ethics has any place in Western traditions, Hargrove shows that environmental concern began to develop among naturalists engaged in exploring wilderness areas. An aesthetic shift away from considering wilderness untamed, threatening, ultimately ugly and in need of human cultivation toward appreciation of natural landscapes underpinned an emerging sense of duty to preserve untouched or especially grand areas. Hargrove's interpretation fits comfortably with Smith's remarks on our motivations for doing public service. Moreover, Smith's views can tie Hargrove's ontological argument to the land ethic, if we emphasize the aesthetic elements of Leopold's essential principle rather than the moral. Thus, Callicott's intuition that Smith lies in line, via Darwin, with Leopold and his own moral theory was correct, but the pathway is different than the one Callicott chose.

The Upshot

Any attention to Smith's moral theory reveals that, contrary to popular belief, Smith would not reduce human motivation or moral evaluation to the rational pursuit of self-interest. By the same token, Smith would most likely not advocate complete reliance on unregulated markets to achieve environmental protection. Thus, resources for an environmental ethic exist in his work. Nonetheless, if the foregoing is plausible, then Callicott's attempt to link the land ethic to this tradition in Western moral philosophy needs significant reformulation. The moral theory of the land ethic seems to represent a significant break with the past and therefore cannot easily invoke the moral sentiments in its defense. This does not make the land ethic wrong or right. It does imply that transforming Western moral, political, and social orientations to encompass greater respect for the integrity of the land requires emphasizing different elements of the Western tradition than those Callicott has highlighted.

Thinking that merely drawing people's attention to the biotic community's interdependencies would be enough to arouse moral concern would be comforting, but Smith's moral theory does not point in this direction. Thinking that the seminal philosopher of commercial society would concur in the land ethic's central ethical principle would be comforting, but close attention to Smith's views, even if one attempts to adjust for the passage of time, indicates that Smith would not consider the land ethic a direct extension of his own science of morals. Nonetheless, he could well agree that moral revulsion attends contemplation of willful and heedless destruction of beauty. If so, then pointing to evolving standards of natural beauty to find in the theory of moral sentiments warrant for environmental ethics is quite sensible. This remains a somewhat indirect justification in that ecosystems or ecological wholes are not seen as having intrinsic value. Rather, their value arises from an emerging appreciation of natural complexity and order that leads us to judge nature beautiful. Still, this is no small gain, in light of arguments that Western philosophy contains almost no basis for environmental ethics.

The project of environmental education remains daunting, but the hopeful message is that although Smith might have seen rude forests and barren oceans, he also recognized that such judgments vary culturally and historically. Adam Smith would not have been surprised that social standards had changed, and so his communitarianism readily accommodates emerging aesthetic and moral judgments. Although environmental educators cannot claim that respected liberal philosophers such as Smith would endorse the land ethic on immediate moral grounds of the kind Callicott offers, they can emphasize his probable revulsion at the frightful ugliness of a society that would destroy nature for the sake of transient comfort and amusement. Such a call to action lacks the normative force Callicott finds in Hume's moral theory, but it may have more political force if indeed it does call on those sentiments given us by the great Director of nature.

Works Cited

Adams, J. (1996). Cost-benefit analysis: The problem, not the solution. *The Ecologist, 26*(1), 2–4.

Attfield, R. (1983). *The ethics of environmental concern.* New York: Columbia University Press.

Callicott, J. B. (1989). *In defense of the land ethic: Essays in environmental philosophy.* Albany NY: SUNY Press.

Callicott, J. B. (1992). Can a theory of moral sentiments support a genuinely normative environmental ethic? *Inquiry, 35*, 183–198.

Clark, H. C. (1992, Spring). Conversation and moderate virtue in Adam Smith's "Theory of Moral Sentiments, *Review of Politics*, 185–210.

Farr, J. (1988). Political science and the enlightenment of enthusiasm." *American Political Science Review, 82* (1), 51–69.

Foreman, D. (1991). *Confessions of an eco-warrior.* New York: Harmony Books.

Green, J. M. (1995). Retrieving the human place in nature. *Environmental Ethics, 17*, 381–396.

Hargrove, E. C. (1989). *Foundations of environmental ethics.* Englewood Cliffs, NJ: Prentice-Hall.

Harpham, E. J. (1984). Liberalism, civic humanism, and the case of Adam Smith. *American Political Science Review, 78*, 764–774.

Holmes, S. (1995). *Passions and constraint: On the theory of liberal democracy.* Chicago: University of Chicago Press.

Hont, I., & Ignatieff, M. (1983). *Wealth and virtue: The shaping of political economy in the Scottish Enlightenment.* Cambridge, England: Cambridge University Press.

Hope, V. M. (1989). *Virtue by consensus: The moral philosophy of Hutcheson, Hume, and Adam Smith.* Oxford: Clarendon Press.

Hume, D. (1957). *An inquiry concerning the principles of morals.* New York: Bobbs-Merrill.

Leopold, Aldo. 1987. *A Sand County Almanac and Sketches Here and There.* Oxford: Oxford University Press. (Original work published 1949.)

McNally, D. (1993). *Against the market: Political economy, market socialism and the Marxist critique.* London: Verso.

Meyerson, A. (1989, Fall). Adam Smith's welfare state. *Policy Review, 50,* 66.

Minowitz, P. (1993). *Profits, priests, and princes: Adam Smith's emancipation of economics from politics and religions.* Stanford, CA: Stanford University Press.

Muller, J. Z. (1993). *Adam Smith in his time and ours: Designing the decent society.* New York: Free Press.

Pack, S. J. (1991). *Capitalism as a moral system.* Brookfield, VT: Edward Elgar.

Partridge, E. (1996). Ecology and nonmoral sentiments. *Environmental Ethics, 18,* 149–163.

Passmore, J. (1974). *Man's responsibility for nature.* London: Duckworth.

Raphael, D. D., & Macfie, A. L. (Eds.) (1976). Introduction. In A. Smith, *The theory of moral sentiments.* Oxford: Clarendon Press.

Smith, A. (1937). *An inquiry into the nature and causes of the "Wealth of Nations,"* E. Carman, (Ed.). New York: Random House/Modern Library. (Original work published 1776.)

Smith, A. (1976). *The theory of moral sentiments,* D. D. Raphael & A. L. MacFie (Eds.). Oxford: Clarendon Press. (Original work published 1790.)

Taliaferro, C. (1988, Fall). The environmental ethics of an ideal observer. *Environmental Ethics, 10,* 233–250.

White, L., Jr. (1967). The historical roots of our ecologic crisis. *Science, 155,* 1203–1207.

Wilson, J. Q. (1989). Adam Smith on business ethics. *California Management Review, 32* (1), 59–72.

Wilson, J. Q. (1993a). The moral sense: Presidential address, American Political Science Association, 1992. *American Political Science Review, 87* (1), 1–11.

Wilson, J. Q. (1993b). *The moral sense.* New York: Free Press.

Winch, D. (1978). *Adam Smith's politics: An essay in historiographic revision.* Cambridge, England: Cambridge University Press.

Zimmerman, M. E., Callicott, J. B., Sessions, G., Warren, K., & Clark, J. (Eds.) (1993). *Environmental philosophy: From animal rights to radical ecology.* Englewood Cliffs, NJ: Prentice Hall.

3

59 to 63

Ecological Science, Philosophy, and Ecological Ethics

ROBERT P. MCINTOSH

E. C. Lindeman (1940) rather boldly titled an article, "Ecology as an Instrument for the Integration of Science and Philosophy." Lindeman wrote: "The ecologist stands in a most advantageous position. He has already acquired the habit of dealing with wholes as well as fractions. To this extent he is a philosopher." Lindeman urged the ecologist to include philosophy "as an integral part of his design," but apparently did not encourage reciprocal interest among philosophers because two decades passed before philosophers gave serious attention to ecology. A major stimulus to this interest among philosophers (ca. 1969) was J. Baird Callicott's (1989a) "hunch" that ecology was a treasure trove of philosophically revolutionary ideas. Over nearly three decades, he has pursued this hunch in elaborating Aldo Leopold's land ethic toward a monistic conception of values predicated on ecological insights. Callicott's hunch came, fortuitously, with the widespread recognition of the environmental crisis in the 1970s. Ecology became an "in" word and was widely adapted outside of its normal scientific sphere, often leaving its content behind (Wali, 1995).

Unlike most nonecologists who seized on ecological verbiage connected with environmental concerns, Callicott has given serious attention to the science. In the shadow of the naturalistic fallacy, he asserted: "The philosophical context of the land ethic and its conceptual foundation is clearly the body of empirical experience and theory which is summed up in the term *ecology*" (Callicott, 1989b). Ecologists had, since its early years, urged its significance for human concerns (Adams, 1935). Some even asserted its relevance to philosophy. Colwell (1970) wrote: "Ecology provides a model to philosophy and to the other human sciences of a new way of viewing the interrelationships between the phenomena of nature. . . . the answer to the value question then, from an ecological point of view is this: human values are found in objectively determined ecological relations within nature." Such assertions by ecologists, naturally, carried little weight in the philosophical community. It was left largely to Professor Callicott to bear the onus of making such an assertion and trying to argue it axiologically.

Professor Callicott initially predicated his philosophical position on the following themes in Aldo Leopold's *Sand County Almanac* (1949):

1. All ethics so far evolved rest upon a single premise: that the individual is a member of a community of interdependent parts.
2. A thing is right when it tends to preserve the integrity, stability and beauty of the biotic community. It is wrong when it tends otherwise.

I intend to examine community, integrity, stability, and some related concepts as they appear in the empirical and theoretical foundations of ecology that Professor Callicott asserted are the basis of a philosophy derived from Leopold's land ethic. Beauty I leave to the eye of the beholder.

Communities, transcending the human community, became explicit in the nineteenth century in the work of Alexander Von Humboldt who recognized vegetation communities based on life form (e.g., trees, shrubs, or herbs). Early nineteenth-century marine biologists, such as Edward Forbes and Victor Hensen, recognized communities of marine animals by identifying species in samples collected by dredge or trawl. The first technical term for a marine community, the oyster bed, *biocoenosis*, was coined by K. Mobius in 1877. Biocoenosis was essentially equivalent to Leopold's biotic community in being restricted to the living components. It was later expanded to *geobiocoenose* in an early effort to join the physical (abiotic) and biotic components of a site (McIntosh, 1985). Community concept permeated protoecology in the nineteenth century and early ecologists designated the study of communities "synecology" as one of two branches of the young science of ecology. Community terminology proliferated, including formation, association, alliance, and society among a host of terms applied to communities of diverse organisms. In 1935 Frederic Clements and Victor Shelford advanced the term *biome* (coined earlier by Clements) to include both plant and animals of a community of a large area. The climax, but by no means the terminus, of community terminology was Arthur Tansley's (1935) *ecosystem*, which incorporated both biotic community and physical attributes of a site and which later was used by some ecologists to displace the biotic community by the physical or abiotic attributes of an area, especially energy and/or nutrients to characterize the system.

Community concept in ecology, in any of its various guises, up to the publication of Leopold's *Sand County Almanac* (1949), was almost entirely predicated upon the metaphor of community as organism or even supra- or superorganism. Allee, Emerson, Park, Park, and Schmidt (1949) stated unequivocally: "The interspecific system has also evolved the characteristics of the organism and may thus be called an ecological supraorganism." This comfortable metaphor allowed the community to be viewed as an integrated, essentially deterministic, aggregation of species that was either stable or developing toward a stable, self-perpetuating entity controlled by climate in the most familiar theory of pre-1950s ecology advanced by Clements (Tobey, 1981). This organismic metaphor largely informed Leopold's view of community (Callicott, 1996) and its organismic intimations fitted comfortably into the natural history tradition familiar to philosophers, historians, and environmentalists (McIntosh, 1998).

The paradigm of organismic theory of community was shattered in the 1950s by what Michael Barbour (1995) described as "something profoundly important among American ecologists." What happened was the resurrection of H. A. Gleason's "individualistic concept" of community, which had been published in 1926 and was roundly attacked and discredited by proponents of the organismic concept. It was published again in 1939 after which it was essentially ignored until 1947, but it was substantiated by an extensive body of research in the 1950s (McIntosh, 1958, 1967, 1975). It is this "fragmentation" of organismic community theory that Callicott (1996) described as "deconstructive ecology" and ecologists commonly describe as a paradigm change (Fiedler, White, & Leidy, 1977; Pickett, Ostfeld, Shachak, & Likens, 1997). By either term the changes amounted to a substantial revision of the empirical basis and theoretical framework of community theory in ecology. Gleason's community was predicated on the premise that species each had distinct physiological and ecological capacities and behaved individualistically in the context of physical environment and the presence of other species. Coupled with the probabilistic distribution of seeds, or other means of dissemination, Gleason held that resulting communities also were individualistic, changing continuously in time and space. This is not to say that individual communities on any site do not exist or that the component species do not interact in complex ways, but it contradicted the traditional view of communities as predictable, integrated, developing, or even evolving, entities. In fact, it argued that communities are not natural, integrated, stable entities and changed the prevailing concept of community from relatively discrete, homogeneous entities to patterns of species heterogeneously distributed on gradients or continua of change. When Callicott described the "now problematic" community concept it was the traditional organismic community that was problematic not the fact of interacting groups of species in a given habitat.

"Deconstructive ecology" was similarly evident in the 1980s following the widespread acceptance in the 1960s and 1970s of a theory of animal community, heavily influenced by mathematical models, commonly termed *niche theory*, and predicated on communities organized by competition among species. This produced hope for a general theory of community as organized entities even constructed according to "assembly rules" (Diamond, 1975). Richardson (1980) saw this as the reemergence of the organismic community concept among animal ecologists and suggested that animals were more prone than plants to form coevolved organismic communities. Moore (1983) expressed confidence in the superorganismic concept of community and described the "revival of the organismal heresy." An ecologist and philosopher joined forces to illustrate the "resilience" of the superorganism writing: "Imposing consistency clearly shows that groups and communities can be organisms in the same sense that individuals are. Furthermore, superorganisms are more than just a theoretical possibility and actually exist in nature" (Wilson & Sober, 1989).

In the face of such confident assertions, animal ecologists "deconstructed" the theory of communities organized by competition and, to a large degree, accepted Gleason's individualistic concept, although a rear-guard action continues to the present (Simberloff, 1980). Animal ecologists still present a spectrum of positions with respect to the individualistic concept of community (McIntosh, 1995) but widespread consensus, if not unanimity, exists that the old balance of nature paradigm is passé, as evidenced in a recent volume on the *Ecological Basis of Conservation* (Pickett et al., 1997). Several authors noted (Talbott, 1997) the new paradigm in the book and the contrasting descriptors for the old paradigm were stable, balance, equilibrium, organism or superorganism, homogeneous, closed, and deterministic; the descriptors for the new paradigm were disturbance, heterogeneous, patchy, flux, stochastic, probabilistic, and contingent. Gleason's concept had a substantial stochastic component. All species had their proper place but they were frequently not in it, contrary to classical balance of nature organismic community lore. However, the resilience of the organismic concept in the ideals of a few is evident in a description of communities as functionally organized units selected by a community-level process of natural selection (Wilson, 1997). The ecological literature, as Professor Callicott (1996) recognizes, is not the firmest substrate on which to build a philosophy. He writes: "But the individualistic dynamic paradigm in deconstructive community ecology seems to undercut two out of three of the land ethic's cardinal values." These are integrity and stability leaving beauty unchallenged. Nevertheless, integrity and stability are all too commonly encountered in the ecological or quasi-ecological literature.

Many criticized the word *community*, always a somewhat loosely used term, and several alternatives were urged as substitutes or supplements. The favorite of these was *ecosystem,* which when coined by Tansley in 1935, incorporated the nonliving (abiotic) and living (biotic) into a single system drawing on physics for the concept of system. Some viewed the rise of the ecosystem concept in ecology as a developing dichotomy in ecology. Traditional ecology concentrated on populations or communities of organisms, the "biosociological" (Hutchinson & Wollack, 1940). Ecosystem ecology concentrated on the transference of energy or matter in biomass, or biogeochemistry. The biological philosophy of Ludwig von Bertalanffy (1952) was transformed as "systems ecology" in the 1960s wave of computer models and differential equations applied to ecosystems. But when the mathematics was criticized, Lazlo (1980) defended it as not a mathematical method, but as a philosophy. Some ecologists agreed. Patten (1975) wrote: "The systems approach is a philosophy and a theory that comes with a formalism and a set of tools. . . . it is not the latter that are important to ecology in the long run, but rather the philosophy and the formalism." Systems ecology appealed to holism, emergence, and energy as numeraire. Some saw the ecosystem as a stable entity that persisted and evolved even though species might come and go. The hopes for a unifying theory of theoretical population ecology or of systems ecology were largely unfulfilled although both are present as residues in ecological studies. Both were, in Callicott's term, "deconstructed" and challenged by proponents of yet another new ecology, which reemphasized organisms and their interactions and urged pluralism in ecology (McIntosh, 1987). Appeals to philosophy by ecologists were increasingly evident and some ecologists and philosophers have ventured philosophical asides about ecology with indeterminate results (Saarinen, 1982).

A problem for philosophers and ecologists is that not only are the concepts of community and ecosystem being deconstructed, new terms are being advanced presumably to clarify the ecological lexicon. Fauth et al. (1996) and Fauth (1997) hopefully, and probably ineffectually, tried to simplify and make operational the "jargon" of community and ecosystem ecology. A recent favorite is *assemblage* variously defined, or not defined, but ostensibly advanced to avoid the semantic difficulties of community concept. Assemblage is sometimes used to avoid the taint of the organismic community or to denote a taxonomic subset of a community, familiarly enough described as a bird, reptile, insect, or other taxon-based community. Assemblage leads naturally enough to assembly, presumably to describe the way an assemblage is assembled. This suggests the mechanical metaphor of assembly line bringing together appropriate parts in proper order to produce a finished product. Such imputed regularity calls up assembly rules, or worse, metarules. These smack of a rule-governed community with the disturbing proviso that the rules may change depending on the sequence of events (McIntosh, 1995). This is daunting enough for ecologists and no doubt more so for philosophers, although it recalls the hoped-for orderliness of the organismic tradition. If ecologists can demonstrate clear rules by which an assemblage is assembled, it may be useful for philosophers but current prospects are not promising. For all its historical faults *community* is a more useful term and *assemblage* adds little clarity to the ecological literature.

Callicott (1996) faces up to "poststructuralist" ecological theory noting that no "poststructuralist ecological theory . . . asserts that organisms are entirely independent of one another." In fact, no prestructuralist concept of the ecological community, except mathematical models thereof, ever asserted that communities were random groupings without interactions among the species. Callicott cites his onetime home community, Stevens Point, Wisconsin, and uses the diverse and changing composition of that human community to provide an analogy with the biotic community and argues that the changing, loosely structured, poorly bounded community of Stevens Point warrants ethical obligations and, similarly, even the deconstructed biotic community is sufficiently defined to "engender analogous environmental duties and obligations." The merit of the philosophical argument I cannot consider, but Professor Callicott's analysis of deconstructed, poststructuralist ecology is well informed and negates the criticism sometimes leveled at ecology that if it cannot be neatly bounded and classified it cannot be science (Sagoff, 1997). Professor Callicott's (1996) concern that deconstructive ecology undercuts two of Leopold's three criteria for valuing the biotic community, integrity and stability, caused him to abandon them and offer "a stab at" an amended, presumably hierarchical, version of a "moral maxim for the land ethic": "A thing is right when it tends to disturb the biotic community only at a normal spatial and temporal scale. It is wrong when it tends otherwise." It remains to be seen if hierarchy theory can provide substance for such a definition.

Callicott, true to his original premise, continues to seek the basis for a land ethic in the "body of empirical experience and theory which is summed up in the term ecology" (Callicott, 1989a). Ecologists have greatly expanded the body of empirical experience, have thoroughly revised their understanding of the organization and operations of communities and ecosystems, but they have not arrived at a consensus about a body of theory to elucidate their new understanding of the way nature works, which frustrates Callicott's efforts to find a rock on which to construct a land ethic. He is not aided by the diverse individuals and groups that have, in the era of the environmental crisis and the expanded versions of conservation biology and public policy, adopted ecological terminology and concepts (Pulliam, 1997). The problems Callicott notes about the definition and boundaries of community and ecosystem are extended to those of the definition of and boundaries of ecological science. Ecology is intrinsically a diffuse science with diverse roots in nineteenth-century biology. It developed relatively unsung until it was thrust into the political and public consciousness in the 1970s confused with environmentalism (McIntosh, 1985). It now suffers the hazard of being transformed from a science to various sociopolitical positions, sometimes termed *ecologism* (McIntosh, 1998) and, perhaps, a greater hazard of being a guide to ethics and philosophy.

Contrary to Callicott's abandoning them, *integrity*, and its sometimes synonym *health*, both terms commonly modified by biotic, ecological, or ecosystem, remain in widespread use in the penumbra of ecological science. They were initiated and have become imbedded in the literatures of conservation, management, law, and aspects of theoretical ecology with connotations well beyond Leopold's purely metaphorical and undefined use of integrity. The use of some characteristic of the biotic community to represent an attribute or state of the environment, or vice versa, goes back to the early years of ecology under the name of indicator. One of the pioneers of ecology, C. Raunkiaer, noted the problems of using a complex of values to measure the environment—the same values of one variable may, in different combinations with others, have different ecological effects. This remains a problem of attempts to develop indices of ecological *integrity* or *health* as they are variously defined. *Indicator* was commonly used in ecology to denote a species that characterized a particular community or environmental attribute. It became familiar in the 1970s as indicators of pollution or environmental degradation with the analogy to the miner's canary to foretell potential disaster. As such, it was purportedly an indicator of a specific state of the community or environment or a measure of risk. Various descriptors of communities or ecosystems were used in the 1970s, the era of recognition of the environmental crisis (e.g. *degraded*, *damaged*, or *stressed*). *Stress ecology* described the impacts of disturbance such as pesticides or radiation that impaired the organization or function of a community or ecosystem in some measurable way (Odum, 1985). *Integrity* remains as the ecosystem or community descriptor of choice for some and is widely advocated, although Callicott no longer gives it credence as a basis for ecological ethics. However, Westra (1994) had offered the

"Principle of Integrity" as an "Environmental Proposal for Ethics" which may leave room for debate about the merits of integrity. Hence, exploring the current use of *ecosystem* or *ecological integrity* seems worthwhile.

Schneider (1992) wrote that integrity is not only a concept with meaning to the scientific community but has been incorporated into U.S. national water quality legislation and into an international agreement between Canada and the United States. In fact, the term *integrity* was absent from most scientific ecological literature until the 1960s, but has increased since then. Its first appearance in Leopold's *Sand County Almanac* (1949) was essentially rhetorical. It appeared in the U.S. Federal Water Pollution Control Act Amendments in 1972. It was used by the aquatic ecologists David Frey (1975) and John Cairns (1975) in a publication of the U.S. Environmental Protection Agency (EPA) entitled *The Integrity of Water* (1975) and was widely used in EPA publications thereafter. According to Regier (1992), "The commitment to integrity was science-forcing and technique-forcing." Regier described *ecosystem integrity* as a "slogan" that "encompasses empirical/scientific and ethical/normative emphases." Regier saw scientific paradigms as changing with the political winds, a view that may not accord with that of most ecologists. He wrote: "If the political changes run deeply, then a once dominant scientific paradigm may be rejected as absolute or reactionary and a new politically relevant paradigm may be legitimated." In many ways, integrity seems to turn more on political change or relevance than on scientific understanding of ecology.

Integrity has been variously defined. Cairns (1975) defined *integrity* as "the maintenance of the community structure and function characteristic of a particular locale or deemed satisfactory to society," a substantially anthropocentric meaning. By contrast, James Karr (1981, 1996) defined *integrity* in biocentric terms as the condition at sites with little or no influence from human actions. Karr (1996) distinguished ecological integrity from ecological health but associated them with two other recent hybrids, *ecological restoration* and *ecological rehabilitation*, respectively. Restoration restores a site to a state before human influence, commonly termed pristine or virgin. Rehabilitation describes a change of a site disturbed by human actions to one that is not offensively degraded or destructive of areas beyond its boundaries. The hybrid discipline ecological engineering undertakes a wide range of activities to improve ecological conditions without purporting to restore a "natural" or pristine state. Integrity has been seized on by lawmakers, diverse conservation agencies, land managers, economists, and some theoretical ecologists as something undeniably good but it is not readily quantified. If Karr's definition is to be used, it suffers from the difficulty of determining a putative prehuman or "reference" state. In the New World at least, *prehuman* is commonly and erroneously taken to mean pre-European; in the Old World, prehuman states are commonly impossible to determine. Karr (1981, 1996) proposed development of Indices of Biotic Integrity, which are multiple measurements of ecological conditions that are aggregated by summation, multiplication, or use of extreme minimum or maximum values into a unit index to be compared with the index of a reference site undisturbed by human actions. Messer (1992) suggests an analogy of such indices with the Dow Jones Industrial Average, dear to the hearts of investors, but the Dow has the advantage of summing the values of stock prices all based on the single value of the dollar as a well-established common denominator, an attribute not shared by ecological measurements, and just what it responds to is described by diverse commentators, such as Louis Rukeyser, in notoriously elusive phraseology.

Noss (1995), however, argued that integrity can be a rigorous measure by appropriate selection of ecological indicators and authenticated this hope by noting that Leopold's undefined use of integrity was quoted "with nothing short of reverence" by biologists and conservationists and, he might have added, philosophers. Reverence is a poor substitute for explicit definition in science. Noss noted the introduction of Gleason's individualistic concept into the discourse that engendered the deconstructive ecology Callicott examined and wondered if aggregations of species are transient whether they can possess integrity. He finds hope for making integrity a useful concept even in the context of ecological ideas of continuously varying nonequilibrium systems. Noss proposed a three-level hierarchical framework for integrity at the Landscape-Regional Level, the Community-Ecosystem Level, and the Species Level and provided lists of measures appropriate to each level.

Integrity is commonly used as a synonym for health of the ecosystem. Karr (1996), however, limits health to the state of systems modified by human actions. Regier (1992) adopted systems health as "relevant" to ecosystem integrity by a series of characteristics including being "productive of goods and opportunities valued by humans." *Goods* is somewhat ambiguous because it could have a material or a nonmaterial meaning. Regier (1992) links integrity, and presumably health, of ecosystems with ecology in the works of the distinguished Odum family of ecologists, notably Eugene and Howard, who are major advocates of systems ecology. The link is manifest in Regier's own adoption of the philosophy of Ludwig von Bertalanffy to characterize the nature of "healthy" organic systems. Bertalanffy's philosophy was heavily cited in the 1960s and 1970s with the meteoric rise of systems ecology (McIntosh, 1985), but has become meshed with other sources of systems ideas and remains as a tenuous link of ecology to philosophy.

Much more explicit definitions and analyses of ecosystem health appear in the writings of David Rapport and Robert Costanza who share an interest in the overlap of the eco twins, ecology and economics. Rapport (1989) advocates the use of metaphor in science and finds ample parallels between human medicine and ecosystem health, which he treats as synonymous with integrity. These parallels are evident in his verbiage; ecosystems may be "sick," have "pathology" but, fortunately, "ecosystem practitioners" and "clinical ecology" exist to attend to such problems. Rapport identifies several "symptoms" in either decreased or increased primary productivity, less efficient nutrient transfer, either decrease or increase in species diversity, instability of populations, increased disease or pest prevalence, loss of larger sized forms of life or occurrence of many contaminants. He advocates methodologies adopted from human medicine to treat these symptoms and to restore, rehabilitate, or even "enhance" ecosystems. Like his sometime collaborator Regier, Rapport gives primacy to systems concepts that were developed early on by Bertalanffy. Although he admits that ecosystem integrity has not been made operational, he finds that indicators of its converse "disintegrity" have achieved an emerging consensus, although identifying such a consensus among ecologists is difficult. Rapport anticipates the emerging practice of "ecosystem medicine" and "indicators of ecosystem integrity that will provide direct measures of the healthiness of natural systems," although the "critical feedback mechanisms" to this end are "much looser at this supraorganismic level."

Robert Costanza (1992), who describes ecosystem health as "a bottom line normative concept," shares the hope of Regier and Rapport for an emerging practice of ecosystem medicine and maintenance of ecosystem health. Costanza anticipates a progression of measurements of indicators of the system, through measures of "end points," to measures of overall ecosystem health. Like Rapport, he argues that ideas from human health are "applicable to evaluating the health of any complex system including ecosystems and economic systems." In Costanza's usage, ecosystem health or integrity "represents a desired end point of environmental management." He reviews several definitions of *ecosystem health* and finds them all wanting in some respect but finds hope in network analysis—the mathematical analysis of interconnections of complex systems. According to Costanza: "Network analysis holds the promise of allowing an integrated, quantitative hierarchical treatment of all complex systems, including ecosystems and combined ecological economic systems." Such sweeping promises have been heard before in systems ecology but did not eventuate.

Relatively unnoticed among the extended publications on ecosystem health is a critique of concepts and indices of ecosystem health by G.W. Suter

(1993) who purports "to examine this bandwagon before it gains momentum." In fact, the momentum may have turned the bandwagon into a juggernaut because it has already produced several symposium volumes and at least one new journal. Suter questions the worth of the metaphor *ecosystem health* to draw on the power of medicine and the sciences of human health. He asserts that the metaphor implies that ecosystems are integrated, homeostatic entities akin to the largely discredited concept of the ecological system as a superorganism, which Clements developed in ecology and which was widely accepted before 1950 and inherent in the models of systems analysis in the 1960s and 1970s. Suter argues that the multiple components of indices are sensitive to the several combining functions (i.e., summing, multiplying) used to calculate them and that a given index value may be produced by different combinations of the components, which may mask ("eclipse") the reason for the value as noted by Raunkiaer many years previously. He says that the value of a given index cannot be related to any biological response. Moreover, indices of health have no relation to scientific theory and are logically tautologies, according to Suter. In Suter's assessment, the metaphor of health creates an illusion that some properties of ecosystems are equivalent to human health and the relative certitude of scientific medicine rather than witch doctors.

Callicott (1995) addressed the concept of ecosystem health as problematic but asserted that it is an important conservation concept and is intelligible to lay persons. Such health metaphors, he wrote, "suggest that certain states of ecosystems are both objective and normative, actual and valuable." He begins at the beginning with the ecosystem concept noting current questions about its ontological status. It is true that some textbooks, as he asserts, have backed away from the systems bandwagon of the 1960s and 1970s but the substance of systems ideas is very evident in current ecology. It is not clear, as Callicott suggests, that the prognosis is good that the "ontological status" of ecosystems can be resolved, that they may be "easily defined in terms of turnover time" according to hierarchy theory. Going from the frying pan of 1970s systems theory into the fire of 1990s hierarchy theory may produce more than "a considerable degree of abstractness." In Callicott's analysis, ecosystem health is a combination of "objective condition of organisms" and "social determination," which allows experts "to set the parameters of ecosystem health" and "the people decide how to use and manage ecosystems."

Callicott turns to the terminology of conservation, which overlaps that of ecology and shares with it some ambiguity. The verbiage of community and ecosystem health and integrity is not derived from or common in most ecology but, as Callicott notes, is confounded in the literature of conservation and management. Some use the terms synonymously, others as distinct. In

Callicott's interpretation, *integrity* refers to community ecology and *health* relates to ecosystem ecology. If it were true, as he asserts, that "ecosystem ecology and community ecology are different sciences" then the distinction might be useful. That most ecologists will accept so categorical a distinction is unlikely, although community ecology and ecosystem ecology, particularly in the form known as systems ecology, clearly have different emphases. Rather than divide ecology to accommodate what Noss (1995) described as "buzz-words," which are not products of or much used in ecology, I would abandon the terms, although such terminological bandwagons are difficult to stop. Sagoff (1997) wrote: "The idea that there are such qualities as the 'health' or 'integrity' of ecosystems and that species are their indicators seems less a refutable proposition of empirical science than a first principle of a certain ecological faith." Whatever the merits of the buzzwords *integrity* and *health* in the literature of public policy and management, they do not derive from or enlighten most scientific ecology. Contrary to the assertion of Regier (1992), that these or other concepts should change with the political winds is not desirable. Ecological science is an essential basis for conservation, management, and public policy related to these activities, and certainly ecologists should be attentive to these needs as they long have been. It does not follow that empirical ecology or theoretical ecology should change with the winds of public policy, although the options may be influenced by policy on funding.

The concepts of *ecosystem health* and *integrity* take on expanded horizons in the recent work of long-time systems theorists Robert Ulanowicz (1997) and Sven Jorgensen (1997). Ulanowicz seizes on Dan Simberloff's ill-documented characterization of ecology as a "sick" science and offers a cure in the guise of "ascendency," which combines a measure of system activity, or "through put," measured as the sum of nonzero energy flows with the "average mutual information" multiplied by the all-too-familiar Shannon-Weaver Information Index. Absent overwhelming external disturbance, an ecosystem "has a propensity to increase in ascendency," according to Ulanowicz. He offers a new version of succession in four stages (i.e., growth, development, maturity, senescence), which except for the omission of immigration is somewhat redolent of Clements's turn-of-the-(twentieth)-century description of succession. Ulanowicz further identifies ascendancy with the "holist" camp he traces from Clements through Lindeman and the Odums and contrasts it with the "probabilists" whose pedigree he traces to Gleason via Simberloff and Williams. Ulanowicz sees (almost) the holists "unfurling the banner of ascendancy" and the probabilists "showing the colors of overhead." In Ulanowicz's unfamiliar lexicon, ascendancy is the ordered functioning of an ecosystem.

Ulanowicz distinguishes ecosystem health, which he incorrectly says has largely occupied U.S. ecologists, from integrity, which he says is the "corresponding watchword in Canada." Whatever the validity of that distinction, health, according to Ulanowicz, is how an ecosystem is functioning at the present time. Integrity encompasses both the past and future of an ecosystem including "its telos" or the direction in which a system is heading. This too is somewhat redolent of Clements's climax and draws in philosophical overtones. According to Ulanowicz "the ecological arena is populated by discrete entities each with its own direction." Ulanowicz claims that the "ascendency hypothesis" had empirical origins in Eugene Odum's (1969) list of "Trends to Be Expected in the Development of an Ecosystem," many of which did not turn out as expected. Nevertheless, according to Ulanowicz, the direction of an ecosystem ("its telos") is not only integral to its integrity, it "can also impart a legitimacy to ethical considerations of how society should interact with the system."

Jorgenson (1997), like Ulanowicz, addresses ecosystem health and integrity. He cites the phrases "take nature's pulse" and "clinical ecology" used by advocates of the concept of ecosystem health. He notes, briefly, some definitions of ecosystem health and integrity and offers as a measure of ecosystem health a "relative energy index" measuring "approximate distance from thermodynamic equilibrium." Jorgenson, like Ulanowicz, cites Odum's (1969) characteristics of developing ecosystems and, like Odum, sees the entire ecosystem as an evolving entity. He endorses Ulanowicz's ascendancy, which he oddly describes as a widely used measure of ecosystem development although it is little known. Jorgenson goes beyond Ulanowicz in propounding an "ecological law of thermodynamics" (ELT), which he describes as a "tentative fourth law of thermodynamics," which states:

> A system that receives a through flow of exergy (high quality energy) will have a propensity to move away from thermodynamic equilibrium, and, if more combinations and processes are offered to utilize the energy flow, the system has the propensity to select the organization that gives the system as much energy as possible.

The ELT, "a thermodynamic translation of Darwin's theory," is described as an evolving "supersystem" with adaptations to cope with abiotic changes in the environment. If, or how, it relates to the traditional superorganism of pre-1950s ecologists is not clear.

The second of Leopold's desiderata for the biotic community, *stability*, has a much longer and more familiar usage in ecology than integrity, but it too has been cursed with myriad meanings. Ecology absorbed the natural history concept of balance of nature and much of its early tradition was predicated on what one of its most eminent pioneers, S. A. Forbes (1880), described as "the ideal balance of nature" and a "tendency toward a just equilibrium." Stability and equilibrium became ideals of many ecologists, sometimes confounded with usages from the physical sciences commonly represented by a ball in a cup or a hill and valley, which made them very evident pictorially, but not so evident ecologically. The easy catchword of stability did not make clear of what or for how long. Stability of a community of short-lived organisms might entail several generations in days or weeks; of long-lived pine trees a single generation might be centuries. Nevertheless, stability of a large-scale community, such as Clements's "climax," was widely accepted and ecologists used stability to characterize a stage of a community perpetuating itself barring disturbance. The premier European limnologist, A. Thienemann, identified as a principle of ecology what came to be called the stability-diversity hypothesis, which held that stability and diversity (in that era meaning the number of species) were causally associated (Hynes, 1970). Species number came to be recognized as an important attribute of a community.

The linkage of stability and species number gave rise to a famous symposium entitled "Diversity and Stability in Ecological Systems" (Brookhaven Symposia in Biology, 1969) predicated on the assumption that "a major means for assuring the continuity of life appears to be the number of species per unit area." Diversity in ecology subsequently took on a more complicated meaning to include number of species and also the proportionate distribution of numbers of individuals among species. St. Thomas Aquinas had anticipated this as a virtue observing that it is better to have a few individuals each of many species than many individuals of one species (Glacken, 1967). A community of ten species with one individual each of nine species and ninety-one of the tenth is very different from a community with ten individuals of each species. Number of species came to be called *richness*, the combination of number of species and proportionate numbers of individuals came to be *diversity*, and the variation of numbers of individuals of species was *evenness*. The aforementioned symposium on diversity and stability was not notably productive because the key terms were not well defined or understood, and neither could be measured to everyone's satisfaction. Species number increased with area, the familiar species-area curve, and diversity gave rise to a plethora of indices, among them an early borrowing from information theory, the Shannon-Weaver Information Index (H'). Ghent (1991) asserted that the continued use of the Shannon-Weaver Index of Diversity (H') was due "more to a hankering after the imagined prestige of association with information theory than to any demonstrable virtue of H' as an index of biological diversity." Nevertheless, H' is still used although less confidently and with fewer decimal places.

Stability, with its halo of traditional balance of nature, was commonly linked with equilibrium in the theoretical ecology of the 1960 and 1970s, usually predicated on mathematical formulations assuming equilibrium conditions with minimal reference to the attributes of populations or communities. Stability, in good biological tradition, reproduced and gave rise to a complex of terms, which Pimm (1984) termed *confusion*, and which Grimm & Wissel (1997) then upgraded to *babel.* Holling (1973) linked "resilience and stability of ecological systems," defining *resilience* as the ability to return to an initial, or reference, state following a disturbance. However, even as stability terminology multiplied, the major reason for interest in it was demolished or, in Professor Callicott's term, deconstructed. The theory of a relationship between diversity and stability was reviewed, and no relationship was found (Goodman, 1975). Goodman was unsparing in his analysis describing models of the theory as "recreational mathematics" and writing that the theory was borne out "neither by experiment, by observation, nor by models." However, Goodman noted the power of the underlying metaphor and wrote, "It is the sort of thing that people like, and want, to believe." He predicted it will retain a "revered position in the popular environmental ethic, where it doubtless will do much good." In fact, the announced demise of stability-diversity theory was premature because it has been resurrected in recent years. A major problem remains the meaning and relation of stability and diversity in ecology although they are widely regarded as undeniably good in conservation and management.

The discussion was continued, and Pimm (1984), who slightly changed the verbiage to "complexity and stability," noted the existence of "several score permutations" of these definitions and offered to resolve at least some of the controversy. He added complexity to diversity of the early discourse, and offered four definitions of complexity and five of stability, stable itself having two variants, local and global. His general definition of *stable* was the return of a system to equilibrium following a perturbation, which assumes equilibrium as the initial state, an issue widely questioned in recent ecology. Pimm added four measures of stability: (1) *resilience*, the rate of return of a variable toward equilibrium; (2) *persistence*, the time the value of a variable lasts; (3) *resistance*, the degree of change of a variable following perturbation; (4) *variability*, the variance of a population measure over time. Fortunately, according to Pimm, "theoretical and even field results fall clearly into the various definitions of stability, although his discussion of these is weighted toward models rather than field results. Pimm complicated the usual criteria of complexity of a community or ecosystem—species number (richness), evenness, the variation of the abundance (number of individuals of species), or the combination of richness and evenness, commonly called *diversity*—by adding "connectance" and "interaction strength." *Connectance* he defined as the number of actual interactions among species divided by the number of possible interactions. *Interaction strength* is the average effect on the number (density) of individuals of one species on the growth rate of another species. He noted that most of the questions about the relation between complexity and stability have not been asked and that questions about connectance and interaction strength probably cannot be readily answered except in models.

Pimm's article added to the complexity of the diversity-stability debate, which was disconcerting to ecologists and probably more so to philosophers looking to ecology for guidance. Grimm (1996) reported this difficulty and commented "the imprecision of the term 'stability' makes it difficult to apply stability concepts"; Grimm and Wissel (1997) offered "An Inventory and Analyses of Terminology and a Guide for Avoiding Confusion." A guide was surely needed because they identified 163 definitions of 70 stability concepts. They provided a list of stability terms, analyzed the essential synonyms, and boiled down the 163 definitions to six overlapping those of Pimm (1984): (1) *constancy*, staying essentially unchanged; (2) *resilience*, returning to a reference state (or dynamic) after a disturbance; (3) *persistence*, lasting through time; (4) *resistance*, staying essentially unchanged in the presence of disturbance; (5) *elasticity*, speed of return to a reference state or dynamic after a disturbance; and (6) *domain of attraction*, the whole range of states from which the reference state (or dynamic) can be reached—essentially Pimm's global. Grimm and Wissel described these six attributes as "stability" properties and concluded, "Stability is not a stability property!" Their "confusion avoidance strategy" is to abandon the term *stability* and to answer three questions each time stability appears: (1) Which of the six stability properties is addressed? (2) What ecological situation is referred to? (3) Is the statement anchored in the researched situation or are there unacceptable generalizations? The crux of the problem Grimm and Wissel (1996) note "is that the term 'stability' has an enormous attraction for ecologists, . . . and politicians, managers and naturalists really love it." The lack of general ecological theory of community or ecosystem development and organization frustrates ecologists, philosophers, and others who hope to build on its theoretical and empirical work. The importation of ideas of equilibrium, information, energy, and system bringing with them a freight of intimations from other disciplines adds to the lamented complexity of ecology and the difficulty of philosophers in developing an ethic predicated on ecology. Ecologists have not helped by confounding definitions and terminology of stability.

The resilience of stability and a number of associated terms is evident in recent assertions that one or another attribute of ecological communities or ecosystems is, in fact, related to stability. Tilman and Downing (1994) describe long-term studies of experimental plots of grassland species to examine what they described as biodiversity and stability. Biodiversity has emerged as a favored term in ecology conservation, management, and legislation and is more inclusive than simple diversity in ecology as it includes genetic and any other differences. In fact, Tilman and Downing reported not on biodiversity or diversity, in its usual ecological meaning, but on the simpler attribute of number of species (richness). They, nevertheless, reported that their findings supported the supposedly defunct diversity-stability hypothesis. Tilman (1996) similarly confounded biodiversity, diversity, and richness in relating community biomass to number of species (richness) although he noted a "weak association" with stability of abundances (number of individuals), the other component of diversity. Karieva (1996) hailed the Tilman experiments by connecting diversity to "sustainability on the prairie," a far cry from Tilman's 3 x 3 m plots even if there are 147 of them. *Sustainability* introduces another buzzword favored by ecologists, managers, and politicians. Sustainability has an appeal like that of stability and in many minds it probably has a connotation not unlike stability.

Tilman used production of plant material (biomass) as the measure of stability, and Johnson, Vogt, Clark, Schmitz, and Vogt (1996) addressed the triumvirate of biodiversity, productivity and stability in examining the diversity-stability hypothesis. Like Tilman, they plotted ecosystem processes, such as production, against simple number of species that is neither diversity or biodiversity, although they recognize the distinction in a table showing species number as distinguished from the familiar diversity index (H'). Johnson et al. catalogued four hypotheses about species diversity whatever it may mean:

1. The *traditional diversity-stability hypothesis*, which implies that loss of any species would decrease the stability (in its traditional hopeful meaning).
2. The *rivet hypothesis*, which allows that a few species, like the rivets on the Titanic, could disappear without problems but the system would be subject to a disastrous failure at some point. The implication of this idea is deplored by some ecologists as implying that loss of some species is insignificant (i.e., they are expendable).
3. The *redundancy hypothesis*, which implies that species as members of certain functional groups can effectively replace other species of the group by expanding their role in the ecosystem. This too allows some species to be expendable.

4. The *idiosyncratic hypothesis* of a null or indeterminate relation between species diversity and ecosystem function.

According to Johnson et al. this array of hypotheses illustrates "a growing appreciation that species diversity can influence the stability and productivity of ecosystems in a variety of ways." An interesting if awkward question, according to Johnson et al., is whether species are the best way to distinguish functional groups in ecosystems. They anticipate that advances in modern cellular biology may make species unreliable as a basis for distinguishing the effect on an ecosystem. This poses the disconcerting prospect of changing the basis of species number or species diversity incorporating relative abundance of individuals, or the ecological properties of species, to some indefinitely diverse mix of phenotypes probably not distinguishable in the field. Such a prospect has little likelihood of serving the purposes of either ecologists or philosophers.

The problem for philosophers in assessing the empirical and theoretical state of ecology, traditional or deconstructed, as a basis for an environmental ethic is that the things that ecology deals with may fulfill the suspicion of biologist J. B. S. Haldane (1928): "Now, my suspicion is that the universe is not only queerer than we can suppose. . . . I suspect that there are more things in heaven and earth than are dreamed of in any philosophy." The latter suspicion, borrowed from Shakespeare, led Haldane to give up on philosophy. That may be a more extreme reaction than that of most ecologists or any philosophers. However, even the hopeful leader of the new theoretical ecology of the 1970s, Robert May (1986), eventually commented "on the ineluctably contingent nature of such rules and patterns as are to be found governing the organization of communities." One of the consequences of the complexity and contingency of ecological phenomena has been increasing appeals to philosophy by ecologists. Oddly, at the same time philosophers have turned to ecology as a source of philosophical insight. Ecology has been criticized by various ecologists and philosophers for its inadequacies as a science. One philosopher suggested "a way that ecology could be of philosophical interest [is] as more than an example of a backward science" (Kiester, 1980). However, Callicott (1986) wrote of "The Metaphysical Implications of Ecology." Being elevated from backwardness to a science with metaphysical implications is almost too much for a science once accused of physics envy. However, ecologist Bernard Patten (1975) anticipated metaphysical hopes for ecology in "holoecology," stating: "The metaphysical themes of a totally unified nature that in biology have been on hold since Darwin drew the organism in sharp relief to its background matrix are now returning for serious development."

One of the difficulties of ecology, however, has been to frame it in the conventional rubrics of conventional philosophy of science (i.e., law, theory, reduction, deduction; McIntosh, 1980). An ecologist offered a plea for "philosophical tools" for ecology including theory reduction and linguistic analysis (Loehle, 1988). A philosopher, K. S. Shrader-Frechette, teamed up with an ecologist, E. McCoy (1990), to question Loehle's philosophical optimism. Another ecologist (Peters, 1991) provided *A Critique for Ecology*, which argued that ecology was in bad shape, failed on predictability but could be improved with some philosophical guidance. Shrader-Frechette and McCoy (1993) again joined forces to analyze *Methods in Ecology: Strategies for Conservation* and claimed that ecology lacked a theoretical framework and was condemned to a natural history approach. More optimistically, ecologists S. T. A. Pickett, J. Kolasa, and C. G. Jones (1994) offered *Ecological Understanding: The Nature of Theory and the Theory of Nature*. The authors describe this volume as a "system of ideas about the philosophy of science by practicing ecologists for practicing ecologists." Some ecologists saw ecology as offering help beyond practicing ecologists. Golley (1993) wrote: "Thus the ecosystem perspective can lead towards an ecological philosophy and from philosophy it can lead to an environmental value system, environmental law and a political agenda."

Despite the promise some ecologists and philosophers see for ecology as a guide to philosophy, some philosophers have found ecology wanting as a science and hence a dubious rock on which to build an ethic. Ecologists have not helped strengthen its case by their tendency for elusive terminology and failure to arrive at consensus about fundamental concepts, let alone theories. They were spared criticism as long as philosophers paid little attention to its "empirical experience and theory." The criticism is not unidirectional; some ecologists have expressed concern about philosophy of science. Pickett et al. (1994) reviewed limitations of the philosophy of science by which ecology is found wanting. The failure of classification that Sagoff (1997) notes and the limitations of ecological theory that Shrader-Frechette and McCoy (1993) recount are considered among its failings as a science. The merits of ecology as the basis of an environmental ethic are unclear if its status as science is questionable. Pickett et al. turned to biologist Ernest Mayr (1982) and philosopher Marjorie Grene (1987). Mayr had asserted that the traditional strictures of philosophy of science must be expanded to incorporate the inherent complexity, historical contingency, and multiple causality that characterize ecological phenomena. Grene called for a new philosophy of science relating to organisms in relation to their environment. "One may contrast this ecological, orientational view of perception, and a fortiori of all our knowledge of the world around us with the anti-biological, anti-ecological hypothetico-deductivism of the older philosophy of science." Perhaps, to paraphrase Callicott, what we need is a deconstruction of philosophy.

Works Cited

Adams, C. C. (1935). The relation of general ecology to human ecology. *Ecology, 16*, 316–335.

Allee, W. C., Emerson, A. E., Park, O., Park T., & Schmidt, K. P. (1949). *Principles of animal ecology.* Philadelphia: Saunders.

Barbour, M. (1995). Ecological fragmentation in the fifties. In W. Cronon (ed.) *Uncommon ground: Toward reinventing nature* (pp. 75–90). New York: Norton.

Bertalanffy, L. von. (1952). *Problems of life: An evaluation of modem biological thought.* New York: Wiley.

Brookhaven Symposia in Biology. (1969). *Diversity and stability in ecological systems*, No. 22. Upton, NY, Author.

Cairns, J., Jr. (1975). Quantification of biological integrity. In *The integrity of water* (pp. 171–188). Proceedings of a Symposium, March 10-12, 1975. Office of Water and Hazardous Materials, U.S. Environmental Protection Agency, Washington, D.C.

Callicott, J. B. (1986). The metaphysical implications of ecology. *Environmental Ethics, 8*, 301–316.

Callicott, J. B. (1989a). Environmental philosophy: Its origins and types. In J. B. Callicott (Ed.), *In defense of the land ethic.* (pp. 1–11). Albany, NY: SUNY Press.

Callicott, J. B. (1989b). Environmental ethics and animal liberation. In J. B. Callicott (Ed.), I*n defense of the land ethic* (pp. 15–38). Albany, NY: SUNY Press.

Callicott, J. B. (1995). A review of some problems with the concept of environmental health. *Ecosystem Health, 1*, 101–112.

Callicott, J. B. (1996). Do deconstructive ecology and sociobiology undermine Leopold's land ethic? *Environmental Ethics, 18*, 353–372.

Clements, F. E. & Shelford, V. E. (1935). *Bio ecology.* New York: Wiley.

Colwell, T. B., Jr. (1970). Some implications of the ecological revolution for the reconstruction of value. In E. Lazlo and J. B. Wilbur (Eds.), *Human values and natural science* (pp. 245–258). New York: Gordon and Breach.

Constanza, R. (1992). Ecological economic issues and considerations in indicator development, selection, and use: Toward an operational definition of system health.

In D. H. McKenzie, D. E. Hyatt, & V. J. McDonald (Eds.), *Ecological indicators* (pp. 1491–1502). New York: Elsevier.

Diamond, J. M. (1975). Assembly of species communities. In M. L. Cody & J. Diamond (Eds.), *Ecology and evolution of communities* (pp. 342–344). Cambridge, MA: Harvard University Press.

Fauth, J. E. (1997). Working toward operational definitions in ecology: Putting the system back into ecosystem. *Bulletin of the Ecological Society of America, 78,* 295–297.

Fauth, J. E., Bernardo, J., Camara, M., Resetarits, W. J., Jr., Buskirk, J. van, & McCollum, S. A. (1996). Simplifying the jargon of community ecology: A conceptual approach. *American Naturalist, 147,* 282–286.

Fiedler, P. L., White, P. S., & Leidy, R. A. (1997). The paradigm shift in ecology and its implications for conservation. In S. T. A. Pickett, R. S. Ostfield, M. Shachak, & G. E. Likens (Eds.), *The ecological basis of conservation: heterogeneity, ecosystems and biodiversity* (pp. 83–92). New York: Chapman and Hall.

Forbes, S. A. (1880). On some interactions of organisms. *Bulletin of the Illinois State Laboratory of Natural History, 1,* 3–17.

Frey, D. G. (1975). Biological integrity of water—an historical approach. In *The integrity of water* (pp. 127–140). Proceedings of a Symposium, March 10–12. Office of Water and Hazardous Materials, U.S. Environmental Protection Agency, Washington, DC.

Ghent, A. W. (1991). Insights into diversity and niche breadth analysis from exact small-sample tests of the equal abundance hypothesis. *American Midland Naturalist, 126,* 213–255.

Glacken, C. J. (1967). *Traces on the Rhodian shore.* Berkeley: University of California Press.

Golley, F. B. (1993). *A history of the ecosystem concept in ecology: More than the sum of its parts.* New Haven: Yale University Press.

Goodman, D. (1975). The theory of diversity-stability relationships in ecology. *Quarterly Review of Biology, 50,* 237–266.

Grene, M. (1987). Perception, interpretation and the sciences: toward a new philosophy of science. In C. McNight & M. Stahedroff (Eds.), *Philosophy and its variety* (pp. 107–129). Belfast, Northern Ireland: Queen's University of Belfast.

Grimm, V. (1996). A down-to-earth assessment of stability concepts in ecology: Dreams, demands, and the real problems. *Senckenbergiana Maritima, 27,* 215–226.

Grimm, V., & Wissel, C. (1997). Babel, or the ecological stability discussions: An inventory and analysis of terminology and a guide for avoiding confusion. *Oecologia, 109,* 323–334.

Haldane, J. B. S. (1928). *Possible worlds, and other papers.* New York: Harper and Brothers.

Holling, C. S. (1973). Resilience and stability of ecological systems. *Annual Review of Ecology and Systematics, 4*, 1–23.

Hutchinson, G. E., & Wollack, A. (1940). Studies on Connecticut lake sediments: Chemical analysis of a core from Linsley Pond, North Branford. *American Journal of Science, 238*, 493–517.

Hynes, H. B. (1970). The ecology of running waters. Toronto: University of Toronto Press.

Johnson, Kris H., Vogt, K. A., Clark, H. J., Schmitz, O. J., & Vogt, D. J. (1996). Biodiversity and the productivity and stability of ecosystems. *Trends in Ecology and Evolution, 11*, 372–377.

Jorgensen, S. E. (1997). Integration of ecosystem theories: A pattern (2nd ed.). Dordrecht: Kluwer.

Kareiva, P. (1996). Diversity and sustainability on the prairie. *Nature, 379*, 673–674.

Karr, J. R. (1981). Assessment of biotic integrity using fish communities. *Fisheries, 6*, 21–27.

Karr, J. R. (1996). Ecological integrity and ecological health are not the same. In P. Schultze (Ed.) *Engineering within ecological constraints* (pp. 97–109). Washington D.C.: National Academy Press.

Kiester, A. R. (1980). Natural kinds, natural history and ecology. *Synthese, 43*, 331–342.

Lazlo, E. (1980). Some reflections on systems theory's critics. *Nature and System, 2*, 49–53.

Leopold, A. (1949). *Sand County Almanac.* New York: Oxford University Press.

Lindeman, E. C. (1940). Ecology: An instrument for the integration of science and philosophy. *Ecological Monographs, 10*, 367–372.

Loehle, C. (1988). Philosophical tools: Potential contributions to ecology. *Oikos, 58*, 115–119.

May, R. M. (1986). Species interactions in ecology. *Science, 231*, 1451–1452.

Mayr, E. (1982). The growth of biological thought: Diversity, evolution, and inheritance. Cambridge, MA: Belknap Press of Harvard University Press.

McIntosh, R. P. (1958). Plant communities. *Science, 128*, 115–120.

McIntosh, R. P. (1967). The continuum concept of vegetation. *Botanical Review, 33*, 130–187.

McIntosh, R. P. (1975). H. A. Gleason—individualistic ecologist 1882–1975: His contributions to ecological theory. *Bulletin of the Torrey Botanical Club, 102*, 253–273,

McIntosh, R. P. (1980). The background and some problems of theoretical ecology. *Synthese, 43*, 195–255.

McIntosh, R. P. (1985). *The background of ecology—concept and theory.* Cambridge, England: Cambridge University Press.

McIntosh, R. P. (1987). Pluralism in ecology. *Annual Review of Ecology and Systematics, 18*, 321–341.

McIntosh, R. P. (1995). H. A. Gleason's individualistic concept and theory of animal communities: A continuing controversy. *Biological Reviews, 70*, 317–357.

McIntosh, R. P. (1998) The myth of community as organism. *Perspectives in Biology and Medicine 41*, 426–438.

Messer, J. J. (1992). Indicators in regional ecological monitoring and risk assessment. In D. McKenzie, H. Hyatt, & J. McDonald (Eds.), *Ecological Indicators* (pp. 135–146). New York: Elsevier.

Moore, P. D. (1983). Revival of the organismal heresy. *Nature, 303*, 132–133.

Noss, R. F. (1995). Ecological integrity and sustainability: Buzzwords in conflict. In L. Westra & J. Lemons (Eds.), *Perspectives on ecological integrity* (pp. 60–76). Dordrecht: Kluwer.

Odum, E. P. (1969). The strategy of ecosystem development. *Science, 164*, 262–270.

Odum, E. P. (1985). Trends expected in stressed ecosystems. *BioScience, 35*, 419–422.

Patten, B. C. (1975). Ecosystem as a coevolutionary unit: A theme for teaching systems ecology. In G. S. Innis (Ed.), *New directions in the analysis of ecological systems*, vol. 5, no. 1 (pp. 1–8). La Jolla, CA: Simulation Council Proceedings.

Peters, R. (1991). *A critique for ecology.* Cambridge, England: Cambridge University Press.

Pickett, S. T. A., Kolasa, J., & Jones, C. G. (1994). *Ecological understanding: The nature of theory and the theory of nature.* San Diego, CA: Academic Press.

Pickett, S. T. A., Ostfeld, R. S., Shachak, M., & Likens, G. E., (Eds.) (1997). *The ecological basis of conservation: Heterogeneity, ecosystems, and biodiversity.* New York: Chapman and Hall.

Pimm, S. L. (1984). The complexity and stability of ecosystems. *Nature, 307*, 321–326.

Pulliam, H. R. (1997). Providing the scientific information that conversation practitioners need. In S. T. A. Pickett, R. S. Ostfield, M. Shackak, & G. E. Likens, (Eds.),. *The ecological basis of conservation: Heterogeneity, ecosystems, and biodiversity* (pp. 16–22). New York: Chapman and Hall.

Price, P. W. (1984). Communities of specialists: Vacant niches in ecological and evolutionary time. In D. R. Strong, Jr., D. Simberloff, L. G. Abele, & A. P. Thistle (Eds.), *Ecological communities: Conceptual issues and the evidence* (pp. 510–523). Princeton, NJ: Princeton University Press.

Rapport, D. (1989). What constitutes ecosystem health? *Perspectives in Biology and Medicine, 33*, 120–132.

Regier, H. A. (1992). Indicators of ecosystem integrity. In D. McKenzie, H. Hyatt, & J. McDonald (Eds.), *Ecological indicators* (pp. 183–200). New York: Elsevier.

Richardson, J. L. (1980). The organismic community: Resilience of an embattled ecological concept. *BioScience, 30*, 465–471.

Saarinen, E. (1982). *Conceptual issues in biology.* Dordrecht: D. Reidel.

Sagoff, M. (1997). Muddle or muddle through? Takings jurisprudence meets the endangered species act. *College of William and Mary Law Review, 38*, 825–993.

Schneider, E. (1992). Monitoring for ecological integrity. In D.H. McKenzie, D. E. Hyatt, & V.J. McDonald (Eds.), *Ecological indicators* (pp. 1403–1419). New York: Elsevier.

Shrader-Frechette, K. S. & McCoy, E. D. (1990). Theory reduction and explanation in ecology. *Oikos, 58*, 109–114.

Shrader-Frechette, K. S. & McCoy, E. D. (1993). *Method in ecology.* Cambridge, England: Cambridge University Press.

Simberloff, D. (1980). A succession of paradigms in ecology: Essentialism to materialism. *Synthese, 43*, 3–39.

Suter, G. W., II. (1993). A critique of ecosystem health concepts and indexes. Environmental Toxicology and Chemistry, *12*, 1533–1539.

Talbott, L. M. (1997). The linkages between ecology and conservation. In S. T. A. Pickett, R. S. Ostfield, M. Shachak, & G. E. Likens (Eds.), *The ecological basis of conservation: Heterogeneity, ecosystems, and biodiversity* (pp. 368–378). New York: Chapman and Hall.

Tansley, A. C. (1935). The use and abuse of vegetational concepts and terms. *Ecology, 16*, 284–307.

Tilman, D. (1996). Biodiversity: Population versus ecosystem stability. *Ecology, 77*, 350–363.

Tilman, D., & Downing, J. A., (1994). Biodiversity and stability in grasslands. *Nature, 367*, 363–365.

Tobey, R. (1981). *Saving the prairies: The life cycle of the founding school of American plant ecology, 1895–1955.* Berkeley: University of California Press.

Ulanowicz, R. E. (1997). *Ecology, the ascendant perspective.* New York: Columbia University Press.

United States Environmental Protection Agency. (1975). *The integrity of water.* Proceedings of a Symposium, March 10–12, 1975. Office of Water and Hazardous Materials, U.S. Environmental Protection Agency, Washington, DC.

Wali, M. (1995). Ecovocabulary: A glossary of our times. *Bulletin of the Ecological Society of America, 76*, 106–111.

Westra, L. (1994). *An environmental proposal for ethics: The principle of integrity.* Boston: Rowman and Littlefield.

Wilson, D. S. (1997). Biological communities as functionally organized units. *Ecology, 78*, 2018–2024.

Wilson, D. S., & Sober, E. (1989). Reviving the super organism. *Journal of Theoretical Biology, 136*, 337–356.

4

Biocentrism, Biological Science, and Ethical Theory

KRISTIN SHRADER-FRECHETTE

Perhaps more than almost anyone else, J. Baird Callicott has been working to extend the frontier of ethical theory in ways that correct the anthropocentric bias of many moral thinkers. He has been creating a new environmental ethics, one grounded in the thought of Aldo Leopold. After highlighting the main contributions of Callicott's seminal book, *In Defense of the Land Ethic*,[1] I shall (1) summarize some of his best ethical insights, and (2) suggest several reasons that community ecology and natural selection, at least at present, may not be able to provide support for all of the environmental and ethical arguments in which Callicott and others enlist their help.

CALLICOTT'S CONTRIBUTIONS

Callicott deserves high praise because although his and Leopold's land/environmental ethic rests on "the ecological concept of a biotic community,"[2] he goes to some length to defend the value of nonhuman species. At least part of his motivation is the widespread species extinctions occurring all over the planet.[3] Many writers, myself included, believe that arguing for the intrinsic value of natural entities and nature as a whole makes sense because this intrinsic value can be grounded in some properties of the entities, such as being living. Some people, however, deny that a particular natural or metaphysical property (e.g., life) is truly good. To counter their denial and the difficulties it brings, Callicott argues that "good and evil, like beauty and ugliness, rest in the final analysis upon feelings or sentiments which are, as it were, projected onto, persons, or actions and affectively 'color' them.'[4] Callicott realizes that his subjectivist position forces him to deny the intrinsic value of nature in any

objective sense. One could respond, however, that one need not define intrinsic or inherent value in a strict way. One could avoid relativism and subjectivism by espousing a philosophical belief in the value of life. Callicott, however, does not take this path. He says his subjectivist axiology allows natural beings to be "valued for themselves."[5] It also escapes relativism, according to Callicott, because sociobiology has achieved a "consensus of feeling" through the "biologization of ethics." Human ethical feelings, he claims, "have been standardized by natural selection."[6]

Callicott also deserves high praise for arguing that genuine ecological education is the main way to reorient people toward a land/environmental ethics. He also shows that land aesthetics can contribute to the effort because it "calls attention to the psychic-spiritual rewards of maintaining the biological integrity and diversity of the rural landscape."[7] Callicott realizes, wisely, as many ethical theorists do not, the truth of the Augustinian insight that one must love something (like the land) to reason well about it and to understand it.

Other Callicott insights include his articulating difficulties with the animal-rights position. Often its proponents do not distinguish human-domestic communities (which include nonhuman animals) from wild biotic communities. Instead, he argues, they say that being a subject of a life (in some sense) is sufficient for being a rights-holder.[8] Apart from whether most animal liberationists fall victim to this error,[9] Callicott's insight is a correct one. This is that the community concept is essential to the notion of moral obligation and that different kinds of communities undergird different moral obligations. The insight is important not only because much of contemporary ethics is erroneously individualistic/atomistic, but also because significant philosophical discussions turn on the necessity of a shared moral community as the basis of duties to community members. Indeed, many of the arguments about rights of future generations focus on whether present and future people can share the same kind of moral community (i.e., have the same conception of the good) as we do. Hence Callicott's point is not only helpful to his own argument but also central to moral philosophy in general.

Likewise, in a significant departure from traditional ethical theory, Callicott insightfully argues that altruism is as fundamental in human nature as egoism. He shows that there are inborn natural sentiments that have society as their natural object.[10] Given the postulated egoism of the two main schools of modern moral philosophy (the deontological and the utilitarian), as Goodpaster recognizes, Callicott's taking the "higher road" of altruism is both refreshing and prophetic. The rational grounds for his doing so are compelling; most arguments (that all actions are done for self-serving reasons) presuppose a tautological definition of *self-serving* and hence are nonfalsifiable, nonempirical, and highly ideological. The psychological and political grounds for endorsing Callicott's grounding moral philosophy on altruism are that positing such a foundation, even if it cannot be proved, is the only possible way of avoiding narcissism. Otherwise, egoism will be a self-fulfilling prophecy for us all.

Finally, although I do not believe that "naturally selected" feelings justify particular ethical stances (see the arguments of the next section), Callicott is insightful when he argues that many moral values originate in the feelings. This insight locates the psychological beginnings of morality in the correct place, just as do many proponents of an ethics of care. Indeed, without feelings such as compassion, whether a principled and rational morality is ever able to develop is questionable. Callicott's emphasis on feelings—as the originators of morality—is important, in part, because it forces moral educators to emphasize the development of the whole person, emotions and intelligence. His emphasis does not allow us merely to nurture his or her ability to engage in rational analysis. Callicott's insight also is significant because it enables environmentalists to begin environmental education at the level of feeling. This is the level of experiences in nature, the level that Holmes Rolston correctly recognizes as crucial.[11]

Problems with Biological Holism and "Community" Definitions

With so much to praise in Callicott, especially his insistence on the importance of altruism and expanding our moral communities, can we find areas in which his views are arguably false? The answer to this question, for me, is "yes." The grounds for this response are, in large part, biological. Two of my concerns include the following. First, no scientifically/biologically coherent notion of "community" is robust enough to ground either contemporary community ecology or environmental ethics. As a consequence, how to safeguard the interests of these communities is not clear. Second, in relying on natural-selection mechanisms to deliver his evolutionary ethics from relativism, Callicott's ethics has lost its normative dimension. Let's examine these points in order.

First, although Leopold's and Callicott's subordination of all creatures to the integrity, beauty, and stability of the biotic community is philosophically defensible, it is biologically problematic. There is, for example, no clear sense in which one can claim, and no ecological consensus, that natural ecosystems proceed toward homeostasis, stability, or some balance, and almost no support for the diversity-stability view MacArthur, Hutchinson, and Commoner hold.[12] Salt marshes and the rocky intertidal are two of the many counterexamples to the diversity-stability view,[13] and mathematically and empirically based counterexamples have multiplied over the past two decades.[14] Even though some laypersons and policymakers appeal to the hypothesis,[15] most scientists either have repudiated it or have cast strong doubt on it.[16]

Doubts about balance and stability have arisen, in part, because ecologists cannot say what it would be, in a non–question-begging way, to hinder some balance, stability, or integrity. This is because communities and ecosystems regularly change and regularly eliminate species. Indeed, change is the norm, and most scientists now believe that biotic "communities" cannot be identified by any specific properties or species that give predictive power over them. Nature does not merely extirpate species or cause them to move elsewhere because their niches are gone. And, if not, then no clear scientific grounds exist for defining and preserving some controversial notion of balance or stability. Hence whether Leopold's and Callicott's appeal to the science of ecology can help environmental ethics in any precise, scientific way is not clear.[17] We cannot say that what happens naturally is good, whereas what happens through human intervention is bad; this would be to solve the problem of defining *balance* or *stability* in a purely stipulative or ad hoc way. Nor can the criterion for what is "natural" or "balanced" be merely that it is wrong for humans to do quickly (e.g., cause lake eutrophication) what nature does more slowly. One would need an argument (given neither by Callicott nor Leopold) that accelerating ecosystemic changes is bad, even if the changes themselves are somehow "natural."

Another conceptual problem besetting environmental appeals to a community-based scientific/ecological balance, wholeness, or integrity is that ecologists must take into account thousands of communities, species, and individuals, as well as the health or balance of ecosystems or the biosphere. How to define (scientifically) the health of a system (as opposed to an individual) is unclear because system health is relative to some specific goal; how to define the system at issue is also unclear. Defining an ecological "whole" to which Callicott and Leopold can refer is especially problematic for at least two reasons. One reason is that contemporary scientists do not accept the views of the biologists (e.g., Clements, Elton, Forbes) Callicott cites to explicate his views. The other reason is that most ecologists have rejected the contemporary variant of Clements's position, the Gaia hypothesis, as unproved metaphor or mere speculation. They admit the scientific facts of interconnectedness and coevolution on a small scale, but they point out that ecosystems and communities, as intact and clearly definable systems, do not persist through time. Hence no clear referent exists for the alleged "dynamic stability" of an ecosystem or community.[18]

Moreover, which (of many) alleged ecological communities whose stability we ought to seek is not clear. One could seek to stabilize (whatever that means) the ecosystem,[19] or the association,[20] or the trophic level, or the biosphere. Optimizing the well-being of one such community typically leads neither to the optimization of another community, nor to that of the biosphere, nor to that of a particular association. If not, then Callicott has no scientific or biocentric basis for choosing a given "whole" as the unit that is to be optimized.[21] Instead one must make a human value judgment to optimize the well-being of a particular community. Admittedly, once one makes a human value judgment about which particular whole one wants to attempt to stabilize or balance, that particular ecological conclusions are valid within certain spatial and temporal scales becomes obvious. Nevertheless, a given ecological conclusion regarding a particular type of balance or stability, for example, typically holds for some "wholes" (e.g., communities), but not for others, and for some spatial and temporal scales, but not for others. Because ecologists cannot optimize the welfare of all the different wholes there is no general "community" level at which ecological problem solving takes place, and thus no unambiguous way to operationalize (biologically) Callicott's and Leopold's views. Because no universal scientific/ecological theory exists to which ethicists can appeal in defining the "whole" about which Leopold and Callicott speak, ecologists are forced to work on a case-by-case basis. Numerous alleged "wholes" (e.g., populations) exhibit density vagueness rather than density dependence, whereas other wholes do not.[22] Also, many ecosystemic or holistic "explanations" are neither falsifiable nor even testable, but arguably "theological ecology."[23] Ecologists simply do not agree on the underlying processes that allegedly structure communities and ecosystems.[24]

A second biological problem with Callicott's grounding environmental ethics on the science of ecology occurs in his arguments against according rights to individual members of the biotic community. He says that safeguarding the rights of each individual is not possible; such a "safeguard" would stop all trophic processes beyond photosynthesis.[25] The biological problem with Callicott's reasoning here is that nature does not respect communities either. We find strong biological evidence (e.g., fossilized pollens) of radical changes in community composition and structure throughout history. These changes in community composition and structure, in turn, suggest that there is no such thing as a stable or balanced community "type" existing through time. Rather communities are definable only stochastically or statistically and the community "types" only appear stable because the time frame of examination is relatively short. Even if climate and environment remained the same, communities could not be classified into balanced or stable "types" on the basis of climate or time.[26] And if not, then the same argument that Callicott uses against Regan can be used against him. Nature does not respect biological communities so, on Callicott's own terms (and this is not an argument that I would make), how can he avoid a stipulative and question-begging argument that humans ought to respect biological communities? The point is not that this objection is correct. Rather, the point is that Callicott's own arguments fall victim to some of the same scientific flaws with which he charges Regan.

Problems with Natural-Selectionist Ethics

Another problem with Callicott's using the science of biology to justify his environmental ethics is that he destroys the normative dimension of his ethics. Callicott reasons, quite correctly, that in relying on a Humean notion of ethics, he is open to the charge of ethical relativism. He responds by postulating that ethical uniformity/unanimity is achieved by means of natural selection. He says "human feelings . . . have been standardized by natural selection" (see note 6). If Callicott is right, then one can neither be morally bound to do something against natural selection or against her genetic makeup, nor praised or blamed for acting in accord with or against natural selection. Hence Callicott has admittedly saved his ethics from relativism, but at the price of its "oughtness" or normative character.

Appealing to natural selection as a way of grounding evolutionary ethics has at least three difficulties. First, arriving at ethical beliefs and actions relies on cognitive and evaluative aims, (anticipating experience, solving problems, etc.). The "evolution-ethics" analogy therefore breaks down because, although evolution does not operate according to ends or aims, ethics does. Also, evolution and natural selection ignore the contribution to reflective self-understanding of ourselves as agents of inquiry, even though this reflective agency is at the core of ethical knowledge.[27] Moreover, the natural selection explanation fails to explain how someone could make the first correct ethical guess or have the first ethical feeling. At best, natural selection could only explain later correct guesses or feelings.[28] A second reason against Callicott's analogy between evolution and ethics is that, in ethics, people select theories and behavior on the basis of hypotheses about the facts and evaluations of them. In evolution, however, each variation arises independently of the adaptive needs of the organism. The facts themselves, not our theories or evaluations of them, guide evolution. Hence evolution is blind to the adaptive needs of the organism, whereas ethics is blind to the facts and operates only on hypotheses about facts.[29] A third difficulty with Callicott's (or any) evolutionary ethics is that he uses it to move from biological theory to realistic ethics. But for this inference to be successful, people must know that the organism (the ethical agent) has an accurate representation (biological theory) of the environment. They do not know this.[30] Indeed, earlier sections of these remarks surveyed some of the current problems with biological and ecological theory. For all these reasons, Callicott's appeal to natural selection appears to create more philosophical problems than it solves.

If Callicott is unable to use natural selection and community ecology to "bail out" environmental ethics, then where do we go from here? Clearly Callicott deserves high praise for showing us much of what is wrong with traditional moral philosophy. Nevertheless, my own preferences are for a metaphysical account that posits intrinsic value in nature itself, an account that deviates only slightly from that of Paul Taylor (see earlier notes). Perhaps ultimately ethicists must rely more on metaphysics and less on biological science if they wish to build (or discover) an environmental ethics. Just as there are no "technological fixes" that will give easy answers to environmental problems, perhaps also there are no "scientific fixes" (such as natural selection) that will give easy answers to ethical problems. Nevertheless Callicott has helped us begin to ask the hard questions.

Answering Callicott

In a reply to the preceding analysis, Callicott states: "I nowhere suggest that ethics and evolution are analogous."[31] Yet Callicott claims: The "conceptual and logical foundations of the land ethic" are a "Darwinian protosociobiological natural history of ethics. . . . Its logic is that natural selection has endowed human beings with an affective moral response to perceived bonds of kinship and community."[32] Value "in the philosophical sense," says Callicott, "is a newly discovered proper object of a specially evolved 'public affection' or moral sense, which all psychologically normal human beings have inherited from a long line of primates."[33] Stating the preceding position (that evolution and natural selection provide the foundations of the land ethic), given in at least three places, and then disavowing its consequences is logically inconsistent for Callicott. It is logically inconsistent, once someone points out the problematic logical consequences of one's position, to affirm the position but deny the logical consequences.

Callicott's response to my scientific and epistemological criticisms of his conception of community have similar logical problems. He says that "if the concept of a human community is coherent and robust enough to support anthropocentric moral obligations . . . then the concept of a biotic community . . . is coherent and robust enough to support ecocentric moral obligations."[34] His claim does not work, however, because of the incompatible properties that Callicott attributes to biotic and human communities. In Callicott's book he says that we humans "remain members of the human community" and that we have moral responsibilities . . . "to respect universal human rights."[35] Yet Callicott also claims: "Not only are other sentient creatures members of the biotic community and subordinate to its integrity, beauty, and stability; so are *we*."[36] Either certain universal human rights have primacy, or the biotic community has primacy. For both to have primacy is impossible. Or as Alice phrased it to the Queen: "One *can't* believe impossible things."[37]

Callicott also denies my charge that his ethics is not normative by claiming that his ethics is normative in the sense in which a body temperature of 98.6° degrees "provides a *norm* against which we measure deviations—fever and hypothermia"; that is, Callicott claims that his ethics (like a norm in science or medicine) is *statistically normative*. My analysis charges, however, that his ethics was not *ethically normative*, namely "one cannot be praised for acting in accord with natural selection."[38] Callicott cannot answer the charge of his denying *ethical norms* by responding that his ethics has *statistical norms*. Statistical norms always tell us what behavior *is* most probable, whereas ethical norms tell us what behavior people *ought* to perform. The two are not the same.

Conclusion

Where does this exchange leave us?[39] At the least, with some agreement. As Callicott correctly states, "ecology does not provide us with objective dynamic norms of ecosystemic health."[40] Furthermore, our exchange suggests that, just as scientific progress comes from a plurality of theories, so also progress in environmental ethics most likely will come from a plurality of philosophical approaches—such as Callicott's work, rooted in a profound grasp of moral theory, and my own work, grounded in biology and philosophy of science. My recommendation for the future is that we take the advice of Ernst Mayr and analyze the key concepts of environmental ethics, concepts such as *evolution*, *community*, and *norm*, as Callicott and I have done. Mayr said that the "spectacular recent progress" in evolutionary theory was not due to improvements in measurement but due to improvements in the clarification of concepts.[41] The same can be said for much of environmental ethics.

Notes

The biological and ethical part of this work was supported by National Science Foundation (NSF) grant BBS–8619533 and NSF grant SBR–95–12133, both from the Ethics and Values Program. The opinions expressed are those of the author and not the NSF.

1. J. Baird Callicott, *In Defense of the Land Ethic: Essays in Environmental Philosophy* (Albany, N.Y.: SUNY Press, 1989).
2. Callicott, *Defense*, 126.
3. Callicott, *Defense*, 154–155.
4. Callicott, *Defense*, 160.
5. Callicott, *Defense*, 161.
6. Callicott, *Defense*, 164.
7. Callicott, *Defense*, 201.
8. Callicott, *Defense*, 47.
9. See Paul Taylor, *Respect for Nature* (Princeton, N.J.: Princeton University Press, 1986). Taylor clearly makes the distinction between natural and domestic communities of animals.
10. Callicott, *Defense*, 84–85, 142–143.
11. See Holmes Rolston, *Philosophy Gone Wild* (Buffalo, N.Y.: Prometheus, 1986), esp. 221–262.
12. P. Taylor, for example, urges us to "preserve ecological integrity" (Taylor, *Respect*, 299), but he never tells us what ecological integrity is, and he denies that ecology can inform environmental ethics (Taylor, *Respect*, 8). See Mark Sagoff, "Fact and Value in Environmental Science," *Environmental Ethics* 7, no. 2 (Summer 1985): 107–110. See also Taylor, *Respect*, 8.
13. M. Sagoff, "Environmental Science and Environmental Law" (College Park, Md.: Center for Philosophy and Public Policy, March 1985), unpublished essay, 109.
14. See R. M. May, *Stability and Complexity in Model Ecosystems* (Princeton, N.J.: Princeton University Press, 1973); R. Levins, "The Qualitative Analysis of Partially Specified Systems," *Annals of the New York Academy of Sciences* 231 (1974): 123–138; J. H. Connell, "Diversity in Tropical Rain Forest and Coral Reefs," *Science* 199 (1978): 1302–1310. See also Sagoff, "Fact," 109, and R. P. McIntosh, *The Background of Ecology: Concept and Theory* (Cambridge, England: Cambridge University Press, 1985), 187–188.
15. See for example, U.S. Congress, Senate, *Congressional Record*, 93rd Congress, First Session, 119 (24 July 1973): 25668; B. Commoner, *The Closing Circle* (New York: Knopf, 1971), 38; N. Myers, *A Wealth of Wild Species* (Boulder, Colo.: Westview Press, 1983).
16. See Sagoff, "Fact," 107. See also R. T. Paine, "A Note on Trophic Complexity and Community Stability," *American Naturalist* 103 (1969): 91–93; R. Lewin, "Fragile Forests Implied by Pleistocene Data," *Science* 226 (1984): 36–37; May, *Stability*; Levins, "Qualitative"; Connell, "Diversity"; Daniel Goodman, "The Theory

of Diversity-Stability Relationships in Ecology," *Quarterly Review of Biology* 50, no. 3 (September 1975): 237–266. Finally see McIntosh, *Background*, p. 142.

17. See the previous note. See also K. S. Shrader-Frechette and E. D. McCoy, *Method in Ecology* (New York: Cambridge University Press, 1993), esp. chap. 2.

18. See Taylor, *Respect*, 45–46, 225–226, 246, 259, 281–282.

19. See R. MacArthur, "Fluctuations of Animal Populations, and a Measure of Community Stability," *Ecology* 36 (1955): 533–536; Goodman, "Theory."

20. Ecologists who follow Clements are more likely to make this claim. See McIntosh, *Background*, 44, 79, 107.

21. McIntosh, *Background*, 228, 252–256. B. Norton, "Environmental Ethics and the Rights of Nonhumans," *Environmental Ethics* 4 (1982): 17–36, raises a similar point. See McIntosh, *Background*, for a discussion of community ecology (pp. 69–146, 263–267), population ecology (pp. 146–193), and ecology (pp. 193–242).

22. See K.S. Shrader-Frechette, "Practical Ecology and Foundations for Environmental Ethics," *Journal of Philosophy* XCII, no. 12 (December 1995): 621–635. McIntosh, *Background*, 126 ff., 157 ff., 181–182 ff., 252; see also K.S. Shrader-Frechette, "Organismic Biology and Ecosystems Ecology," in *Current Issues of Teleology*, ed. N. Rescher (Pittsburgh, Pa.: University of Pittsburgh Center for the Philosophy of Science, 1986), 77–92. For information on density dependence, see D. Strong, "Density Vagueness: Abiding the Variance in the Demography of Real Populations," in *Community Ecology*, ed. J. Diamond and T. Case (New York: Harper and Row, 1986).

23. McIntosh, *Background*, 193.

24. See Shrader-Frechette and McCoy, *Method*, and Shrader-Frechette, "Practical." What pattern of excellence is it that an ecosystem maximizes? Ecologists cannot answer the question. See M. Cody and J. Diamond (eds.), *Ecology and the Evolution of Communities* (Cambridge, Mass.: Harvard University Press, 1975); D. Strong et al. (eds.), *Ecological Communities* (Princeton, N.J.: Princeton University Press, 1984). See especially M. Gilpin and J. Diamond, "Are Species Co-occurrences . . . ?" in Diamond and Case, *Community*, 298–315. See also D. S. Simberloff, "Competition Theory, Hypothesis Testing, and Other Community Ecological Buzzwords," *American Naturalist* 122 (1983): 626–635. If there is no accepted natural process followed by ecosystems that are "maximizing excellence", then how did competition become so entrenched in "explaining" ecosystemic processes? One scientist claims that competition has survived as a hypothesis merely because it fits in with our notions of homeostasis and the balance of nature. (R. Lewin, "Santa Rosalia Was a Goat," *Science* 221 [12 August 1983]: 636–639.) If this is so, and I think that it is, then ecology is in the midst of a revolution, Kuhnian or not, to overthrow entrenched and untestable competition.

25. Callicott, *Defense*, 43, 51.

26. See Strong, "Density," and Strong et al., *Ecological*. See also M. Davis, "Climate Instability, Time Lags, and Community Disequilibrium," and R. Graham, "Response of Mammalian Communities to Environmental Changes During the Late Quaternary" in Diamond and Case, *Community*, 269–284, 300–313.

27. C. Hookway, "Naturalism, Fallibilism, and Evolutionary Epistemology," in C. Hookway, ed., *Minds, Machines, and Evolution* (Cambridge, England: Cambridge University Press), 13–15, makes similar points. See also A. O'Hear, "Has the Theory of Evolution Any Relevance to Philosophy?" *Ratio* 29 (1987): 27–29.

28. P. Skagestad, "Taking Evolution Seriously: Critical Comments on D. T. Campbell's Evolutionary," *The Monist* 61, no. 4 (October 1978): 615, makes a similar point.
29. Skagestad, "Evolution," 617, makes a similar point.
30. O'Hear, "Theory," 25, makes a similar point.
31. J. B. Callicott, "Reply," *Between the Species: A Journal of Ethics* 6, no. 4 (1990): 193–194.
32. Callicott, *Defense*, 83.
33. Callicott, *Defense*, 86.
34. Callicott, "Reply."
35. Callicott, *Defense*, 93.
36. Callicott, *Defense*, 92.
37. Lewis Carroll, *The Annotated Alice: Alice's Adventures in Wonderland and Through the Looking Glass* (New York: Bramhall House, 1960), 251.
38. K. S. Shrader-Frechette, "Biological Holism and the Evolution of Ethics," *Between the Species: A Journal of Ethics* 6, no. 4 (1990): 185–192. See Shrader-Frechette, "Practical."
39. See Shrader-Frechette, "Practical": Shrader-Frechette and McCoy, *Method*; K. S. Shrader-Frechette, "Ethics and the Environment," *World Health Forum* 12, no. 3 (1991): 311–321; K. S. Shrader-Frechette, "Sustainability and Environmental Ethics," in *The Notion of Sustainability and Its Normative Implications*, ed. Gunnar Skirbekk (Oslo: Scandinavian University Press, 1994), 57–79.
40. Callicott, "Reply."
41. Ernst Mayr, *Toward a New Philosophy of Biology* (Cambridge, Mass.: Harvard University Press, 1988), vi.

18. [illegible] "Taking the [illegible] Seriously: Critical Comments on [illegible]

19. [illegible] makes a similar point.

20. [illegible] similar point.

21. [illegible] *Journal of Ethics* [illegible]

22. Callicott, [illegible]

23. [illegible]

24. [illegible]

25. Callicott, [illegible]

26. Callicott, [illegible]

27. [illegible]

28. [illegible] and the Evolution of [illegible] 192. See also [illegible]

29. [illegible] and the Environment [illegible] Environmental [illegible] University Press, 1997), 75.

30. Callicott, "Reply."

31. [illegible] *Philosophy of Biology* (Cambridge, Mass.: Harvard University Press, 1988), [illegible]

Part II

Intrinsic Value in Nature

5

Callicott on Intrinsic Value and Moral Standing in Environmental Ethics

WENDY DONNER

INTRODUCTION

One central question in environmental ethics is, what obligations do humans have to the environment? The environment is taken to include individuals such as nonhuman animals and plants as well as holistic elements such as species, ecosystems, and the entire biotic community. To determine our obligations, we first need to determine what sorts of things have value in themselves or moral standing. In this essay I critically explore J. Baird Callicott's position on these questions, which are prominent in recent debates in environmental ethics.

Callicott argues that it is necessary to get beyond the limitations of extensionist approaches to the environment. According to Callicott, extensionists such as Tom Regan and Peter Singer want to extend moral concern to other sentient animals. Callicott argues that such theories are not an environmental ethic but an ethic for managing the environment for the use of sentient beings. The rest of nature would then be "mere means" for the ends of sentient creatures. Callicott argues that these theories will not suffice as an ethics of the environment because they do not allow for "direct moral consideration of plants and all the many animals that may not be either sentient, or . . . 'subjects of a life.' "[1]

The disagreements between the extensionists and the holistic theorists are apparent in Callicott's rather harsh critiques of Regan and Singer and their countercritiques. Callicott argues that Singer and Regan ignore the "ecological order of nature," which is "premised on one fundamental principle—all life . . . depends on death."[2] Thus according to the land ethic, "To the extent that the animal liberation/animal rights ethics condemn the taking of life (as a violation of the rights of a subject of a life or the infliction of pain on a sentient being), they are irreconcilably at odds with the ecological 'facts of life.' "[3]

Holism: The Land Ethic

Holism proposes to center environmental theory directly on the environment where ecologists claim the center belongs. However, as a theory the land ethic has evoked a lot of controversy and more than a little hostility and can seem to be quite jarring and unsettling in its challenge to the conceptual frameworks of more traditional theories of value and of the environment. What this means is that the ethic does not take human or even animal individualistic interests or rights as its focus, but rather it centers itself squarely on the environmental ecosystems as the prime bearer of inherent value.

Callicott justifies his theory on the basis of both ecological and evolutionary Darwinian principles and considerations. Although eschewing traditional extensionalism that we have seen he rejects, Callicott nonetheless claims, "All contemporary forms of life thus are represented to be kin, relatives, members of one extended family. And all are equally members in good standing of one society or community, the biotic community or global ecosystem."[4] This theory holds as the prime bearer of value the biotic community and other wholes such as ecosystems and entire species, as opposed to individual members of such species. Thus when we are making practical decisions about the environment, we ought to follow the fundamental rule of doing that which will "enhance the diversity, integrity, beauty and stability of the biotic community."[5] He also argues that nonhuman species as a whole have intrinsic value, and we may be obligated to sacrifice individual members of sentient species to save an endangered but nonsentient plant species from extinction.[6] But Callicott also attempts to respond to criticisms of untempered holism and tries to embrace both holism and individualism. His theory "provides moral standing for both environmental individuals and for the environment as a whole."[7]

Callicott appeals to a Humean-Darwinian account of human feelings of benevolence to explain how we can feel sympathy for both individuals and wholes. Darwinian natural selection has operated to select for those members of the human community who had both more intense and wide-ranging feelings of sympathy. This provides a basis in our feelings for our placing value on the natural environment. He says that both individuals and wholes as such can be "the objects of certain special, naturally selected moral sentiments."[8] Intrinsic value on this model is a bivalent concept grounded on both subjective and objective factors.

> Intrinsic value is, as it were, "projected" onto appropriate objects by virtue of certain naturally selected and inherited intentional feelings, some of which . . . simply have social wholes as their natural objects. . . . Wholes may thus have intrinsic value no less problematically than individuals.[9]

This account of intrinsic value is not without problems of its own, a point to which I shortly turn. But Callicott is at pains to emphasize that in his view the land ethic can accommodate standing for individuals. This is because many of the harshest criticisms of the theory question its commitment to individual members of species and argue that it has inhumane consequences. According to Callicott our moral sentiments can move back and forth between holistic and individualistic objects. However he wavers on this point and in the end where he puts the primary weight is not clear.

Callicott is painfully aware of the difficulties engendered by this focus on whole systems. The land ethic has been attacked for its failure to make value distinctions among different species based on their place on the evolutionary scale. According to this theory Homo sapiens as a species is no more or less valuable than any other species, and its value as a species must be assessed in terms of its impact on the environment. Because this impact has often led to massive environmental destruction, this would seem to place our species low on the value scale and leave little room for pity when war and famine strike. It seems that ecocentrism also requires a cold-hearted attitude toward individual animals of other sentient species if their species overpopulate or are domesticated by humans. This conflicts with the concerns of those who regard individual animals as having moral standing or rights independent of such ecological considerations and who would not be willing to abandon regard for members of domesticated or overpopulated species.

Ecocentrism and Inherent Value

I now turn my attention to some of the deeper theoretical issues concerning value. When we shift our focus from an individualist, consciousness-based framework to one centered on wholes and species and ecosystems, we subtly change the very question we ask about value. Traditional moral theory asks the question of what sorts of beings have moral standing, that is, what beings ought or deserve to be taken into account for their own sakes when we make moral decisions. In the traditional framework, consciousness is taken to be the bottom line because it matters to a conscious being what is done to it. It sounds absurd to say that it matters to a rock or to the Amazon rain forest what is done to it. Now we ask what sorts of things, conscious or otherwise, have inherent or intrinsic value because it does not seem absurd to ask whether rain forests have value in themselves. On the other hand, although the question is not absurd, it is still a requirement that the case be made that rain forests or other nonconscious things have such value.

How shall we define intrinsic or inherent value? Callicott provides the following: "Something is intrinsically valuable if it is valuable *in* and *for* itself—if its value . . . is independent of any use or function it may have in relation to something or someone else . . . an intrinsically valuable entity is said to be an 'end-in-itself,' not just a 'means' to another's ends."[10] Callicott proposes in the place of this strong sense of inherent value a second version that can be called weak inherent value. In this weak sense, an appreciative consciousness is necessary to project value on to a nonconscious object, but the object of appreciation is valued for itself, for properties of its own. As Callicott puts it:

> I concede that . . . the *source* of all value is human consciousness, but it by no means follows that the *locus* of all value is consciousness itself or a mode of consciousness like reason, pleasure, or knowledge. In other words, something may be valuable only because someone values it, but it may also be valued for itself, not for the sake of any subjective experience . . . it may afford the valuer.[11]

But Callicott has conceded a great deal by this move without gaining too much ground for his argument. For at the very least we may say that he has conceded that there is a hierarchy of value, with conscious beings, contrary to the claims of his theory, having greater value than nonconscious valuable things. For whereas nonhuman species (the immediate subject of his discussion) or ecosystems may be the locus but not source of value, conscious beings are *both* the source and locus of value, and this confers greater status. Conscious beings not only have the very features that make them valuable in themselves, but they also have the capacities to value and appreciate themselves and others. They project value onto others as well as existing as value unto themselves. The fact that they have this double-impact value is not a small point. If the light requires to be shined or projected by others, if value is conferred by others, then this places the value of the appreciated object on rather shaky ground. The traditional framework avoids this unhappy consequence by insisting that a being's moral considerability is not conferred by others although others may acknowledge or refuse to acknowledge their stature. Thus slaves had the same moral considerability as their owners even though their rights were not recognized and were grossly violated. The great moral revolution and theoretical breakthrough of individual human rights consisted in the insight that all humans were beings who had moral value grounded on features of human nature, and they had this value, based on consciousness, regardless whether others recognized it. Callicott's projectivist account of intrinsic value moves away from this.

But the account is suspect beyond this because if we give up the strong sense of intrinsic value, that is, value independent of consciousness, how value resides in the object that is valued is not clear. Value is conferred by the lighting up or projection of a human valuer. The features of the object exist whether human projectors react with them. The value on this account seems rather more subjectively based in the human valuer than Callicott allows; in what sense value resides in the object is not clear. At most it resides in a relation between valuing consciousness and features of the object. What resides in the object are natural features that can be picked out by conscious valuers as providing triggers of value. These natural features are good-making properties but they remain as neutral natural properties unless and until value is placed upon them by appreciative valuers. Value remains with consciousness.

Callicott's analysis of value does not establish the conclusion that ecosystems and species are the primary bearers of value. In fact, his analysis shows just the opposite; because the value of nonhuman species is projected or conferred by individual conscious beings, then these latter would seem to be the prime focus of value. However, his argument does help to establish a weaker conclusion that a human-centered moral theorist with serious environmental concerns would welcome. That is, his argument gives grounds for showing why conscious individual humans should show respect for the natural environment for its own features and not for the sake of any human ends.

This much is welcome. But there are other problems with and constraints on the theory. An environmental ethics needs to give human moral agents some clear guidance in cases of conflict between the very different elements of focus of the theory. Callicott tries to balance concern for wholes, ecosystems, and entire species and their diversity against individual members of human and nonhuman species. Such balancings mirror the complexities and difficulties of the extreme extensionist whom Callicott ridicules, and yet his theory does little better and has some additional headaches. Many such balancings are hard enough from within the traditional framework. But the land ethic calls for a balancing of entirely different orders of things—human and nonhuman individuals versus species and systems—and it calls for the balancing to be done in a new framework, one not dependent on wise or enlightened long-term preservation of nature for human use, but dependent, for example, on a high valuing of endangered species that may be low on the evolutionary scale. When we try to balance such entirely different orders of things, and the values placed on these elements are out of line with our traditional valuings, the weighing problem becomes acute.

A moral theory must do more than give general guidelines. If we need to weigh the value of or interests of different elements such as sentient individuals and nonconscious wholes, we need a theory that has either one fundamental principle to resolve conflicts, or, if there is more than one, we need principles that set out clearly how we prioritize the interests of or value of these different elements consistently. Callicott's theory does not furnish this.

Recall as well that the theory has unsettling results, so the balancings required by this theory may be unclear or they may be horrifying. Callicott's Humean-Darwinian analysis of human sympathy as naturally selected softens the impact. Our sympathy for fellow humans may deflect our judgment that the human population should be reduced drastically. In particular, this sympathy grounds the particular ties we feel with our immediate kin and community.

The Darwinian substructure Callicott appeals to brings out another troubling feature of the land ethic. The social sentiments that give us special ties to our immediate kin and community coexist with opposite feelings for outsiders—aggression and rage against those from beyond the tribe. Sympathy and aggression are partners in the evolutionary play. And this parallels the ambivalence that ecocentrists and deep ecologists feel toward animals and hunting. Animals are supposed to be killed with respect, but this does not seem to apply to sport hunting, which is celebrated by these theorists.

Thus the land ethic has serious problems: an analysis of value that does not support the value claims of the theory, a lack of guidance on how weighings of very different elements are to be carried out, and decisions that are troubling at best, horrifying at worst.

Notes

1. J. Baird Callicott, "The Search for an Environmental Ethic," in *Matters of Life and Death: New Introductory Essays in Moral Philosophy*, ed. Tom Regan, 3d ed. (New York: McGraw-Hill, 1993), 350.
2. Callicott, "Search," 351–352.
3. Callicott, "Search," 352.
4. Callicott, "Search," 364.

5. Callicott, "Search," 365–366.
6. J. Baird Callicott, "On the Intrinsic Value of Nonhuman Species," in *In Defense of the Land Ethic: Essays in Environmental Philosophy*, ed. J. Baird Callicott (Albany, N.Y.: SUNY Press, 1989), pp. 129–155.
7. Callicott, "Search," 364.
8. Callicott, "Search," 364.
9. Callicott, "Search," 364–365.
10. Callicott, "Intrinsic," 131.
11. Callicott, "Intrinsic,"133.

6

Naturalizing Callicott

HOLMES ROLSTON III

Philosophy, the "love of wisdom," becomes troublesome when friends and truth conflict. Aristotle responded to Plato that, especially when considering the good, both are dear, but our duty "requires us to honor truth above our friends."[1] Baird Callicott is a longtime friend whose philosophy I much respect, but the truth is dearer than Callicott. Because Callicott is also a Platonist scholar, he will remember the Aristotelian duty to prefer truth to friendship. Callicott, although a dear friend, is I fear, a doubtful guide at rather critical turning points and has gotten himself lost. He cannot find values in nature, not intrinsically. Indeed, at times he cannot find nature at all, not original nature, only a nature commingled with culture. So, paradoxically, we need to get Callicott, although he thinks of himself as a naturalist, really naturalized.

A pity, too, that he loses his way, because he and I travel together over much of the landscape of environmental philosophy and policy, unfamiliar terrain that he and I (and increasingly many others) have been exploring for a quarter of a century. I cannot follow him in his arguments (1) about nature and culture, or (2) about intrinsic natural value. All good scholars know that an attack on argument differs from an attack on persons. Because Callicott earlier took it upon himself to "deconstruct Rolston,"[2] perhaps now it is my turn to "reconstruct Callicott."

NATURE AND CULTURE

Callicott is anxious not to be a dualist, especially not a Cartesian dualist, which he thinks characterizes Enlightenment and modern thought and is one of the causes of environmental crises. Thinkers who distinguish between nature and culture are such dualists, working in the legacy of matter and mind because they think that the cultures produced deliberately by human minds are something different from the productions of spontaneous nature, the latter resulting from the self-organizing causal processes of energetic matter.

Callicott desires a new concept of nature that includes culture. "The modern picture of nature is false and its historical tenure has been pernicious. A new dynamic and systemic postmodern concept of nature, which includes rather than excludes human beings, is presently taking shape."[3] He would probably say he wishes to naturalize culture. Callicott puts this provocatively: "We are animals ourselves, large omnivorous primates, very precocious to be sure, but just big monkeys, nevertheless. We are therefore a part of nature, not set apart from it. Chicago is no less a phenomenon of nature than is the Great Barrier Reef."[4] That ought to cure us from the "sharp dichotomy between man and nature," which has too long been a feature alike of religion and philosophy, "both wellsprings of the Western intellectual heritage."[5]

If one is a metaphysical naturalist, as Callicott seems to be, then whatever is, is natural. In this respect he does not differ from many modernists, who are often also metaphysical naturalists, as some ancient thinkers also were. In this sense, the word *natural* has no contrast class, at least none occupied by any existing thing. Other metaphysicians might hold, for example, that the supernatural exists, contrasting with the natural. There might be supernatural things going on in Chicago, in the churches. But Callicott is not entering this debate. He wants to claim, as a helpful insight in environmental ethics, that humans are natural, their culture (exemplified in Chicago) is quite natural (as much as the Great Barrier Reef). Realizing this "might even help to disseminate broadly an ecological world view and an associated environmental ethics."[6]

The trouble is that, outside of metaphysics, a word becomes useless if it has no contrast class. Naturalizing everything delimits nothing. In environmental ethics it seems rather necessary to mark off what happens in wild spontaneous nature from what happens as a result of humans in their cultural activities, that is, in significant measure at least, to set humans apart from nature. Otherwise, we are not going to get any helpful analysis, such as might guide human conduct, by inquiring whether *x* is natural because any and all cultural activities will be natural activities as well (setting aside any supernatural events).

Naturalizing everything overnaturalizes too much. The products of Chicago industries, such as compact disks and Styrofoam cups, are natural just as much as coral reefs with their polyps and fishes. Corporate executives deciding to break the standards of the Clean Water Act and polluting Lake Michigan, are behaving in accord with nature as much as those deciding to meet or exceed the standards to preserve the integrity of the waters. Or as clams feeding underwater off the coast of Australia.

Callicott is sensitive to this problem, and he separates desirable from undesirable human behaviors by asking whether they are healthy, not whether they are natural (because they are all natural). Doubtless that will give some of the needed guidance. But whose health do we seek? Our human health? Callicott replies, rather, that we seek ecosystem health, assuming that this supports human health, which is ordinarily a quite reasonable assumption. Health is not just a skin-in matter; it is a skin-out matter. One cannot be healthy in a sick environment. Aldo Leopold wrote of our "responsibility for the health of the land."[7]

But human health might also permit or even require some rather radically transformed natural systems, making tall grass prairies into cornfields and short grass prairies into wheat fields. The prairies of the Midwest can be quite healthy ecosystems even if the whooping cranes go extinct. "An ecological system is healthy and free from 'distress syndrome' if it is stable and sustainable—that is, if it is active and it maintains its organization and autonomy over time and is resilient to stress."[8] Yes, that sounds plausible and desirable, but where is the place for cultural alterations of landscapes?

A disanalogy exists between humans wishing bodily health and landscape health. A person prefers bodily natural health. We repair breakdowns, but we do not rebuild the healthy body. We only go to doctors when we are sick. By contrast, we do not want entirely natural ecosystems, healthy though they might be, and nothing more. If we are to have any culture at all, especially a modern culture, we must transform wild nature into rebuilt environments. We constantly labor to make something better (judged by our cultural standards) out of wild nature, not just healing something sick. We do not revise our bodies as we revise wild nature.

A flourishing culture requires revamping much of wild nature. However, if this goes too far, then the natural system can collapse. We have to identify a pristine biological integrity, wild healthy environments, present ideally in wilderness areas, hopefully in protected areas, and contrast that with a culturally modified biological health, which we will try to maintain all over the landscape. But all this requires the distinction between nature and culture that Callicott has denied us.

Ought there to be any prairies saved for what they are in themselves, with a flourishing population of whooping cranes, preserved as healthy nature apart from its healthy support of culture and agriculture? This question cannot be addressed without specifying in more detail whose health is involved; and, sometimes at least, the health of wild natural ecosystems and their members may be at stake, not just that of humans in their cultures. This again requires the forbidden distinguishing of nature from culture. The most we could do might be to include the cranes and the wild prairies somewhere in our desires for quality of life. But if we had some other desires, the cranes and the prairies could go, assuming we kept the healthy cornfields and wheat fields.

Nature differs from culture, and vice versa, in ways we need to specify. The problem is that, anxious not to be a dualist, Callicott is not discriminating enough to see that although humans evolve out of nature and its processes, they significantly evolve *out of* it. That can confuse him and others into saying that humans are just natural because they are products of various natural laws and events operating through evolutionary history, and because their origins were natural, they continue to be natural. But that is to fall into a "nothing but" fallacy (more accurately, the genetic fallacy), which confuses what a thing now essentially is with what its historical origins once were. It cannot take emergence seriously. Environmental philosophy needs to see the difference in being human, and only after we get clear about that, do we also want to see the senses in which, although evolved out of it, culture has to remain in relative harmony with nature.

Humans superimpose cultures on the wild nature out of which they once emerged with radical innovations, leading to the contrast we regularly make in ordinary language, between the natural and the artifacted, between a clam in the Great Barrier Reef and a Styrofoam cup in Chicago. The difference in ordinary language is catching something significant, something of which we need to take account (regardless of whether one is a metaphysical naturalist). Culture does introduce emergent novelties not previously present in wild nature.

Information in wild nature travels intergenerationally on genes; information in culture travels neurally as persons are educated into transmissible cultures. Although the higher animals can learn limited behaviors from parents and conspecifics, animals do not form cumulative transmissible cultures. In nature, the coping skills are coded on chromosomes. In culture, the skills are coded in craftsman's traditions, religious rituals, or technology manuals. Information acquired during an organism's lifetime is not transmitted genetically; the essence of culture is acquired information transmitted to the next generation.

Information transfer in culture can be several orders of magnitude faster and overleap genetic lines. A typical couple may have only two or three children who inherit their genetic information. But those children are educated by taking classes from dozens of teachers, by reading hundreds of books, using libraries with tens of thousands of books, written by authors to whom they are genetically quite unrelated and who may have been dead for centuries. The children learn from television programs with information coming from all over the world. A human being develops typically in one of some ten thousand cultures, inheriting a heritage that is historically conditioned, perpetuated by language, conventionally established, using symbols with locally effective meanings. Cultures may exchange ideas; sometimes people are reared at the crossroads of cultures; well-educated persons choose and criticize their cultures.

Animals are what they are genetically, instinctively, and environmentally without any options in what they shall be at all, even if they do make some limited choices. Humans have myriad lifestyle options, evidenced by their cultures, and each human makes daily decisions that affect his or her character. The highly deliberative character of human actions is without real precedent in nature, even though animals may have some precursor options in what they shall do. Natural selection pressures are relaxed in culture. As a result of their reflective deliberations, humans help each other out compassionately with charity, affirmative action, or Head Start programs. They study medicine to cure their bodily diseases. The determinants of animal and plant behavior, much less the determinants of climate or nutrient recycling, are never anthropological, political, economic, technological, scientific, philosophical, ethical, or religious. Little or nothing in wild nature approaches all this. If we are going to evaluate what natural and cultural values we want to treasure, we must appreciate and criticize human affairs with insight into their radically different character.

We might want, for instance, to insist, as I will in the argument to follow, contra Callicott, that intrinsic wild values exist that are not human values. Just because the human presence is so radically different, humans ought sometimes to draw back and let nature be. If so, we will have to debate whether all values are anthropocentric (human-centered), as Callicott thinks not, or anthropogenic (human-generated), as Callicott thinks—or at least used to think. But all this is quite outside the capacity of plants and animals. Humans can and ought see outside their own sector; they can relate their species self-interest to other natural values. And only humans have conscience enough to do this; indeed, it seems likely that only humans have conscience at all.

These contrasts between nature and culture were not always as bold as they now are. Once upon a time, culture evolved out of nature. The early hunter-gatherers had transmissible cultures but, sometimes, were not much different in their ecological effects from the wild predators and omnivores among whom they moved. Cultural discoveries are cumulatively transmissible; we would expect early cultures to have limited technologies. As culture grows, more and more power accumulates to rebuild and alter nature, more skills and information are transmitted. A few aboriginal peoples may remain today, with low-power technologies, although even they probably have accumulated rather complex cultures. But we now do not and cannot live in such a twilight society. Any society that we envision must be scientifically sophisticated, technologically advanced, globally oriented, as well as (we hope) just and charitable, caring for universal human rights and for biospheric values. This society will try to fit itself in intelligently with the ecosystemic processes on which it is superposed. But they are not going to be helped in doing so by thinking of themselves as nothing but precocious monkeys in a Chicago jungle. Overnaturalizing human affairs is not the answer.

Intrinsic Natural Value

Although Callicott is resolute about not being a dualist and separating humans from nature, he nevertheless makes a rather striking separation between humans and plants or animals. According to his value theory, nature comes to have intrinsic value only on human encounter and habitation. At least that has been his characteristic claim, although as we see later, he sometimes modifies it to include some related vertebrates. This first connects humans with nature and that seems promising, but, alas, this also prevents disconnecting nature from humans so that it can have any intrinsic value on its own—and that is disconcerting. Nature only comes to have such value when humans take it up into their experience.

Suddenly, the dichotomy comes back with a vengeance. Only humans produce value; wild nature is intrinsically valueless without humans. All it has without humans is the potential to be evaluated by humans, who, if and when they appear, may incline, sometimes, to value nature in noninstrumental ways. Maybe there is no metaphysical difference of substance or process; human activities and those in wild nature are equally natural. But there is an axiological difference of value; only humans can value anything in this way. That is quite separatist. Maybe we humans are metaphysically different after all, in process if not in substance, if we have such a remarkably different capacity.

Callicott is quite clear about our unique value-ability. All intrinsic value is "grounded in human feelings" but is "projected" onto the natural object that "excites" the value. "Intrinsic value ultimately depends upon human valuers." "Value depends upon human sentiments."[9] We humans can and ought *place* such value on natural things, at times, but there is no value already *in place* before we come. Intrinsic value is our construct, interactively with nature, but not something discovered that was there before we came. "There can be no value apart from an evaluator, . . . all value is as it were in the eye of the beholder [and] . . . therefore, is humanly dependent."[10] Such value is "anthropogenic."[11]

> The *source* of all value is human consciousness, but it by no means follows that the locus of all value is consciousness itself. . . . An intrinsically valuable thing on this reading is valuable *for* its own sake, *for* itself, but it is not valuable *in* itself, i.e. completely independently of any consciousness, since no value can in principle . . . be altogether independent of a valuing consciousness. . . . Value is, as it were, projected onto natural objects or events by the subjective feelings of observers. If all consciousness were annihilated at a stroke, there would be no good and evil, no beauty and ugliness, no right and wrong; only impassive phenomena would remain.[12]

This, Callicott says, is a "truncated sense" of value where "'intrinsic value' retains only half its traditional meaning." At the same time, "value is, to be sure, humanly conferred, but not necessarily homocentric."[13]

The word *project* here needs analysis. Motion picture projectors project an image when light travels from the projector to the screen, but we are not here to think of a value-bestowing ray. Nothing travels from the human valuer to the natural object. Rather, humans value trees somewhat like they color them green. The greenness of the tree is in my head, but it looks as though the tree is green. Out there are only electromagnetic waves of 550 nanometers. The greenness is projected, manufactured in my head and apparently hung onto the tree. Dogs, with black and white vision, project no greenness onto the same tree. I have no options about the greenness; I do have options about the valuing—to some extent. I can see the tree as board-feet of timber or a poem (Joyce Kilmer). I can value it as an instrument to satisfy my desires or I can see it as having intrinsic value.

In all this nothing travels from the human to the tree. The "projection" is better called a "translation." The "value conferring" does not transmit anything to the tree, and in that sense the value never really gets outside of the human head. The tree is *sending* and the human is *receiving*. The human is not really doing any sending, nor the tree any receiving. The incoming signals from the tree are "translated" as green, and so the tree appears green. In one sense this is an illusion; in another it is not. There is no experience of green in the tree, but there is ample reality (radiation) out there, behind and exciting my experience. My coloring the tree green is mapping what is really there, although my mind is translating as it maps. My "finding" of intrinsic value in nature is to be modeled after my "finding" green. (Green insects, camouflaged on the leaves, are protected from predators who, although they have no experience of green, have other sense modalities that catch electromagnetic signals and distinguish wavelengths).

To say that a natural *x* is valuable means that *x* is able to be valued if and when (human) valuers come along, but *x* has this property whether humans (or other valuers) ever arrive. To say that something is intrinsically valuable means that it is of such kind that were valuers to arrive they might value it intrinsically rather than instrumentally. The trilobites that went extinct before humans evolved were (potentially) intrinsically valuable. Undiscovered species on Earth now or on uninhabited planets are intrinsically valuable in this potential sense.

By this account no actual value ownership is autonomous to the valued and valuable wildflower; there is a value ignition when humans come. The object plays its necessary part, although this is not sufficient without the subject. Out there, apart from humans, there is only "a range of *potential* values in nature actualizable upon interaction with consciousness."[14]

Notice that, although anthropogenic, value is not necessarily anthropocentric. Value is not self-regarding, or even human-regarding, merely, although it is human-generated (anthropogenic). It is not centered on human well-being, although it is still tethered to human experience. Sometimes humans value nature instrumentally, as when they want soil to grow crops. Sometimes humans value nature intrinsically, as when they save endangered lemurs, refusing to convert a lemur forest sanctuary into cropland. But this is always humans doing the valuation: anthropocentric if the decision is for croplands, but still anthropogenic if the decision is for lemurs. Wild nature is value free and only becomes valuable when humans evaluate it. Also, humans err; they can (and often do) value wildflowers and lemurs insufficiently; they fail to appreciate how they can and ought to value these things in themselves.

This compromise account is certainly to be welcomed over less enlightened humanistic accounts. It affords enormously more environmental respect and protection than weaker theories. Only human beings value (evaluate) natural things; but it does not follow that when human beings do value (evaluate) things, they conclude that only humans have value. Man is the only measurer of things, but man does not have to make himself the only measure he uses. If we do, we will miss much richness in natural values. Still, values in nature have in fact been "truncated," and that is unwelcome; we may still be missing much of the richness of value in nature. This is not yet a genuinely biological or ecological theory of value, but residually a psychological one, which has to keep these humanistic bridges connecting with people as it enters the terrain of environmental ethics. Surely it is anomalous to have the philosopher who values the "land" so much, who urges a "land ethic" so intensely, finding nothing of value in the "land" at all, until we humans place or project it there.

The problem is first one of language. Callicott may use the language of valuing nature for itself, but this is misleading; value is always and only relational with humans one of the relata. Despite the language of value projection and conferral, if we try to take the term *intrinsic* seriously, this cannot refer to anything the object gains, to something within ("intra") the object because the human subject does not really project anything to the natural object. We have only this "truncated sense" of intrinsic. All the attributes under consideration are objectively there before humans come, but the attribution of value is subjective. The object causally affects the subject, who is excited by the incoming data and translates this as value, after which the object appears as having value (and color). But nothing is really added intrinsically to the object at all; everything in the object remains what it before was. Despite the language that humans are the source of value that they locate in the natural object, no value is really located there. The only new event is that these properties are registered in—translated into felt values by—the perceptual apparatus of the beholder.

The term *intrinsic*, although claimed in a truncated sense for this view, is misleading. What is really meant is better specified by the term *extrinsic*,[15] the *ex* indicating the external, anthropogenic coagulation of the value, which is not *in*, *intrinsic*, internal to the nonsentient organism, even though this value, once generated, is apparently conferred on the organism. This value is noncontributory in the sense that it is not used in some human reference frame, that is, not possessed in a rebuilt environment. The value is accepted, reflected, enjoyed just as it is. Still, human consciousness realizes this value in the organism, which the organism did not have before, but that on encounter with humans, it does come to have extrinsically.

The value-generating event is something like the light in a refrigerator—it is only on when the door is opened. Values in flora and nonsentient fauna are only "on" when humans are perceiving them and otherwise "off." This is said to be the ignition, or projection, of value, hitherto only potentially present. There are only "*potential* instrumental and intrinsic values in nature . . . awaiting actualization by a conscious physical subject."[16] But is not this like looking for time in the clock that measures it, looking for a birthday party in the camera that photographs it? I seem to be assuming that, among all the phenomena in the universe, only one sort of thing, psychological interest, produces actual value intrinsically, although I recognize that myriad things present in the world before, during, or after the presence of (human) valuers can excite such value. Actual value was not lost when the various species of trilobites went extinct, nor is value lost now when unknown species in tropical forests go extinct, bulldozed away unawares to humans.

But this leaves us with an uneasy concern that, for all this seemingly generous talk about caring for others, about our placing value there, because it is only we who can place value anywhere, humans really do remain at the center of concern; their concern is central to having any value at all. Their concern is all that matters, and being concerned for animals or plants, or species or ecosystems that really do not matter in themselves is not always going to be easy. We are more likely to be concerned only if they matter to and for us, which places humans right back at the center. Nature is actually valuable only when it pleases, as well as serves, us. That seems to be the ultimate truth, even though we penultimately have placed intrinsic value on nature and take our pleasure enjoying these natural things for what they are in themselves. Without us there is no such pleasure taken in anything. What is value–able, able to value things, is people; nature is able to be valued only if such able people are there to do such valuing. Nature is not value–able—able to generate values—on its own, nor do plants and most animals have any such value–ability, on their own. Callicott has not really gotten his values naturalized, not yet.

Callicott does enter a caveat about whether such valuing is done by humans only. Anyone who observes animals will soon see that man is not the only measure, or measurer, of things. Those lemurs, on which we chose to confer intrinsic value, may appreciate our favor; they will also take a dim view of any such anthropogenic theory, no matter how generous, because lemurs, all by themselves, value insects and fruits instrumentally as food to eat. They do not behave as if these were anthropogenic values at all. They were doing these things before any humans came to Madagascar approximately 1,500 years ago. The value of the food they eat is not "humanly conferred." Lemurs cannot reflect on value theory, of course; they cannot self-consciously evaluate their value theory, but they can behaviorally demonstrate what they value. And humans, who can reflect on value theory, ought to be able to see that the lemurs are not valuing anthropogenically at all. They have their own ends. There is autonomous intrinsic value, not just anthropogenic intrinsic value.

Callicott has come to accept this point increasingly over the years; indeed he recognized this possibility from the start.[17] Value, he now says is not always "anthropogenic"; it may sometimes be "vertebragenic, since nonhuman animals, all vertebrates at the very least, are conscious and therefore may be said, in the widest sense of the term, to value things."[18] Well, that is a help because at least the vertebrates (including the lemurs) share in our ability to value things. But how do these fellow vertebrates value things? They value things instrumentally, no doubt, because they seek other plants and insects for food; they value water to drink, their dens for shelter, and so on.

Do they value anything intrinsically? Callicott does not address this question, but perhaps he would say (and I would agree) that a vertebrate animal values its own life intrinsically. The lemur defends its life as a good of its own; it desires to live its own life. Such life is valued without further contributory reference, even if boa constrictors in turn make use of lemurs for food. Perhaps the lemur can value its young intrinsically because the mother lemur puts herself at risk to bear young and values the ongoing species line.

Do these nonhuman vertebrates have our human capacity to place intrinsic value on other individual plants and animals, on species, or ecosystems other than themselves? Presumably not. No lemur is ever going to become concerned about valuing boa constrictors for what they are in themselves or saving that species line. Any vertabragenic value is going to be vertebracentric for just that species and no further. So it is humans alone who have this remarkable ability to value intrinsically something other than themselves.

Meanwhile, the vertebrates comprise a very small fraction of the animals, much less of the living things. What are we to say of the insects, or the worms, or the trees, or the wildflowers? Bees cannot value honey unless we can find enough neurons in them to provide consciousness. Plants cannot value their seeds, or the lives they defend, because they have no vertebrae or neurons at all. When we run out of psychological experience, value is over. That still leaves most of the world valueless because the vertebrates are only about 4 percent of the described species. Indeed, because most as yet undescribed species are not vertebrates and because the numbers of individuals in vertebrate species is typically much lower than the numbers of individuals in invertebrate species, or in plant species, real valuers form only some minuscule fraction of the living organisms on Earth. Nearly everything on Earth is still quite valueless, unless and until these humans come along and place intrinsic value there. As Callicott admits, until humans do this, "there simply is no inherent or intrinsic value in nature."[19]

All this seems to fall short of valuing what an ecosystem is in itself, a healthy, lively place whether or not we humans are around, full of animals and plants, including vertebrates, who are defending their own lives for what they are in themselves, each with their own modes of coping, only a few of whom have the capacity for consciously evaluating what they are doing. A powerful emotion when leaving culture to return to nature is the sense of entrance into a natural place flourishing independently of any human presence. The forces by which natural systems run are not human forces; they are the biological and physical forces that have generated the world. Wild creatures are selected for their fitness in the places they inhabit; the wilderness is a complex tapestry of values with each living thing defending itself, with vital needs, and the whole system a network in which goods are circulated round and integrated into other goods through both conflict and complementarity. The natural history that envelops us is of value, not only because we humans place value there, but because value is there regardless of whether we value it.

By now we begin to suspect that the anthropogenic account of intrinsic value is a strained saving of what is really an inadequate paradigm, that of the subjectivity of value conferral. A thoroughgoing value theory in environmental ethics is more radical than this; it fully values the objective roots of value with or without their fruits in subjectivity. Sometimes to be radical is also to be simpler. The anthropogenic theory of intrinsic value insists on the subjectivity of value conferral while trying hard to preserve the object with all its properties. It admits that the exciting object is necessary for generating value.

A simpler, less anthropically based, more biocentric theory holds that some values are objectively there, discovered rather than generated by the subjectivist valuer. A fully objective environmental ethics can quite enjoy a "translator" when subjective appreciators of value appear. It can value such appreciation (experienced respect) more highly than untranslated objective value. Value appreciates (increases) with humans. But such an ethic does not insist on a human translator for value to be present throughout 99 percent of the creation. That commits a fallacy of the misplaced location of values. It has not yet naturalized value.

Trees may not be colored without a perceiver, but they do exist per se. Is their value like their color or their existence? Trees have their norms and needs, defenses, programs; these are factors in their existence, and so value, coupling with existence defended, is not an analog of color after all. Trees do appear to be green and perhaps we do not want to call the electromagnetic waves actually there "greenness." Trees are also valuable in themselves, able to value themselves; they stand on their own. By contrast with "greenness," we do want to say that "treeness" is objectively there, the tree with its life project defended. We want to call this valuable regardless of what "seems" to us. Some values are already there, discovered not generated by the valuer because the first project here is really the natural object, nature's project; the principal projecting is nature creating formed integrity. Beside this, the human projecting of value is an epiphenomenon. The theory of anthropogenic intrinsic value needs to give place to a theory of autonomous intrinsic value.

Natural selection picks out whatever traits an organism has that are valuable to it, relative to its survival. When natural selection has been at work gathering these traits into an organism, that organism is able to value on the basis of those traits. It is a valuing organism, even if the organism is not a sentient valuer, much less vertebrate, much less a human evaluator. And those traits, although picked out by natural selection, are innate in the organism, that is, stored in its genes. Dissociating the idea of value from natural selection is difficult.

Any sentient, psychogenic, vertebragenic, or anthropogenic theory of value has got to argue away all such natural selection as not dealing with "real" value at all, but mere function. Those arguments are, in the end, more likely to be stipulations than real arguments. If you stipulate that valuing must be felt valuing, that there must be somebody there, some subject of a life, then trees are not able to value; their leaves and thorns are no good to them and that is so by your definition. But what someone advocating a "land ethic," with its focus on members of biotic communities, wishes to examine is whether that definition, faced with the facts of biology, is plausible. Perhaps the sentientist definition covers correctly but narrowly certain kinds of higher animal valuing, namely that done by humans and their vertebrate relatives, and omits all the rest.

Callicott seems to be misled by thinking that all relationships can be modeled after a particular reading of quantum theory in which the observer interacts with what is observed. From this he draws sweeping conclusions:

> Mass and motion, color and flavor, good and evil, beauty and ugliness, all alike, are equally potentialities which are actualized in relation to us or to similarly constituted organisms. . . . *No* properties in nature are strictly intrinsic, that is, ontologically objective and independent of consciousness. Borrowing now from the vocabulary of quantum theory, we may assert, rather, that values are virtual. Virtual value is an ontological category encompassing all values. Within its purview fall the entire spectrum of instrumental and inherent values. . . . Inherent value is a virtual value in nature actualized upon interaction with consciousness.[20]

That is implausible. Yes, the tree is not experienced as colored green until interacting with consciousness. But the tree is photosynthesizing. The activity and the energy captured and stored metabolically is valuable to the tree quite ontologically objectively and independently of any consciousness, human or otherwise. There is nothing virtual about that. Perhaps the food in the refrigerator is not colored until we open the door; perhaps it is not tasty until we consume it. But the energy stored in the potatoes was there in the dark, and it was first put into the potato underground because it was of value to the plant, whether any humans or other conscious evaluators even came on scene.

Callicott holds that in a modern scientific perspective, a tree's goodness is not more objective than its greenness. If one "grants that there are independent ('free-standing') objects and correspondingly independent subjects, and primary qualities and secondary qualities . . . all the argument in the world to the effect that goodness is more objective than greenness is going to look like a magic show, brought off with smoke and mirrors."[21] Callicott wishes to be postmodern rather than modern, of course, but he continues the same line of argument. "After thinking very hard, during the mid-1980s, about the ontology of value finally I came reluctantly to the conclusion that intrinsic value cannot exist objectively."[22] There is "no 'truth' or 'falsity' to value judgments, since there are no objective or intrinsic values to which value judgments may or may not correspond."[23]

But my reply is that photosynthesis is indeed more objective than greenness, and that this is exactly what modern science teaches, not to be undone by some postmodern smoke and mirrors. Photosynthesis is quite true, and quite valuable to the tree, and all this quite objective. Quantum theory does not make photosynthesis subjective in the slightest. What is good for a tree (nitrogen, carbon dioxide, water) is observer-independent. This leads at once to the fact that the good of the tree (whether it is injured or healthy) is equally objective. The tree's defense of its own life, its coping based on DNA coding, is quite objective (even if, no doubt, there is some observer construction in the theories and instruments by which all this is known). The sequoia tree has, after all, been there 2,000 years, whether any green-experiencing humans were around. *Sequoia sempervirens*, the species line, has been around several million years, with each of its individual sequoia trees defending a good of their kind.

Those who value wild nature, having discovered the intrinsic natural values that we have been defending, wish to preserve natural processes as well as natural products. Humans can and ought to see outside their own sector and affirm nonanthropogenic, noncultural values. Only humans have the cognitive power to erect cultures that destroy wild nature. Humans must, and ought to, destroy wilderness when they build their cultures; neither agricultural nor urban lands can be wilderness. At the same time, only humans have conscience. That conscience emerges for the building of culture to relate humans to other humans with justice and love, but it also emerges—so environmental ethicists are now arguing—for the relating of humans to nature, to the larger community of life on the planet. That relationship, governed by conscience (and also by pragmatic self-interest), requires a harmonious blending of nature and culture, where this is possible. The same conscience also generates a duty that respects wild nature at some times and places for values present there independently of humans.

So the problem with Callicott, repeatedly, is to get his environmental philosophy really naturalized. He so resolutely opposes dichotomizing humans and nature that he cannot find any integrity for nature on its own. He remains, for a would-be naturalist, surprisingly humanistic—with people projecting their values onto nature, with people managing their landscapes. No doubt this is indeed required; it is half the truth in environmental ethics. But it is not the whole truth.

Notes

1. *Nicomachean Ethics*, 1096a.
2. J. Baird Callicott, "Rolston on Intrinsic Value: A Deconstruction," *Environmental Ethics* 14 (1992): 129–143.
3. J. Baird Callicott, "La Nature est morte, vive la nature!" *Hastings Center Report* 22, no. 5 (September/October 1992): 16–23, citation on p. 16.
4. Callicott, "Nature," 17.
5. Callicott, "Nature," 16.
6. Callicott, "Nature," 16.
7. Aldo Leopold, *A Sand County Almanac* (New York: Oxford University Press, 1968), 221.
8. Robert Costanza, Bryan G. Norton, and Benjamin D. Haskell, *Ecosystem Health: Now Goals for Environmental Management* (Washington, D.C.: Island Press, 1992), 9.
9. J. Baird Callicott, "Non-anthropocentric Value Theory and Environmental Ethics," *American Philosophical Quarterly* 21 (1984): 299–309, citation on p. 305.
10. J. Baird Callicott, *In Defense of the Land Ethic* (Albany, N.Y.: SUNY Press, 1989), 26. From "Animal Liberation: A Triangular Affair," *Environmental Ethics* 2 (1980): 311–338.
11. Callicott, "Rolston," 132.
12. Callicott, *In Defense*, 133–134, 147. From "On the Intrinsic Value of Nonhuman Species," in *The Preservation of Species*, ed. Bryan G. Norton (Princeton, N.J.: Princeton University Press, 1986), 138–172.
13. Callicott, *In Defense*, 133–134, 151. For "truncated intrinsic value," see also Callicott, "Rolston," 132.
14. Callicott, "Rolston," 129.
15. This is the traditional terminology. "No objective existent has strictly intrinsic value; all values in objects are extrinsic only. . . . The goodness of good objects consists in the possibility *of* their leading to some realization of directly experienced goodness." C. I. Lewis, *An Analysis of Knowledge and Valuation* (LaSalle, Ind.: Open Court, 1946), 307. All that nonsentient organisms offer is the standing possibility of valuation; they do not have intrinsic value, nor do they gain it by human conferral.

16. Callicott, "Rolston," 140.
17. Callicott, *In Defense*, 26.
18. Callicott, "Rolston," 132, 138.
19. Callicott, *In Defense*, 160.
20. Callicott, *In Defense*, 169–170.
21. Callicott, "Rolston," 138.
22. Callicott, "Rolston," 132.
23. Callicott, *In Defense*, 164.

7

Epistemology and Environmental Values

Bryan Norton

Gifford Pinchot, the first official U.S. Forester, wrote: "There are just two things on this material earth—people and natural resources."[1] This philosophy apparently implies that all things other than people have only *instrumental* value. To counter thinking such as Pinchot's, some philosophers have argued that some nonhuman natural objects have value, sometimes called *intrinsic* value, independent of human use of them. Within environmental ethics, this question has taken on a life of its own.

At first it seemed that Pinchot's anthropocentric instrumentalism and "nonanthropocentrism"—the view that nature has value extrinsic to human consciousness—represented mutually exclusive and exhaustive choices as foundations of environmental ethics.[2] Accordingly, the proposed anti-Pinchotist theory argues that objective values must be located outside the conscious subject, in the world "out there." I argue, however, against this value realism, contending that it is thoroughly dependent on the Cartesian dualistic distinction between subject and object, and only makes sense in the Cartesian, modern paradigm. The nonanthropocentric theory, therefore, shares with Pinchot's view, which it hopes to replace, a common Cartesian assumption, that the world is necessarily organized into conscious subject and external object.

As we undertake the task of replacing the modernist worldview, we must keep in mind the possibility that Pinchot's errors are best criticized without appeal to nonanthropocentric, intrinsic values, that a truly environmental ethic will require a wholly new environmental philosophy—not merely an extension of traditional ethical categories to nonhumans—and that the task of understanding environmental values must proceed simultaneously with the construction of a postmodern and post-Cartesian paradigm in epistemology and metaphysics. This more radical approach involves questioning the common epistemological assumptions of both Pinchotism and its nonanthropocentric competitors. The positing of values "out there" in the world does not, that is, constitute the only alternative to Pinchot's narrow instrumentalism. Furthermore, I believe that preoccupation with the search for intrinsic value has diverted philosophers from the more important and creative work of constructing a new, ecological worldview, a new vocabulary, and a more adequate style of thinking about human roles in the natural world.

Why Intrinsic Value?

We might ask how philosophers explain their preoccupation with the hypothesis of intrinsic value in nature. Experts offer two explanations.

The first explanation rests on a commitment to respond to the need of environmentalists to claim that their values and goals are "objective," epistemologically, and that these values rest on more than subjective preferences. Independent values in nature are necessary, according to J. Baird Callicott, because they will rescue environmental management "from reduction to cost-benefit analyses in which valued natural aesthetic, religious, and epistemic experiences are shadow priced and weighed against the usually overwhelming material and economic benefits of development and exploitation."[3] Only if they are working to save nature for its own sake, according to this line of reasoning, will environmentalists be pursuing objectively worthy goals. The implication, then, is that if goals sought by environmentalists are supported only by their subjective, personal, or culture-bound tastes, environmentalists will be no better than their opponents who block actions to redress environmental harms because of their preferences for rapid development and monetary profits.

Callicott explains the second reason by citing early essays in environmental ethics, and he quotes Tom Regan, who concluded that, "The development of what can properly be called an environmental ethic requires that we postulate inherent value in nature."[4] Callicott apparently accepts Regan's requirement and paraphrases his reasoning: "Otherwise, according to Regan, a putative environmental ethic would collapse into a 'management' ethic."[5] Regan and his challenge to create a "real" and independent environmental ethic therefore sets Callicott's agenda: to create an independent discipline of environmental ethics by offering a theory of inherent (or intrinsic) value. I argue later that this second philosophical reason is parasitic on the first. I turn first, however, to a discussion of independent value in nature as a support for epistemic objectivity of environmental policy goals: do attributions of human-independent values in nature support, or help to support, claims that the goals of environmentalists are objectively justified?

Before examining the epistemological status of attributions of independent value to natural objects, distinguishing between two importantly different theories regarding that value is necessary.[6] Some advocates of independent value in nature believe that nature is valuable in the strong, "intrinsic" sense that natural objects have value entirely independent of human consciousness. According to this theory, the value in nature existed prior to human consciousness, and it will continue to exist even after human consciousness disappears. Other theorists adopt a less heroic version of the hypothesis, accepting that valuing is a conscious activity and that value, therefore, will be only "inherent" in nature. According to the inherentists, nature has value that is independent of the values and goals of human valuers—it is not merely instrumental to human ends—but this value is attributed by conscious valuers, either human or otherwise.

We cannot overestimate the importance of this distinction between autonomous intrinsic value and attributed inherent value; in particular, advocates of inherent value believe that attributions of value to nature are culture-specific and always projected within a culture-laden milieu, whereas the main motivation for defenders of intrinsic value is to avoid cultural relativity by insisting that natural value is independent of human consciousness.

Callicott has most actively discussed the difference between strongly and weakly independent values and has been the most-cited advocate of the latter view. Callicott offers a theory of "inherent value" by which he means value that natural objects have "*for themselves*, quite independently from the satisfying aesthetic, religious, or epistemic experiences they may occasion in nature aesthetes, nature worshipers, or natural scientists."[7] Callicott also calls this a "truncated" theory of intrinsic value: "An intrinsically valuable thing on this reading is valuable *for* its own sake, *for itself,* but is not valuable in itself, that is, completely independent of human [or other valuing] consciousness."[8] Like Holmes Rolston III, one of the most respected advocates of strongly independent values in nature, he assumes the importance of an independent environmental ethic and, like Rolston, he assumes that such an ethic will require development of a nonanthropocentric, noninstrumental theory of value. He says that although "the central theoretical problem for environmental ethics [is] the construction of a coherent and persuasive theory of intrinsic or inherent value in nature, intrinsic or inherent value in nature in the strict, objective sense of the terms must by definition be abandoned if one assumes a subjectivist axiology. Nevertheless, in an important sense, consistently with this axiology, persons and other natural beings may be valued *for themselves* as well as for the utility they afford those who value them."[9]

What the two positions, so different in their outcome and in their implications for environmental values, have in common is a belief that the following, apparently quite different, questions are philosophically equivalent:

1. The epistemic question: Can environmentalists claim that their goals and the value claims that support them are epistemically justifiable, that they are more than merely subjective preferences?
2. The locational question: Can environmentalists' values be located "out there" in the world itself, independent of human consciousness?

Defenders of independent value in nature are clearly unified by a commitment to a particular conception of objectivity. According to this conception: For any characteristic *Z*, *Z* can be objectively attributed to an object *x*, only if subject *S* "finds," or "locates," *Z* in *x*; both *x* and *Z* must, that is, exist independently of human consciousness.

Because they share this basic criteriological assumption, the positions of Callicott and Rolston fall in direct opposition to each other: Rolston believes, and Callicott denies, that achieving "objectivity" for environmental values is possible, according to this locational criterion. Callicott, for example, states the issue as follows: "The very sense of the hypothesis that inherent or intrinsic value exists in nature seems to be that value *inheres* in natural objects as an intrinsic characteristic, that is, as part of the constitution of things. To assert that something is inherently or intrinsically valuable seems, indeed, to entail that its value is objective."[10] Callicott, however, believes that there are "insurmountable logical impediments to axiological objectivism,"[11] leaving him in apparently unavoidable skepticism about inherent values.

Inherent Value

What are we to say about Callicott's theory of "inherent" or "truncated intrinsic" value? Might we, recognizing the epistemological pitfalls of strongly intrinsic value, claim that value properties merely "inhere" in objects themselves? Callicott has in his recent writings acknowledged that any sense of value "out there" is associated with a modernist worldview, and he only refers to inherent value since 1982 to characterize objectivity *within the modernist paradigm*.[12] Because Callicott claims also to reject the modernist paradigm, it seems to follow that he should reject the very distinction between value "in here" and value "out there." If Callicott is a true postmodernist, his assertions that nature has inherent value would make as much sense as a modern physician's assertion that the bodily humours tend to collect in the feet.

Nevertheless, Callicott has several reasons for retaining an interest in defending the view that natural objects have inherent value, provided the objects are described within a Cartesian, representational view of the world. First, Callicott has noted that probably most environmentalists remain broadly Cartesian in their worldview and for these environmentalists the subject-object dichotomy remains live.[13] Second, Callicott, who is best known as the leading philosophical exponent of Aldo Leopold's land ethic, believes Leopold interpreted the world in a Cartesian way and, consequently, he interprets Leopold's comment that he perceives "philosophical" value in nature as a commitment to value in nature independent of human utility.[14] Thus, most of what Callicott has written on inherent and intrinsic value is to be understood hypothetically—they are his views only insofar as his views are described within an inadequate Cartesian conceptual framework, and only insofar as he is interpreting Leopold's version of the land ethic.

Most important, however, Callicott wishes to establish inherent value because of his professional agenda, because of his desire to respond to Regan's challenge to construct a noninstrumentalist and nonanthropocentric environmental ethic. Callicott's other motivations, and their pursuit, must be understood against the backdrop of this agenda. He thinks that offering a theory of intrinsic or inherent value for nonphilosophical environmentalists is important because Leopold's land ethic remains the locus classicus for a noninstrumentalist environmental ethic, which he assumes to be grounded in a modernist worldview.

In fact, I believe Leopold saw, perhaps only dimly, past Cartesian modernism, and that his land ethic was based, both ethically (as Callicott perceives) and epistemologically on a broadly Darwinian worldview (as Callicott denies). My evidence for this conclusion, which I have presented in detail elsewhere, is that Leopold applied the pragmatic definition of truth when discussing "philosophical" questions to which he despaired of answers.[15]

However, Callicott's inherent value project fails, even within a Cartesian framework, to achieve its original goals—the desire to help environmentalists claim moral authority in policy debates and the desire to establish a noninstrumental environmental ethic. For the practical problems of environmentalists, Callicott's subjectivist theory represents a total capitulation because, as he notes, his theory that nature has value for itself, but not in itself, falls short of objectivity as demanded by the Cartesians. Callicott admits that the values in nature are ascribed or attributed from within human consciousness, and follows modernists in characterizing them as therefore "subjective." But, he says: "In the literature of normal metaethics the line of subjectivism is that it necessarily degenerates into a morally intolerable radical relativism." What one fears is that Callicott, straddling the Cartesian and the post-Cartesian worldviews, has pitched out the baby of objectivity while saving the bathwater of Cartesian representationalism in the form of the assumption that the epistemological objectivity of values requires placement of them "out there."

Callicott, as solace to environmentalists, proceeds to explain that environmental values are not radically relative because, following Hume's subjectivist axiology supported by Darwinian theory, one can assert that the urge toward altruism is genetically fixed, that core values exist that all humans follow unless they are "morally abnormal." These Humean moral sentiments can be cognitively redirected to include in the moral community our ecological fellow travelers, the species and ecosystems that are our more distant zoological cousins, which are then valued for their own sake and not merely anthropocentrically and instrumentally. The science of ecology can teach us to broaden the applicability of value attributed "for itself." Thus Callicott opts for a Darwinian, naturalistic ethic within a Humean subjective framework of knowledge and understanding. We value nature, according to this theory, as a parent values the child. The value is attributed by the parent, but it is noninstrumental value, value that does not depend on the child's contribution to any selfish goal of the parent. It is not, however, located independently in the child and is therefore not objective in the locational sense. Splitting hairs, Callicott says that this value is anthropo*genic*, but not anthropo*centric*.[16] This value, Callicott argues, is adequate to establish that an ethic of nature exists, and not just an ethic for the management of nature.

The irony now becomes clear: Callicott has denied the objectivity of environmental values in any sense that would be helpful in environmental policy because environmental values are the subjective judgments of those environmentalists who are convinced that we need an ecological ethic. Nevertheless, Callicott has responded to Regan's challenge: such so-called inherent values are independent in a weaker sense than Regan meant to require; environmental ethics can therefore claim "independence" in this weaker sense. I doubt that active environmentalists care, or should care, about whether environmental ethics is independent in either of these senses, but I especially doubt whether they should care about independence in Callicott's weak sense that results in value subjectivism. Worse, it seems all but evident that the original worry for a nonanthropocentric environmental ethic was based on the discredited Cartesian assumption that epistemic objectivity requires location outside of (i.e., independent of) human consciousness. As Callicott realizes,

Regan's original challenge presupposed an exclusive dichotomy between anthropocentric and instrumental on the one hand, and nonanthropocentric and noninstrumental-independent on the other. Once that dichotomy collapses and noninstrumental, anthropocentric values are recognized, Callicott's argument for noninstrumental but anthropogenic values is at best splitting hairs.

It seems much more reasonable, and simpler, to follow Eugene Hargrove, who believes the distinction between instrumental and noninstrumental value is important, but who views it as a distinction between two types of anthropocentric values. Or, more radically, to follow me in the path of John Dewey, who rejected the idea of intrinsic value altogether. I do not follow Dewey in therefore assimilating all value to instrumental value. I simply choose to stop using the tainted terms. This more radical course obviates the need for hairsplitting and, more important, points the way toward an exploration of anthropocentric, noninstrumental values within a relational theory of perception. Once we reject the Cartesian, locational criterion of objectivity, seeking alternative epistemological justifications for environmental goals and values is possible.

Callicott, unfortunately, never poses the question: Must attributions of epistemically objective value locate that value in "the constitution of things"? If not, and there is another route to epistemic objectivity, then we can abandon the "out there," locational sense of objectivity associated with Cartesian modernism and undertake the task of reconstructing objectivity from within a relational theory of perception.

Callicott's Post-Cartesian Theories of Value

Callicott, as we have already noted, explicitly recognizes the Cartesian roots of his inherent value theory—his hairsplitting, subjectivist theory of nonanthropocentric but not anthropogenic noninstrumental value—and has discussed two post-Cartesian alternatives to it. He examines what he takes to represent the core idea of deep ecology, and he sketches his own view, which he describes as a "quantum theoretical axiology" for human values, but one that "is intended to be as conservatively and uncontroversially stated as possible."[17]

Let us first examine Callicott's treatment of the more speculative theory of deep ecology with which he flirts. Quoting Fritjof Capra, the popular expositor of the new physics, Callicott describes a theory that not only questions the fact-value distinction (because it rests on the Cartesian subject-object distinction), but that also "goes further. 'A basic oneness in the universe' is also implied which include[s] the observer [the 'I'] in an essential way." Callicott acknowledges that this theory is attractive because it is at least analogically consonant and structurally isomorphic with an "ecological" approach to understanding reality. Adopting Alan Watts's view that "the world is your body," this approach denies "the conventional separation between self and world" and implies that harms to the rain forest are harms to our extended selves. In this sense the rain forest is intrinsically valuable if the self is. Deep ecology therefore provides, according to Callicott, one simple and direct solution to "the central axiological problem of environmental ethics, the problem of intrinsic value in nature."

Callicott, however, finally abandons this project, because he perceives that the hidden assumption, that the self is intrinsically valuable, has its genesis in the supposedly rejected Cartesian dichotomy between subject and object. Nevertheless, ever true to the agenda of providing an independent environmental value based on nonanthropocentrism, Callicott seems to offer encouragement based on a theory that even he recognizes is theoretically inadequate: "Since nature is the self fully extended and diffused, and the self, complementarily, is nature concentrated and focused in one of the interactions, the 'knots' of the web of life or in the trajectory of one of the world lines in the four dimensional space-time continuum, nature is intrinsically valuable *to the extent* that the self is intrinsically valuable."[18]

Callicott's own theory, which he prefers because it does not implicitly reify a moral ego, nevertheless finds noninstrumental value in nature because "a fully consistent contemporary environmental ethic . . . requires a theory of the noninstrumental value of nature which is neither subjectivist nor objectivist." The new axiology must "not rest, either explicitly or implicitly, upon Descartes' obsolete bifurcation."[19]

From this promising beginning, Callicott proceeds to put the best face possible upon our failure to achieve locational objectivity in science or ethics: "Mass and motion, color and flavor, good and evil, beauty and ugliness, all alike, are equally potentialities which are actualized in relation to us or to similarly constituted organisms."[20] But Callicott, still yearning for Cartesian objectivity, is finally forced to admit that "the difference between this account of inherent value and the account of axiological subjectivism is less practical than theoretical," and he offers a post-Cartesian substitute for true Cartesian objectivity: "inherent value is virtual value in nature actualized upon interaction with consciousness."[21] The original problem, that nothing short of epistemic objectivity will be helpful to environmental activists, has been pushed into the background by a defensive argument that environmental ethics is as objective as science, or, as Callicott phrases it: "Physics and ethics are, in other words, equally descriptive of nature."[22]

What unifies Callicott's position, his guiding light in any worldview from Cartesianism to deep ecology, is a commitment to the ideal of Cartesian objectivity. Realizing that such objectivity is a will-o'-the-wisp, Callicott strives for various substitutes. Eloquent in his defense of Darwinism in ethics, Callicott refuses to repudiate the ideal of Cartesian, locational objectivity and adopt a naturalistic, post-foundational epistemology. The only motivation for this hesitancy, as far as I tell, is a commitment to a rather phony sense of nonanthropocentric noninstrumental value as a subject matter for environmental ethics.

Environmental ethics needs a revolution; the field needs to focus on a whole new set of post-Cartesian questions, and these questions will carry practitioners of environmental ethics far beyond the confines of their "independent" (i.e., nonanthropocentric, noninstrumental) ethical theory. Environmental ethics will only achieve adolescence when it escapes Cartesian structures. Perhaps by the time the field reaches maturity and is able to begin creating a new, post-Cartesian worldview, the emphasis will be less on protection of conceptual independence in the form of distinctively philosophical concepts such as "nonanthropocentrism" or "noninstrumentalism," and more effort will be expended in breaking down barriers among disciplines. Environmental philosophy, because it should be the study of how to formulate and understand limits inherent in the world as we encounter it, should be the focus of a postdisciplinary debate regarding how to achieve an adequate conceptualization of our world.

Notes

1. Gifford Pinchot, *Breaking New Ground* (1947; reprint, Washington, D.C.: Island Press, 1990) 325.
2. See for example, Tom Regan, "The Nature and Possibility of an Environmental Ethic," *Environmental Ethics* 3, no. 1 (1981): 19–34.
3. J. Baird Callicott, *In Defense of the Land Ethic: Essays in Environmental Philosophy* (Albany, N.Y.: SUNY Press, 1989), 163.
4. See J. Baird Callicott, "Whither Conservation Ethics?" *Conservation Biology* 4 (1990): 157. The quotation is from Regan's essay, "The Nature and Possibility of an Environmental Ethic," 34.
5. Regan, "Nature," 34.
6. Considerable confusion surrounds the use of terms such as *intrinsic*, *inherent*, and so forth, but I will follow the definitions of Callicott. See Callicott, *In Defense,* 161.

7. Callicott, *In Defense*, 163.
8. Callicott, *In Defense*, 133.
9. Callicott, *In Defense*, 160–61.
10. Callicott, *In Defense*, 160.
11. Callicott, *In Defense*, 159.
12. See for example, "Just the Facts, Ma'am," *Environmental Professional* 9 (1987): 279–288, and "Rolston on Intrinsic Value: A Deconstruction," *Environmental Ethics* 14 (1992): 129–143.
13. Personal communication, May 1991.
14. See Aldo Leopold, *A Sand County Almanac* (New York: Oxford University Press, 1949), and J. Baird Callicott, "The Philosophical Value of Wildlife," in *Valuing Wildlife: Economic and Social Perspectives,* ed. Daniel J. Decker and Guy R. Goff (Boulder, Colo.: Westview Press, 1987), 214–221.
15. See Bryan G. Norton, "The Constancy of Leopold's Land Ethic," *Conservation Biology* 2 (1988): 93–102.
16. Callicott, "Rolston on Intrinsic Value," 132.
17. Callicott, *In Defense*, p. 170.
18. These quotes summarize Callicott's argument in *In Defense*, 171–174.
19. Callicott, *In Defense*, 166.
20. Callicott, *In Defense*, 169.
21. Callicott, *In Defense*, 170.
22. Callicott, *In Defense*, 170.

Part III

Metaphysics and Metaethics

8

Environmental Ethics without a Metaphysics

EUGENE C. HARGROVE

J. Baird Callicott has written that the moral pluralism of Christopher Stone, Peter Wenz, and myself, among others, "severs ethical theory from moral philosophy, from the metaphysical foundations in which ethical theory is, whether we are conscious of it or not, grounded."[1] In contrast, Callicott has concluded that his own environmental ethics "involves one metaphysics of morals: one concept of the nature of morality (as rooted in moral sentiments), one concept of human nature (that we are social animals voyaging with fellow creatures in the odyssey of evolution), one moral psychology (that we respond in subtly shaded ways to the fellow members of our multiple, diverse, tiered communities per se)."[2] Although in this particular instance Callicott appears to be equating "metaphysics" with a "metaphysics of morals," he has discussed metaphysics more broadly, specifically criticizing the "fundamental ontology of atomic materialism."[3] The object of that discussion is to develop a metaphysics of interrelatedness, reminiscent of Whiteheadian prehension, but in the context of the self-realization of deep ecology.[4]

In this essay, I examine the role of metaphysics in environmental ethics, using the positions of Callicott and myself as points of contrast. This approach is useful because we are each working in the context of specific traditional philosophers—Hume in the case of Callicott and Aristotle in my own case. I argue that despite such monistic beginnings, maintaining a single, monistic metaphysical framework is not possible. Although environmental ethics must deal with metaphysical considerations, these considerations are so diverse that they are unlikely to form a single metaphysics.

The term *metaphysics* was assigned to some of the writings of Aristotle by Andronicus of Rhodes sometime around 70 B.C. to distinguish Aristotle's general views about the nature of being from his more scientific writings, to which the term *physics* is applied.[5] The term *ontology* is closely associated with *metaphysics* and in some circumstances can be used in place of it. This term, literally the "logic of being," was coined by Aristotle in the context of his First Principles and was introduced into philosophy as a general term, unrelated to Aristotle's philosophy, by Christian Wolf in the eighteenth century.[6] I prefer to talk or write ontologically rather than metaphysically because I am not engaged in system building. A metaphysics is usually an elaborate system developed through reason that provides a rational explanation of the nature of being or existence. This general overview of being or existence then becomes the perspective from which more specific matters are viewed, for example, science or physics. By claiming that I am dealing with ontological matters, I am presenting my metaphysical deliberations in a more modest fashion so that no one will expect that I am trying to build a grand metaphysical artifact.

Despite my modesty in metaphysical matters, however, I do hold that metaphysical or ontological problems are found in the history of philosophy that create difficulties in environmental ethics. In the first chapter of my book *Foundations of Environmental Ethics*, I argue that metaphysical considerations made thinking in environmental terms difficult for philosophers in the ancient and modern periods and that the training philosophers receive in the history of philosophy produces attitudes in contemporary philosophy that continue to inhibit research in the field of environmental ethics and the acceptance of that research in mainstream philosophy.[7] From the standpoint of ancient philosophy, I hold that the idea that the real is indestructible and unchanging made thinking in terms of the preservation of nature difficult. I argue that a proper perspective did not arise until the end of the eighteenth century with the appearance of uniformitarian geology, according to which nature changes slowly in accordance with physical and chemical processes. This conception of nature approaches the tenacity of the Greek conception because although change is normally very slow, it recognizes that nature can also be fragile when it is susceptible to catastrophic change.[8] From the perspective of modern philosophy, I hold that epistemological concerns about the existence of the external world made playing a role in the history of ideas that produced contemporary environmental thought impossible for philosophy and philosophers.[9]

John Cobb Jr., writing in 1970, noted that the then very early environmental ethics literature tended to appear most successfully as a blend of ordinary language analysis and Aristotelian philosophy. Ordinary language analysis, he writes, "is largely in the empirical tradition, but the paradoxical results of empiricism have lost their hold. Since the common language corresponds roughly with Aristotelian philosophy, the results of its analysis tend to return to that tradition."[10] My work tends to follow this pattern (a mix of Wittgenstein and Aristotle) even though I frequently borrow elements of such diverse philosophers as Descartes, Sartre, Hume, Whitehead, and Moore. Aristotle can easily be converted into a uniformitarian geologist and Wittgenstein deals with the epistemological tangles that form the problem of the existence of the external world.

As an Aristotelian, I do not see the point of developing a metaphysics for my environmental ethic. Aristotle himself considered his ethics and metaphysics to be distinct and largely unrelated subjects. To be sure, similarities exist between them. For example, just as the actualized properties of a physical object are between two opposites (warm is somewhere between hot and cold) so are the dispositions making up moral character somewhere between two opposites called vices (courage is somewhere between cowardliness and foolhardiness). Substantive connections, however, only arise in his discussions of God, where humans are said to be hybrid unmoved movers. For the most part, metaphysics just does not come up. My focus on Aristotle concerns his conception of moral character as based in dispositions shaped by habit, his analysis of moral deliberation and wish, and his notion of moral perception and the relativity of social standards (in which the virtues of Athens and Sparta fall at different points between their respective extremes or vices). I am able to disregard Aristotle's metaphysics because he did not formally connect it to his ethics.

As a Wittgensteinian, I consider matters of existence to be tied to language (or in his terminology, language games). Metaphysical speculation quickly advances beyond what he calls the "limits of language," rendering the resulting metaphysical claims useless as part of a practical ethic that is supposed to be accepted, followed, or used by ordinary people. As Wittgenstein once noted, the difference between what children trained as idealists and those trained as realists might say is only "one of battle cry."[11] Hume makes much the same point when he notes that the "universal and primary opinion of all men [that the external universe exists independent of perception and mind] is soon destroyed by the slightest philosophy."[12] Wittgenstein, who spent several years as an elementary school teacher, points out in *On Certainty* the difficulties a child would have trying to question the existence of the external world.[13] A child who endlessly interrupted the teacher to cast doubt on the existence of the world would eventually be classified as having a psychological problem or a learning disability. According to Wittgenstein, under normal circumstances, "A child learns there are reliable and unreliable informants much later than it learns facts which are told to it. It doesn't learn at all that the mountain has existed for a long time; that is, the question whether it is so doesn't arise at all.

It swallows this consequence down, so to speak, together with what it learns."[14] What is swallowed could be called a people's metaphysics carried forward by the evolution of our societal beliefs and the evolution of our language.

Strawson has called this kind of metaphysics "descriptive metaphysics" in contrast to "revisionary metaphysics." According to Strawson, "Descriptive metaphysics is content to describe the actual structure of our thought about the world, revisionary metaphysics is concerned to produce a better structure."[15] As far as environmental ethics is concerned, I recommend that it rely on descriptive metaphysics for the most part, turning to the revisionary only when a conflict with common perception arises. This position is consistent with Strawson's view because he holds that "revisionary metaphysics is at the service of descriptive metaphysics."[16] An emphasis on the revisionary at the expense of the descriptive will, in my view, simply make environmental ethics completely esoteric and unusable. The rejection of nonanthropocentric intrinsic value by the general public and by people in environmental affairs is a good example of this problem.

In *Foundations of Environmental Ethics*, I present an ontological argument for the preservation of natural beauty. It includes a number of elements of metaphysical positions from various periods in the history of Western philosophy. Central to my position is a type of thinking that is characteristic of the modern period called *representational thinking*. Hume presents this type of thinking as the principle of resemblance, giving as an example the fact that "a picture naturally leads our thoughts to the original."[17] This type of thinking in philosophy, via Descartes's *Meditations*, created the problem of the existence of the external world. Outside of philosophy, however, it contributed to a love for natural beauty as ordinary people moved from the aesthetic appreciation of landscape paintings, the representation, to actual landscapes, the objects of that representation. This development was accompanied by a belief that natural beauty is superior to artistic beauty because a natural landscape is more real than a painting of it. This belief can be found in Hume's *Enquiry*, where he writes that, "All the colors of poetry, however splendid, can never paint natural objects in such a manner as to make the description be taken for a real landscape. The most likely thought is still inferior to the dullest sensation,"[18] and in Moore's *Principia Ethica*, where he writes that a belief that beautiful qualities exist in nature enhances the beauty of a representation of those qualities.[19] The basic intuitions of Americans who bought paintings in the nineteenth century and philosophers such as Hume and Moore suggest that existence is an aesthetic predicate of nature, much as existence is a predicate of God in the ontological argument. In addition, I use Descartes's remarks about the indifference of God in creating the world, which he borrowed from William of Ockham and others, to distinguish artistic creativity in the production of artistic beauty and indifferent creativity in the production of natural beauty through geological uniformitarianism and biological evolution. Finally, I argue, in terms of Sartre, that "existence precedes essence" applies as well or better to nature as it does to God or human beings.

Although each borrowed element normally carries considerable baggage with it, I overlooked these problems. For example, while embracing Moore's thought experiments about the value of existence, I avoided reference to his claims that value exists in nature nonnaturally. I used Hume's critique of the inferiority of imagination without accepting his view that only sensations exist. I placed Descartes and Sartre side by side despite the fact that Sartre would obviously reject Descartes's view of world as a conglomeration of mind and matter, created by God and reenacted moment by moment. I disregarded the inconsistencies between the metaphysics of these philosophers not because I was intent on creating a bad metaphysics, but rather because I was primarily concerned with showing that there is a history of ideas consistent with the history of Western philosophy on the one hand, and with the intuitions of ordinary people who care about the environment on the other. The various elements were borrowed to help explain the view that was being developed, not to create a metaphysical system. As I note in my book, the position I develop is historically contingent on specific assumptions made at various times by philosophers, most of which would not be accepted today and that would not have been accepted by non-Westerners at any time.[20]

In comparing my view with Callicott's primary position (the Hume-Darwin-Leopold-Midgley version [21]), I do not find that I have misrepresented traditional historical metaphysical positions any more or less than he has, and that if he has a metaphysics, perhaps I do too. Callicott, as I have already noted, states that he has a metaphysics in the sense that he has a "metaphysics of morals."[22] According to Kant, who as far as I know is responsible for the phrase, a "metaphysics of morals" is an understanding of the rational part of ethics. Metaphysics as opposed to logic is not "merely formal" and "is restricted to definite objects of the understanding." This understanding is focused on "morality," the rational a priori part, rather than "practical anthropology," the empirical part. [23] Callicott's metaphysics of morals, nevertheless, is focused on Hume's "moral sentiments," which are empirical, roughly the equivalent to the inclinations in Kantian terminology, making Callicott's usage of the phrase metaphorical at best, and therefore not a true "metaphysics of morals" or metaphysics at all from Kant's perspective. Granted, Hume's "moral sentiments" are real insofar as they are sensations or impressions, but the establishment of that ontological status does not constitute a metaphysics.

Nor is the matter improved with the introduction of Darwin's work. According to Callicott, quoting Flew, Hume's ethics "might almost seem to demand an evolutionary background," and, furthermore, "Darwin's theory provides a very plausible explanation, namely, that moral sentiments are fixed in human nature, like all other traits, by natural selection."[24] Although I agree with Callicott that a connection exists between Hume's philosophy and Darwin's biology, I am not sure that Hume himself would have approved of Darwin's work. Darwin's primary problem was that he was unable to give an account of how evolution worked because he had no knowledge of what genetic material is or how it interacts. Hume might have accused Darwin of developing a theory, to paraphrase from an analogous context pertaining to the existence of God, "too bold to ever carry conviction" because it does not have an impression corresponding to both the cause and the effect in evolution, and he might have added, as he did about the role of God in metaphysics, that, "We are got into fairy land, long ere we have reached the last steps of our theory."[25] Or on the other hand, he might have applauded Darwin's restraint in trying to find a "secret power" or "necessary connection," citing his own reluctance to go too far in developing an account of moral sentiments. If the first, Hume would have been accusing Darwin of engaging in improper metaphysical speculation; if the second, he would have been praising him for not engaging in metaphysical speculation. In either case no legitimate metaphysics is produced from a Humean standpoint.

A further difficulty arises in establishing a Hume-Darwin-Leopold-Midgley metaphysical connection when we turn to Leopold. Although Leopold usually does not write about metaphysics, it is not entirely missing from *Sand County Almanac*, as this passage about the possible extinction of the ruffed grouse makes clear:

> It is easy to say that the loss is all in our mind's eye, but is there any sober ecologist who will agree? He knows full well that there has been an ecological death, the significance of which is inexpressible in terms of contemporary science. A philosopher has called this imponderable essence the *numenon* of material things. It stands in contradistinction to *phenomenon*, which is ponderable and predictable, even to the tossings and turnings of the remotest star.
>
> The grouse is the numenon of the north woods, the blue jay of the hickory groves, the whisky-jack of the muskegs, the piñonero of the juniper foothills. Ornithological texts do not record these facts. I suppose they are new to science, however obvious to the discern-

> ing scientist. Be that as it may, I here record the discovery of the numenon of the Sierra Madre: the Thick-billed Parrot.[26]

This strange confusion of the term *noumenon* with what might be called *indicator species* or *keystone species* results from Leopold's reading of Ouspensky's *Tertium Organum,* where, for example, Ouspensky writes:

> The positivist finds himself in the presence of nature almost in the position of a savage in a library of rare and valuable books. For a savage a book is a *thing* of definite size and weight. However long he may ask himself what purpose this strange thing serves, he will never discover the truth from its appearance, and *the contents of the book* will remain for him the *incomprehensible noumenon.* In like manner the contents of nature are incomprehensible to the positivistic scientist. But if a man *knows* of the existence of the contents of the book—the *noumenon* of life—if he knows that a mysterious meaning is hidden under visible phenomenon, there is a possibility that in the long run he will discover the contents. For success in this it is necessary to grasp the idea of the inner contents, i.e., the meaning of the thing itself.[27]

The term *noumenon* is used in this passage in a Kantian sense, but Ouspensky intimates that although noumena are incomprehensible from a Kantian standpoint, bridging the gap between noumena and phenomena, presumably turning selected noumena into phenomena, may be possible for those who are not positivists. Because Leopold had read Ouspensky's definition of *noumenon* as the inner meaning of phenomena, Leopold's rather jolting use of the term is excusable. It does not, however, forward the development of a Leopoldian metaphysics or contribute to Callicott's thesis.

To forward his thesis, Callicott must rely on passages in "The Land Ethic," where Leopold refers to "our intellectual emphasis, loyalties, affections, and convictions,"[28] and "love, respect, and admiration for land, and a high regard for its value."[29] There are, nevertheless, problems here as well. Whereas love, respect, loyalties, affections, convictions, and so on can be tied to Hume's notion of moral sentiments, they could also be associated with Aristotelian virtues, connecting Leopold's views with my environmental ethics as well. In both cases we have connections to biology—Darwin's biology and Aristotle's biology, respectively. For example, the circles of moral considerability from self to family to citizens is equally present in Aristotelian philosophy and Darwinian-dependent philosophy. The differences between Callicott and myself actually pertain to whether virtues or moral sentiments are primarily the product of social evolution or biological evolution. In biological terms, Callicott takes his stand with the sociobiologists, tying moral behavior most fundamentally to biology,[30] whereas I side with the ethologists, who hold that the behavior is a mix of biological and social factors involving, for example, imprinting and malprinting. This issue, once again, is not a metaphysical issue.

Similar problems can also be found in Midgley's contribution to Callicott's thesis. Callicott claims that "Mary Midgley's suggested animal welfare ethic and Aldo Leopold's seminal environmental ethic . . . share a common, fundamentally Humean understanding of ethics as grounded in altruistic feelings."[31] In addition, he claims that both Midgley and Leopold share a "biosocial moral theory," according to which animals of various kinds fall into different communities,[32] corresponding to circles of moral considerability as discussed earlier. With regard to the first claim, although Midgley does have some Humean tendencies, to which Callicott correctly calls attention, she is not very committed to the view he is developing. In *Beast and Man,* where she discusses the role of feeling in ethical choice, she writes:

> Because these conflicts are so complex, it is unrealistic to suggest that they are solved by the pulling and hauling of mere unordered feeling. (There is something comic in Hume's picture of Reason as the slave of the Passions—how is it supposed to know which of them to obey? Slaves have a bad time in such circumstances.)[33]

With regard to the second claim, once again, the same conclusions follow equally well from an Aristotelian position, which denied evolution altogether. In addition, Midgley's inclusion in Callicott's thesis is problematic because he does not take into account the fact that Midgley is a critic of sociobiology. Her book, *Man and Beast*, for example, is primarily a criticism of the ethical conclusions that are normally drawn from sociobiology that relies heavily on ethological research for factual arguments.[34] The key issues with regard to Callicott's discussion are, insofar as I can determine, concerned with the role of contemporary biology, not metaphysics.

In the foregoing comments on Callicott's thesis about a connection between Hume, Darwin, Leopold, and Midgley, I in no way wish to suggest that I do not find Callicott's analysis to be an insightful and important contribution to environmental ethics literature. I think that Callicott has brought together an important set of interrelationships among the various scholars and scientists discussed. Worth noting, perhaps, is that as editor of *Environmental Ethics* I provided Callicott an opportunity to develop and present many of his ideas about these connections. I am not sure, however, that the historical figures Callicott focuses on were completely aware of these connections or that the specific contributions of each are tied to any particular metaphysical position. Darwin, for example, attempted to write about metaphysics in some sense in his secret notebooks, material that Callicott does not try to work into his theoretical schema.[35] If Callicott wants to claim that he is presenting a history of ideas that helps explain his position and that consciously or unconsciously helped bring us to our current views on these matters, I applaud his analysis. I object to it only if he wishes to claim that a specific metaphysics begins in Hume's conception of moral sentiments and is carried forward through Darwin into Leopold, Midgley, and the sociobiologists.

By way of confession, I probably should admit that as a graduate student I was trained that metaphysics is dead, and that although I have stridently objected to the positivists' analysis of ethics, aesthetics, and religion, I find their view of metaphysics on target. As far as I am concerned, A. J. Ayer is correct in arguing that many metaphysical statements do not make much sense or have any practical application, for example, "the Absolute enters into, but is itself incapable of, evolution and progress."[36] Nevertheless, anyone who thinks about environmental ethics at all, specifically with regard to nature preservation, can hardly fail to note that existence is a key concept of indisputable importance in the literature. Because of its importance, metaphysical discussion occasionally must arise. However, given that environmental ethics is supposed to be a philosophy of moral action, and not just an esoteric subfield within professional philosophy, it is not possible, or at least not wise, for environmental ethicists to wander too far into metaphysical territory. As I noted earlier, nonanthropocentrists' efforts to establish the independent existence of intrinsic value in the world have contributed to the view among environmental professionals that environmental philosophy has no practical value. It is for this reason that I recommend that environmental ethicists engage in metaphysical speculation sparingly and only when necessary for explaining or conceptualizing some matter that cannot be dealt with in any other way. Although metaphysical system building is most likely dead, the results of that activity form a vast library of material that can on appropriate occasions be brought to bear fruitfully in environmental ethics.

Callicott objects to this kind of eclecticism because he fears that it will lead to the abandonment of the "philosopher's love of theoretical unity, coherency, and self-consistency" and to "incommensurability."[37] The debate about theoretical unity is a very old one. It can be found in the writings of Plato (who tries to unite all knowledge within one theory) and Aristotle (who approaches each subject on its own terms). I do not believe that any clear winner in this argument has emerged, although, of course, philosophers continue to be expected to make their positions as coherent and consistent as possible. Callicott simply takes a particular side in this debate, aligning himself with Plato. Callicott objects to the use of various philosophical positions in making moral decisions because he holds that following "Kant here, Bentham and Mill there, Singer and/or Regan yonder, and Leopold at the frontier"[38] is inconsistent because all the baggage of each philosopher, everything that each has said about ethics, metaphysics, epistemology, and every other field within philosophy, is drawn into and attaches itself to the decision-making process, producing irreconcilable conflicts and contradictions. He seemingly holds that mixing the views of historical figures produces conflicts because the views of each are generally incommensurable. This problem is overcome, apparently in Callicott's view, by sticking to one philosopher or to a series of philosophers who have been shown to hold the same view, for example, a belief in Humean moral sentiments as a metaphysics of morals for environmental ethics. I hold, to the contrary, that if it is Callicott's view, it is itself inconsistent in that it permits him to jettison unwanted baggage from selected historical positions while denying similar license to others. In truth, Callicott selectively pulls elements from the views of others without confronting all of the inconsistencies and incommensurability of their total philosophical work. If he can do it, then others should be permitted to do so as well without criticism.

I want to close by making two points about incommensurability. First, concern about it in science has proved to be very useful in helping to highlight areas where particular theories do not work very well, which in turn has led in some cases to the development of new theories that are better able to deal with those troublesome areas. The same can be said about philosophy. New philosophical positions are frequently developed specifically to deal with problems that proved intractable in terms of the previous theories. Unlike science, however, philosophy has not gone on to discard the earlier theories because they continue to be effective in dealing with matters closely related to the inspirations that brought them about and have never been completely discredited by the philosophical theories and movements that displaced them. In this sense, incommensurability is more of a fact of life in philosophy than Callicott seems to be willing to acknowledge.

Second, incommensurability is much less of a problem than Callicott suggests. On the one hand, seemingly incompatible views sometimes promote the same conclusions. For example, although Sagoff in *The Economy of the Earth* [39] works within an Kantian conception of virtue and I rely on an Aristotelian conception, ultimately our views are similar. On the other hand, transplanting particular views of particular philosophers without bringing along everything else that they said is often possible. Whitehead, for example, operates, in a medical sense, on a number of rationalist and empiricist philosophies, removes vital elements, puts them together in innovative ways in his own process philosophy, and discards the rest.[40] The severed transplants survive quite well within the new system.

The malleability of descriptive metaphysics—the kind, which I have claimed, is most useful in environmental ethics—is likewise very impressive. Consider, for example, Wittgenstein's criticism of Moore's common-sense metaphysics.[41] In "A Defense of Common Sense," Moore makes long lists of statements that he "knows" to be true with Cartesian certainty. Together these statements form a descriptive metaphysics in a Strawsonian sense, as discussed earlier, including, for instance, in Wittgensteinian paraphrase, "My body has never been far from the surface of the Earth," and "The Earth existed long before my birth and will continue to exist long after my death." With regard to the first statement, Wittgenstein imagines "Moore being captured by a wild tribe and their expressing the suspicion that he has come from somewhere between the earth and the moon. Moore tells them that he knows, etc. but he can't give them the grounds for his certainty because they have fantastic ideas of human ability to fly and know nothing about physics."[42] With regard to the second statement, Wittgenstein imagines Moore arguing with a king who has been brought up to believe that the Earth came into existence when he was born and will cease to exist when he dies.[43] In each case, Moore is unable to convince the wild natives or king that what he "knows" is better than what they "know." Concerning the king, Wittgenstein writes: "I do not say that Moore could not convert the king to his view, but it would be a conversion of a special kind; the king would be brought to look at the world in a different way."[44] He adds elsewhere that this conversion would be "through a kind of persuasion."[45] Persuasion rather than argument is needed because Moore's opponents in debate quite possibly might hold all of the other statements in common with him. Moore and the king could carefully check off each of Moore's common-sense statements and discover that they only disagree about the time that the Earth came into existence. Essentially, even though the king believes that he lives in a different kind of world than Moore, this belief has no affect on the other beliefs that they have in common. Likewise, the claim that people's bodies are never far from the surface of the Earth was true when Moore asserted it; today with space travel and trips to the moon, the opposite is true, and this change in truth value has made little or no difference with regard to the truth status of Moore's other statements. The negative of one of Moore's beliefs about the existence of the world has so little effect because the beliefs are simply a list, not a set of logically interrelated statements. They are a description of a worldview, not a metaphysical system. Although negating one of the statements may change the worldview, doing so does not cause a ripple of changes throughout the description. The rest of the statements, in Wittgenstein's terminology, simply "stand fast."[46]

The impact of selective borrowing of the kind that I employed in developing my position and that I claim Callicott employed in developing his position appears as minimal as those involved in making small changes in Moore's list. If a metaphysical fragment from one view is introduced into another metaphysical view, the worldview implied by the metaphysical system adjusts with little or no need to reinterpret the main body of statements comprising that system. Incommensurability becomes an issue only if someone insists that large numbers of other statements must also be introduced with the new fragment, thereby creating irreconcilable conflicts. Imagine the difficulty Whitehead would have had when he borrowed a little here and there from Leibnitz and Berkeley if philosophical conventions forced him to introduce everything else both philosophers said as true. Normally the problems involved in borrowing from historical figures are somewhere in between no impact and a great deal. Someone writes an essay showing that some element of someone's philosophy is of great benefit to environmental ethics literature. Half a dozen opponents then reply that the borrowing of that element also requires the acceptance of some other element held by the historical figure, which then makes the borrowing worthless. The original author then responds that he or she is dealing with the views of the historical figure in question normatively rather than descriptively, and that it is irrelevant what the philosopher actually thought. In truth, the issue is not whether the element can be borrowed but whether it retains the distinctive character it had when it was first formulated (continues to belong to the original philosopher) or is transformed into something new during the borrowing, that is, is permitted to make peace with the other elements in the new system into which it is placed.

In my view, Callicott's borrowing of moral sentiments from Hume's philosophy is not the borrowing of Hume's metaphysics, merely a borrowing of an element of Hume's ethics. To borrow Hume's metaphysics, he would need to borrow Hume's claim that nothing but sensations or impressions exist, a position that would leave him with little more to say, until he left Hume's skeptical philosophy and returned to inquiries based on custom and habit, "renouncing all speculations which lie not within common life and practice."[47] When Callicott moves on to Darwin, Leopold, and Midgley, he continues to make no reference to, nor in any way makes use of, Hume's metaphysical claims about impressions. The connection between Hume and Darwin works for Callicott because Darwin actually read Hume and used him. The connection with Leopold is more dubious in that Leopold uses some discordant, and slightly misunderstood, terminology from Kantian metaphysics and makes no mention of Hume. A connection with Aristotelian virtue ethics is equally workable and, given the Kantian metaphysics, a Kantian approach following Sagoff might be the best historically correct approach. (After all, Leopold is frequently said to have a categorical imperative of his own.) Concerning the Midgley connection, although she exhibits some elements of Hume (with some scorn, noted earlier), her circles of moral considerability are also compatible with Aristotle's, and her interpretation of Darwinian evolution leads to ethology rather than sociobiology. Callicott makes sense not because he is bringing forward a monistic metaphysics, but because he is doing interesting history of ideas (creatively discarding unnecessary baggage) and because his views ultimately align with Strawsonian-style descriptive metaphysics.

Whatever approach is taken metaphysically, nearly everyone, except perhaps the Whiteheadian process philosophers, end up with a view of reality that is readily intelligible to ordinary people who know nothing of the history of philosophy and know nothing of what metaphysics is. And even the process philosophers hide the more difficult elements of their philosophy when writing about environmental ethics. Books by established process philosophers can frequently be read in whole or in part without the readers knowing anything about process philosophy. If there is to be a metaphysics in environmental ethics, it currently is, and probably has to be, an ordinary language conception of existence, adjusted now and then to cover some specific issue, for example, the sense in which species exist. To go beyond such minor adjustments risks the possibility, to paraphrase Hume, that people will conclude that we are "got into fairy land." The recommendation that we stay within the limits of ordinary life, custom, and discourse, which I find not only in Hume, but also in Wittgenstein and Aristotle, seems to preclude the creation of a traditional metaphysical system tailored especially for environmental ethics.

Notes

1. J. Baird Callicott, "The Case against Moral Pluralism," *Environmental Ethics* 12 (1990): 113.
2. Callicott, "Case," 123–24.
3. J. Baird Callicott, "The Metaphysical Implications of Ecology," *Environmental Ethics* 8 (1986): 303.
4. Callicott, "Metaphysical," 313–16; see Alfred North Whitehead, *Science and the Modern World: Lowell Lectures*, (1925; reprint, New York: Free Press, 1967).
5. Dagobert D. Runes, ed., *Dictionary of Philosophy* (Totowa, N.J.: Littlefield, Adams, 1960), 196.

6. Runes, *Dictionary*, 219.
7. Eugene C. Hargrove, *Foundations of Environmental Ethics* (Englewood Cliffs, N.J.: Prentice Hall, 1989), chap. 1.
8. Hargrove, *Foundations*, 33.
9. Hargrove, *Foundations*, 35–37.
10. John B. Cobb Jr., *Is It Too Late? A Theology of Ecology* (1972; reprint, Denton, Texas: Environmental Ethics Books, 1995), 56.
11. Ludwig Wittgenstein, *Zettel* (Berkeley: University of California Press, 1970), 74e, para. 413–414.
12. David Hume, *An Enquiry Concerning Human Understanding* (Edinburgh, 1748), sec. 12, pt. 1.
13 Ludwig Wittgenstein, *On Certainty* (New York: Harper and Row, 1969), 40e–41e, secs. 310–317.
14. Wittgenstein, *On Certainty*, 21e, sec. 143.
15. P. F. Strawson, *Individuals: An Essay in Descriptive Metaphysics* (Garden City, N.Y.: Doubleday, 1959), xiii.
16. Strawson, *Individuals*, xiii.
17. Hume, *Enquiry*, sec. 3.
18. Hume, *Enquiry*, sec. 2.
19. G. E. Moore, *Principia Ethica* (Cambridge, England: Cambridge University Press, 1965), 193.
20. Hargrove, *Foundations*, 200, cf. 185.
21. See J. Baird Callicott, "Animal Liberation and Environmental Ethics: Back Together Again," in J. Baird Callicott, ed., *In Defense of the Land Ethic: Essays in Environmental Philosophy* (Albany, N.Y.: SUNY Press, 1989), 52–56.
22. Callicott, "Case," 123–124.
23. Immanuel Kant, *Fundamental Principles of the Metaphysics of Morals*, trans. Thomas K. Abbott (Indianapolis, Ind.: Bobbs-Merrill, 1949), 4.
24. Callicott, "Hume's Is/Ought Dichotomy and the Relation of Ecology to Leopold's Land Ethic," in Callicott, *Defense*, 118–119; Anthony Flew, *Evolutionary Ethics* (London: Macmillan, 1967), 59.
25 Hume, *Enquiry*, sec. 7, pt. 2.
26. Leopold, *A Sand County Almanac and Sketches Here and There* (London: Oxford University Press, 1949), 138.
27. P. D. Ouspensky, *Tertium Organum* (New York: Random House, 1920), 129.
28. Leopold, *Sand County Almanac*, 210.
29. Leopold, *Sand County Almanac*, 223. Leopold continues, "By value I of course mean something far broader than economic value; I mean value in a philosophical sense."
30. Callicott, *Defense*, 11.
31. Callicott, "Animal Liberation and Environmental Ethics," in Callicott, *Defense*, 54.
32. Callicott, "Animal," 56.
33. Mary Midgley, *Beast and Man: The Roots of Human Nature* (New York: New American Library, 1978), 184.
34. Midgley, *Beast*, 184. For example, she writes, "it seems unrealistic to talk as though the tendency to rescue people were something that could be carried by a single gene" (p. 134).

35. See Paul H. Barrett, ed., *Metaphysics, Materialism, and the Evolution of Mind: Early Writings of Charles Darwin*, (Chicago: Chicago University Press, 1974).
36. Alfred Jules Ayer, *Language, Truth, and Logic* (New York: Dover Publications, 1946), 36. Attributed to F. H. Bradley in *Appearance and Reality*.
37. Callicott, "Animal," p. 50.
38. Callicott, "Case," 124.
39. Mark Sagoff, *The Economy of the Earth* (Cambridge, England: Cambridge University Press, 1988).
40. See Alfred North Whitehead, *Science and the Modem World* (New York: Free Press, 1967), where insights of Leibnitz, Berkeley, Hume, and many others are put together in a new way.
41. G. E. Moore, "A Defense of Common Sense" and "Proof of an External World," in *Philosophical Papers* (New York: Collier Books, 1959); Wittgenstein, *On Certainty*.
42. Wittgenstein, *On Certainty*, 34e–35e, para. 264.
43. Wittgenstein, *On Certainty*, 14e, para. 92.
44. Wittgenstein, *On Certainty*.
45. Wittgenstein, *On Certainty*, 34e, para. 262.
46. Wittgenstein, *On Certainty*, 18e, para. 116.
47. Hume, *Enquiry*, sec. 5, pt. 1.

9

Philosophy of Nature or Natural Philosophy? Science and Philosophy in Callicott's Metaphysics

CATHERINE LARRÈRE

Environmental philosophy began in the 1970s, in the anglophone academic philosophical community, by arguing for a new, an environmental ethic,[1] and Callicott, among others, participated in the quest for intrinsic value in nature. He elaborated a theory of his own and helped clarify the various positions.[2] His main contribution, however, was to show that environmental ethics not only required a metaethics, but a metaphysics before all else. In two articles, "Intrinsic Value, Quantum Theory, and Environmental Ethics" and "The Metaphysical Implications of Ecology,"[3] he argued that contemporary science (physics as well as ecology) "enfolded, involved, or engaged" a new ontology that was able to put into question the self-centered individualism on which modern ethics (Utilitarian as well as Kantian) was based. Such an ontology would provide a new basis for a new ethics, the "holistic" ethics environmentalism required.

Such an idea—grounded in the "metaphysical implications of ecology"—can be traced back to Aldo Leopold, who in a 1938 essay, according to Callicott's presentation, "expresses his belief that ecology contains the seeds of a philosophy of nature with the potential to supplant the reigning mechanical worldview epitomized by engineering."[4] This suggested a contrast between the "mechanical worldview" and a new, contemporary one that Callicott put in full historical perspective in subsequent essays.[5] He traced "a genealogy of science"—the development of "natural philosophy" from the Greek pre-Socratics to contemporary physics through successive paradigms.

First came the systematization of Greek atomism in modern Classical Mechanics, followed by the emergence of a new, nonmechanical paradigm. He could then state that if "the mechanical worldview was to be held responsible for the environmental crisis" (the untoward consequences of the mechanical handling of nature), and if this required us "to develop an environmental ethics to temper its effects," the possibility of such an ethics was to be grounded in "the emerging postmodern paradigm" because it promised "to be much more hospitable to such an enterprise."[6] This was Callicott's contribution to a "reconstructive"—as opposed to "deconstructive"—postmodernity.

Such an attempt to find in contemporary science a positive alternative to modernity can be compared to similar approaches in the so-called European poststructuralist philosophy, the French one especially. Deleuze announces a new "philosophy of nature," Prigogine and Stengers assert that such a prospect is possible, for René Thom and those who follow him, it is not only possible, but necessary.[7] In the late 1950s and early 1960s, Merleau-Ponty, in the course he was giving in the Collège de France, had already argued that contemporary science (physics and biology) had undermined modern ontology (the distinction between subject and object) so that envisioning a new philosophy of nature was possible, a philosophy of nature that would not be hostile to science, as had been the case in the romantic German *Naturphilosophie*, but one that could reunite science and philosophy.[8]

The European reconstruction of natural philosophy did not arise (contrary to Callicott's) directly from environmental concern, but from other motivations.[9] However, it very often leads to calling into question modern technological ambition,[10] and to developing an ecological awareness.[11] Prigogine and Stengers ask for a "new alliance" with—and due respect for—Nature. Michel Serres, whose conception of contemporary science owes much to Prigogine's contribution to contemporary thermodynamics, developed his own environmental philosophy and ethics in *Le contrat naturel.* Deleuze stated that he shared his project of a philosophy of nature with Guattari, with whom he had been writing some of his most famous books.[12] Guattari had anticipated the outcome of such a collaboration in *Les trois écologies*, a blueprint of sorts for ecological activism.[13]

Hence comparing Callicott's philosophy of nature with these French approaches to assess Callicott's originality and relevance to his own concern and project does not seem arbitrary. But must we speak of "philosophy of nature" or of "natural philosophy"? Callicott refers to both. He begins by inquiring about "philosophy of nature" and goes on reviewing the development of "natural philosophy" from the Greek pre-Socratics to contemporary science.

All the "philosophies of nature" do not appear in "natural philosophy." A philosophy of nature, some claim, aims at "settling the relationships between science and philosophy,"[14] but, in doing so, one can oppose them—as in the romantic claim that philosophy, through poetry, knows better about nature than science—as well positively connect them. Until the end of the eighteenth century at least, the phrase *natural philosophy* was synonymous with physics.[15] So "natural philosophy" was distinguished from "moral philosophy." This dyad implied a certain unity between science and philosophy ("natural" and "moral" philosophy being two branches of the same general inquiry). So to use anew the phrase "natural philosophy" implies that we still can be in quest of a similar unity. To what extent can it be similar? This is the point I want to examine.

The "metaphysical implications of ecology" may mean two things. First I argue that it means that science "enfolds" an ontology; and thus that Callicott's idea of "natural philosophy," compared to poststructuralist approaches to philosophy of nature, provides a surer path toward the reconstruction of a postmodern concept of nature. Second, I argue that it means that natural philosophy is able to inform a moral philosophy. However, I argue further, this entails difficulties—because actual sciences are not identical to natural philosophy—which require philosophy to be more independent of science than Callicott thinks it should be.

Natural Philosophy and Philosophy of Nature

To argue his thesis of the "metaphysical implications" of contemporary science, Callicott presents the "general conceptual notions" that can be "abstracted" from ecology and from what he calls the "new physics" (relativity and quantum theory). Three of them are especially notable:

1. Quantum physics, and more especially Heisenberg's principle of uncertainty, undermines the subject-object dichotomy, which is at the core of modern metaphysics (from Descartes to Kant).
2. Contemporary ontology is relational, "relations are 'prior' to the things related" and entities are not independent objects, but internally related, they are "knots in the web of life."[16]
3. Contemporary ontology is holistic. It is not "possible to conceive of an entity in isolation from its milieu."[17]

Thus summarized, these are the basic assumptions of any constructive alternative to modernity, based on contemporary science. Merleau-Ponty, Serres,

Thom, Prigogine, Stengers—everybody agrees that nature can no longer be seen, as it was in the modern era, as made of "*partes extra partes*," discrete entities connected by external relations. The emphasis is upon internality, upon the interactions between the entities and their milieu. Prigogine and Stengers speak of "entities intrinsically constituted by their irreversible interaction with the world."[18] Serres criticizes the archaism of those who think that sciences deal with objects, or substances, when relations and interactions prevail.[19]

Most important is the common interpretation of measure and observation in quantum physics. When Rolston writes that with relativity and quantum physics, "the subjectivists have won all the chips," that "subjectivity has eaten up everything,"[20] he echoes a rather prevailing interpretation, a pessimistic or negative one. Heisenberg's principle, according to such an interpretation, leads to skepticism, inserts irrationality—literally so: nonmeasurability—at the very root of physical reality. So to avoid jeopardizing the whole set of science and rationality one can say that quantum physics is true for a specific level, or field, of reality, but not for all of it (as Rolston does[21]). One can also see quantum physics not as a new development of physics but as an attempt to undermine physics, an irrational assault against scientific rationality. This is what Karl Popper does when he compares quantum physics to "brainwashing" and says that Bohr's teaching of quantum physics was aimed at saying that it was impossible to understand.[22]

Prigogine and Stengers have criticized this subjectivist interpretation, showing that it went together with a very positivist conception. The subjectivist interpretation, which founds quantum mechanics on the perturbations caused by observation, implies that an objective state of the system exists where, independently of an intruding observer, the set of parameters have defined values. Hence, in such an interpretation, the classical idea of physics (the separation between reality and the observer) stays untouched; there is just a positivist interdiction of interfering with reality. According to their interpretation we need not, in a negative way, refrain from interfering, but, positively, choose how we interfere. We are not outside of the reality, we are part of the same process, of the same reality. Knowledge is not a subjective observation, it is a physical process.[23]

Knowledge is a physical process, Callicott explains also, answering Rolston. Once increasingly sophisticated experimental techniques "permitted investigation into smaller and smaller levels of phenomena," he writes, "it became increasingly evident that to make an observation energy must be exchanged between the object of observation and the observer." Energy is information, and information, energy. Far from subjectivizing everything, it "physicalizes knowledge and consciousness. The *res cogitans* collapses into the *res extensa*."[24] We are part of the world, and knowledge is not an immaterial gaze; it belongs to physics. To know is to actualize a physical relationship. "Relativity turns the observer into a physical being," Prigogine and Stengers write, and they go on speaking of "a world populated with observers, that is nature."[25] For Merleau-Ponty, the philosophy that can correspond to quantum physics should be a more "realistic" philosophy than the transcendental one. The current physicist cannot be thought of as an universal "*cogito*" but as a body-subject—a located, situated, incarnated being.[26] Thus contemporary physics converges with phenomenological, prescientific experience. Nature is not an external object; we are incorporated in it, we have a cobelonging relationship with other natural beings.

Therefore the subject-object dichotomy collapses and with it a related idea of science, which reduces it to a way of handling, manipulating, or instrumentalizing a world made of external objects, or things. To see science as manipulation (or technology) is to see it from the standpoint of applied science, to reduce it to predictive knowledge. As Thom explains, "to predict is not to explain," least of all to understand.[27] He wants to promote a science whose main aim is not to grasp nature, but to understand, which seeks not so much "how it works" as "what is it." Such a science cannot be accused of "ignoring being," it can be seen as "natural philosophy," a way of answering questions, rather than an opportunistic manipulation of the world.

Hence the common core of every reconstructive alternative to modernity based on contemporary science is to consider science as natural philosophy. But on such a basis, different philosophies of nature can be constructed. The prevailing idea, in poststructuralist philosophy, is that after the demise of the metaphysics of substance, the new ontology is one of process and event, the emphasis is on singularities and becoming. Such an ontology can be drawn directly from science, as Prigogine and Stengers do with nonlinear thermodynamics, far from equilibrium systems, dissipative structures, chaos theory—nature driven by the irreversible flow of time, an unpredictable nature. Others, such as Merleau-Ponty or Deleuze, look for a proper metaphysics corresponding to the new sciences: Merleau-Ponty turns to Schelling, and Deleuze looks toward Bergson and Whitehead to understand what an event is.[28]

This is not what Callicott does. Although he mentions that an ontology of internal relations could stir some philosophical memories from "nineteenth- and early-twentieth-century German and English idealism,"[29] he is not in quest of a specific metaphysics for postmodernity. Relying on "natural philosophy," he is not in need of a "philosophy of nature." According to Merleau-Ponty, science is able to give us "negative philosophical discoveries." It can undermine some philosophical prejudices, such as the substantive ontology of the moderns. But it is not a philosophy; science finds ways to bypass traditional philosophical concepts, but it does not construct the right concepts. This is, for Merleau-Ponty, the proper task of philosophy.[30] As soon as Callicott assumes that the "abstractive distillates" of ecology or physics are ontological concepts, he assumes that science involves a positive ontology. It could be said however that to elaborate such an ontology from "distillates" is the philosopher's work, rather than the scientist's.[31]

No one region of science can provide a metaphysical framework for natural philosophy. Objecting to Prigogine and Stengers's new synthesis—organization emerging from instability, in a spontaneous process, without being directed by any cause[32]—Thom reminds them that "it is difficult to concede that concepts coming from a particular science [thermodynamics and dissipative structures] might have, by themselves, an interdisciplinary value."[33] Stretching a part of the knowledge to adjust all knowledge does not lead to false claims but rather to vague, fuzzy, void claims.[34] Ontology, in such a case, is not drawn from a genuine examination of science. Rather, science is instrumentalized to answer preexisting philosophical questions.[35] To avoid such abuse of science, Thom asserts the foundational part of mathematics and physics, whose "fundamental concepts inform our whole worldview."[36] This does not imply a reductive unity of sciences. Thom borrows from Husserl the concept of "regional ontology"—by which Husserl meant several areas of "natural" phenomenology—and applies it to different scientific fields, each of which has its own set of fundamental concepts, which organize the field.[37]

Although he is in no way ignorant of the importance of thermodynamics, especially in the development of ecology, and although much of his relational ontology is elaborated with concepts coming from thermodynamics, Callicott does not share Prigogine and Stengers's assessment of its unifying importance.[38] First, he certainly does not think that the prevailing feature of contemporary ontology is the general instability of an erratically becoming world (but rather the holistic embeddedness of internally related entities). Second, to equate ecology with its thermodynamic elaboration is to treat it as a mere extension of physics, whereas Callicott clearly considers it as a biological science (which includes other forms of ecology, such as community ecology). Moreover, doing so, he maintains the distinction of sciences, their ontological regional disparity, which he presents as a "hierarchy," thus safeguarding the foundational part of fundamental physics (relativity and quantum theory).

Clearly, Callicott does not confuse the metaphysical implications of ecology with those of physics. He speaks of "converging" features of "the mutually reinforcing holistic metaphysical interpretation of quantum theory and ecology."[39] To be located in quantum physics gives these "metaphysical implications" a foundational ground, whereas ecology—that is, the complexity and the locality of "middle-level," the Earth we are living on—offers the empirical level where these implications can be experienced. That is answer enough to Rolston's objection that the implications of quantum theory are relevant only for the microscopic or macroscopic level, but not for everyday experienced middle level. In both cases, the observer is part of the situation he is observing. With it, he has an internal relationship that knowledge actualizes.

But this does not mean that these implications are identical. Callicott borrows from quantum physics the word *complementarity*,[40] which Bohr used to mean the multiplicity of perspectives, which makes a reductive unity of nature impossible and renders the use of metaphors necessary.[41] The two main characteristics of the new ontology (i.e., being relational and holistic) are closely linked, as can be seen in their negative counterpart: There is no independent, self-sufficient entity. Ecology is as relational as the new physics. But according to whether one follows the implications (mostly relational) of physics or those (mostly holistic) of ecology, the moral implications of each—the relational and the holistic—are not identical. In both cases, to be sure, the new ontology means the end of the egoistic, atomistic, self-interested individual of modern ethics (and economics[42]). But the relational self is the extended or realized self of deep ecology, whereas the holistic self is the embedded self that Aldo Leopold's land ethic requires.[43]

Therefore, the "emerging worldview" is far from being completed. The affirmative interpretation of quantum physics provides us with the certainty that it is a new beginning, not a threat to scientific rationality. Contemporary physics—"the sciences at the base of the hierarchy of sciences"—and ecology—"those at the apex"[44]—provide convergent metaphysical implications. But much remains in between, and Callicott refuses to stretch these regional aspects to the whole area in between. Hence one could expect the supporters of a unifying philosophy of process to be quite audacious in foreseeing what the new worldview is going to be, although Callicott would not go beyond the sketchy delineation of a possible global worldview. Just the contrary happens. Although Callicott boldly draws a fully reconstructive postmodern worldview,[45] French poststructuralist philosophers deny that such an anticipation is possible.

With Plato's *Timeus*, Descartes's *World*, Hegel's *Naturphilosophie*, Bergson's *Creative Evolution*, Whitehead's "process philosophy"—a long philosophical tradition of grand sweeping descriptions of Nature as a whole exists. Deleuze acknowledges it, but makes no attempt to give a new one. He "would rather suggest the void possibility (or the current impossibility) to complete such a description."[46] So one can argue that Deleuze, although he announces the need of a new philosophy of nature, eventually assumes the impossibility of any philosophy of nature. His reference to nature is mainly negative. Nature, with Deleuze, has no limits, it is equated to being, and its main feature is diversity.[47] Thus the classical subject is undermined, collapsed in the infinite fragmentation of being that he (or she) is part of. Then a new subjectivity can be reconstructed (that is, what the concept of "fold" (*pli*) provides) but not nature.[48] Announcing his project of thinking about a "new philosophy of nature" (sort of; he actually says: "*une sorte de philosophie de la nature*"), Deleuze says one such is necessary because "any difference between nature and artifact is being blurred."[49] Does that mean that everything must be seen from the standpoint of nature? We can doubt it, judging from the sample of a possible philosophy of nature Guattari gives in his *Les trois écologies.* These three ecologies are the mental, the social, and the natural or "environmental" ones. The first two are the most important; in them the word *ecology* is only a metaphor because it does not mean that minds or societies are parts of nature, but that these are not closed areas, that they are related to ongoing transformations. About ecology properly speaking, which he calls *environmental ecology*, he has very little to say, just enough so that we understand that nature is no longer nature—that which is so by itself—but something thoroughly man-made, an artifact. Soon, he says, we will have no longer to preserve nature, but to "fix" it. Guattari's is a "mechanistic" ecology, an increasingly artificial ecology.[50] Hence it is nature that collapsed in artifact once the difference was blurred, not the other way around. Thus Deleuze and Guattari participate in the hypermodern achievement of modernity: the tale of the end of nature.[51]

The basic requirement of a philosophy of nature is that nature exist. And what a contemporary "philosophy of nature" (sort of) forbids, natural philosophy permits. Callicott's confidence in an "emerging worldview" relies on such grounds. He anticipates the unified natural philosophy by assuming that a unity is to be found between the four fundamental forces that physicists have identified, "the strong and the weak nuclear forces, electromagnetism and gravity."[52] This is still to be done. One could say that each of the two sciences on which Callicott grounds his certainty of a unified emerging worldview are far from being actually unified. People have been striving at quantifying gravitational forces for twenty or thirty years. Ecology is split between a thermodynamic and a Darwinian synthesis.[53] Moreover it leaves open the question of how such an unity can be achieved once we take for granted that postmodern unity cannot be reductionist as modern unity was and that the problem of unifying disparate regional ontologies must be addressed.

Regardless, Callicott's optimism mostly rests on analogies, from the existence of the modern paradigm to the predictable existence of a postmodern one. Two postulates make such an analogical process possible. The first one is the continuity of natural philosophy, from the Greeks to contemporary science. We have been dealing with it up until now, and we can conclude that because a positive alternative to modernity envisions science not as a way of successfully manipulating nature, but in a more contemplative way, then science can be understood as natural philosophy, continuous with the pre-Socratic questions about the world.[54]

Callicott's second postulate concerns the priority of natural philosophy in relation to moral philosophy. And, as we are going to see, it faces serious objections.

NATURAL PHILOSOPHY AND MORAL PHILOSOPHY: THE ETHICAL IMPLICATIONS OF METAPHYSICS

To identify the unity of a historical era, especially the modern one, Callicott uses the concept of paradigm, which comes from Thomas Kuhn, but he uses it in a broader sense. Callicott's paradigm neither means only the unity of the regional area of one peculiar science, nor even the natural sciences unified in a whole model of nature, but rather the whole culture of an era. Callicott's insistent idea is that the reigning paradigm—"the general conceptual model of nature"—informs all the aspects of culture and practices, technology as well as ethics.[55] Hence all the branches of philosophy adjusted themselves in the modern era to the reigning paradigm in natural philosophy. According to this thesis, natural philosophy is prior to moral philosophy, moral philosophy is "ancillary" to natural philosophy.[56]

Therefore, although the emerging new paradigm will be different from the modern one, natural philosophy will remain prior to moral philosophy, will still "inform" moral philosophy. With the demise of classical atomism, the moral individualism characteristic of the modern era will collapse, and a new moral pattern, adjusted to the new model of nature, will be elaborated. One can thus go from the metaphysical to the moral implications of contemporary science. From one paradigm to the other, the content changes, but the general structure remains. Our vision of nature shapes, or informs, our whole worldview, ethics included.

Michel Serres quite disagrees with such an idea. Although he announces a "new synthesis" and wants to "construct a philosophy that is compatible with this new world," and hence is in search of an enlightened alternative to modernity, he does not think that the new synthesis can achieve a unity similar to that of the modern era.[57] It is not so much the content that is going to change as the pattern itself, the relationship between natural science and ethics.[58] Not unlike Prigogine and Stengers, Serres thinks that postmodernity should bridge the gap that modernity deepened between science and the humanities because modernity, as a model of rationality, relied on science alone, and left the humanities (poetry) outside of the rational area, moral as well as natural. Modern enlightenment, according to Serres, promoted natural science as the sole model of rationality, and this led to subsequent positivism and scientism. So a new philosophy, consistent with the contemporary world, should not subordinate moral philosophy to natural philosophy, but should acknowledge the independent contribution of humanities to rationality. Hence we have a much more heterogeneous picture of our worldview, which leaves room for poetic construction as well as for scientific computation. Serres strongly rejects the positivist, scientist idea according to which "there exists no reasoned activity nor any valid ethics outside of the sciences."[59]

Outside of natural philosophy, is no valid ethics possible? Callicott certainly does not think so. That would be absurd. "Since science is Western in provenance,"[60] that would mean that outside the Western world, no valid ethics exist. Callicott certainly did not write his *Multicultural Survey of Ecological Ethics from the Mediterranean Basin to the Australian Outback* with such an idea! To elaborate an ethic of social sentiments, as he does, means that sentiments (i.e., ethics) are prior to the more or less rational representations that may inform them. If natural philosophy is prior to moral philosophy, ethics is certainly prior to natural philosophy. How then can we understand the way these three elements (ethics, natural philosophy, moral philosophy) can fit together?

Human specificity, states Serres, in a very anti-Kantian way, is not to invent ethics—or morality—but to invent science (i.e., objectivity): He writes, "Humanity begins with things; animals don't have things."[61] Human beings, he says, are political animals; like other animals, they are able to live socially, to live together. This implies ethics, and valuing, which is a subjective process, something that happens between "me" and "you," with "us." This is common to human as well as animal societies: "All animals enter into contracts among themselves that are purely social, empty, based exclusively on the concept of us."[62] The real change comes when we escape the subjective commerce, and posit an exterior third person pronoun, a neutral "it," an object. And this is science: the objectification of nature. So the Other, contrary to what Levinas says, is not the human person, or God envisioned in someone else's gaze, the Other is Nature, nature rationally constructed as an object.[63]

Therefore we can understand that if ethics and religions are coextensive with societies, philosophy is not. Societies can exist without philosophies. But when philosophy begins, or appears, it necessarily begins as natural philosophy. It begins with the objectivity of nature. It is no historical contingency that philosophy began in Greece with physics.

Callicott's additional claim to such a statement about the priority of natural philosophy is that natural philosophy entails, or involves a moral philosophy. Contemporary science, despite its Western origin, is now universal. It is so because it is widespread everywhere in the world. Hence the universality of contemporary science is that of a network, rather than a totalizing, abstract universality. Science is universal insofar as it extends, or expands, itself, insofar as it is endorsed by different people as a common language. So it coexists with different cultures, more or less harmoniously so. "Citizens of Iran," writes Callicott, "watching a fundamentalist ayatollah fulminate against Western ideas and values on TV receive contradictory messages."[64] Indeed. D'Alembert, a philosopher very representative of the Enlightenment spirit, thought that to get rid of religious obscurantism, not much had to be done. Just teach a young man mathematics, he said, and the rest (the demise of naive religious ideas) will follow. No such thing happened. We cannot be as naive as Enlightenment thinkers used to be.

The moral implications of natural philosophy are no more causal than they are merely logical.[65] But there are moral implications nonetheless. Science is not neutral; Callicott says that the messages are "contradictory"; something in Western science contradicts Islamic fundamentalism. So he claims that the contemporary worldview involves an environmental moral philosophy, and, because it is universal (i.e., universally endorsed), "it is intended to serve as a standard for evaluating the others."[66] So, because moral philosophy is ancillary to the universal natural philosophy, there is a preeminence of moral philosophy upon regional ethics. Finally, the solution to environmental crisis is to be found in contemporary natural philosophy, in our ability to understand it and to draw carefully its moral implications.

Philosophers have already objected to such a confidence in natural sciences. Husserl, in his 1935 lecture about the European crisis, denounces the naturalistic blindness of the "objectivist" sciences.[67] They are blind to the spirit. It has been naively believed that natural sciences could embrace the whole of being, whereas the power and place of subjectivity, without which science would not be possible, can be acknowledged only in its own realm, not on the "objective" science field. For Husserl, therefore, the solution to the European crisis was to state the universal ambition of European spirituality. Such a criticism presumes precisely what is put into question: the separation between spirit and matter, man and nature, subject and object. As soon as we assume that a philosophy of nature is possible, which includes humanity, we cannot go on flatly saying, as Serres does, echoing Husserl, "The hard sciences go their way without man, thereby risking becoming inhuman."[68]

The reverse objection is also possible. The problem with "hard" sciences is not that they are too naturalistic, but that they are thoroughly social. Science is a social product; it is made collectively inside a social community; it needs funding, and it must be socially justified and legitimized. So the hierarchy of sciences echoes or reflects social values. The Enlightenment has fully achieved its aim, natural sciences have won *über alles.* Science, Serres writes, "recruits the best intellects, the most efficient technical and financial means. As a result science finds itself in a dominant position, at the top of the heap, as we say, single-handedly preparing the future and in a position to occupy more and more territory."[69] So one could say that when Callicott submits moral philosophy to natural philosophy, (i.e., philosophy to science) he obeys a social hierarchy, he submits himself to the social power of science, to say nothing of the importance of science in U.S. hegemony over the world.

Of course, such a way of arguing, if it is the only one, is thoroughly deconstructive; everything is an issue of power and nothing else. But Serres does not say so. For him, science is both socially constructed and objective. It is a social structure, united by contract, like every social institution, but it has a specificity, it is able to single out an object. It is not ignorant of being; science is a human community that is able to have a commerce with nature.[70] But considering science as pure natural philosophy, overlooking its social dimension, would be delusional. In this sense, too, natural sciences are human sciences. Natural philosophy might be prior to moral philosophy, but social—moral—ideas are able to permeate human sciences. Of course, Callicott is not unaware of such things. Does he not speak of "deconstructive ecology"?[71] Even brave new natural sciences cannot escape the decadent perversity of postmodern philosophy.

Taking or not taking into account this double aspect of actual sciences greatly affects the way a new paradigm can be envisioned. According to Thom any paradigm begins with a limited, successful regional scientific field. Then such successes are extended on a wider range.[72] This can be done in two ways. The extension may be the result of stretching the original concepts by giving them a more universal span than they originally have, transforming them into what Thom names "fuzzy concepts." The extension may also be more cautiously effected, safeguarding the regional diversity, by proceeding with controlled analogies or metaphors. Actually Callicott proceeds both ways.

The original field is that of ecology where he discovers the "seeds" of a new philosophy of nature (i.e., a certain sort of unity), both relational and holistic.

He validates this original piece of ontology by checking its convergence with fundamental, foundational new physics. Then he draws the moral implications of such a metaphysics. This first step, which is very close to Leopold's own progress from ecology to ethics, is followed by a second one, when a whole postmodern paradigm is elaborated. But even at this extended level, Callicott generally gives two characteristics of the new worldview. On the one hand, it is said to be "organicist"; on the other, a systemic holism is asserted.[73] This duality bears testimony to the double process from which the paradigmatic extension proceeds.

To name the characteristic ecological unity, one can use the term *ecosystem*. Tansley coined the word explicitly to avoid the "abuse" of terms. But following Leopold, who did not speak of ecosystem, Callicott does not choose it as the cornerstone of his paradigmatic construction.[74] Rather, he compares ecological unity either to an organism, or to a community. This refers to two successive steps in the development of ecology. It was first *organicist* with Clements; then Charles Elton spoke of "biotic communities." "The ecological paradigm thus began to change from nature-as-organism to nature-as-society."[75]

Although conceiving of ecological unities as organisms is outdated, Callicott holds on, more often than not, to such a unifying conception. Indeed, it could be very convenient for his extensive purpose. A successful concept in nineteenth-century biology (it made possible a science of life that classical mechanism precluded), the concept of organism could be judged reliable for further extensions in other areas of natural sciences (such as ecology) and it could as well bridge the gap between natural and moral philosophy.[76] Very long ago, already, the living being, or the body, has been the model of moral unity and of the hierarchical subordination of the parts to the whole.[77] Hence it could represent the very extension of science, the priority of natural philosophy upon moral philosophy. Unfortunately, things are just the other way around. To support his once organicistic assumptions, Leopold, Callicott must acknowledge, had to rely on a not very "palatable" philosophy, the mystic philosophy of a Russian philosopher, P. D. Ouspensky.[78] One could then argue that the organicistic unity of nature, as well as the mechanistic one, is a reductionist unity, but that contrary to the latter, it goes with an irrational metaphysics.[79]

Elton's community concept applies to ecology, and natural life, a social concept. So we learn that the relation between natural and social sciences (and moral philosophy) is not a one-way street. Felicitously so. Because it is a metaphor, applying the community concept to a new area does not lead to reductionist extension. The analogy must be cautiously applied by taking into account the differences. For instance, notes Callicott, the community concept originally is monospecific (for animal as well as human societies), whereas biotic communities are multispecific and even include nonliving entities.[80] Moreover, as he also notices, it is more liable to and suitable for moral implications.[81]

Above all, these are different moral implications. The organism analogy is not designed to address the same moral problems as the community metaphor. Kant, while stating that humanity only was capable of moral actions because moral behavior originates out of free will, developed practical philosophy as a problem of causality. He distinguished between the causality that rules natural phenomena (the linear, transitive, mechanical causality, where every cause has its effect, and every effect its cause) and moral causality, from which moral actions proceed (the expressive causality, the *causa sui* concept[82]). Thus the free act has its own origin in itself; it is not the effect of a cause. At the same time, or about, Hume and Smith, grounding ethics in sentiments, had tried to explain morality as a causal process. But they referred to the linear, transitive concept of efficient causes, and so Kant's argument of moral autonomy (morality proceeds from nothing exterior to morality) could be maintained. The organismic moral metaphor can be understood as an attempt to find a biological, natural, analogue to moral autonomy. Actually, an organism may be said to be autopoiêtic, actively and creatively maintaining its existing conditions, not independently of, but inspite of, the exterior circumstances.[83] The level of organic unity thus indicates the level of moral autonomy.

But what if autonomy is no longer a moral requirement? Environmental philosophers could find some interest in knowing that the concept of autonomy, as an inescapable moral requirement, is being contested in the traditional field of anthropocentric ethics. The American philosopher Stanley Cavell closely examines in *The Claim of Reason*[84] the moral autonomy thesis, in its different versions, especially the grammatical, or syntactical, formulation that prohibits drawing moral (imperative) conclusions from factual (descriptive) premises because, experts generally argue, a gap exists between the two moments, which makes inference impossible.[85] There is no such gap, concludes Cavell, simply because no inference is drawn. When one says to another, "You ought to do that" after having explained the reasons why (the factual data), such a statement is not inferential according to Cavell. To derive an imperative from an imperative is tautological; one has to give the specific reasons why to do so, that is, to provide the context inside which such an act will have a meaning, will improve the moral condition of the agent. Morality deals with the description of a world, of a community in which we can live together, in which we can share value, and assume responsibility. Morality requires expertise, not only good will.

This is good news. If Cavell is right (and I think he is), we have been relieved of one of the most irritating metaethical mysteries: How can we know that an action is a moral one?[86] We have no longer to answer such a question but to sort out different descriptions of values as well as facts. The requirement that the biotic community concept must meet is, "Can it provide a relevant context for ethical action?" And, in my opinion, Callicott's examination of the biotic community as an ethical community meets such a requirement. Therefore, as Callicott wrote, "Physics and ethics are . . . equally descriptive of nature."[87]

This can be said, if by physics is meant "natural philosophy," and not the knowledge of the means our ends require. It means that something philosophical must be said of nature, whereas since the modern era—and especially since Kant—the only relevant question for philosophy has been man. According to Kant, nothing philosophical can be said about nature, nature being nothing but means for human ends. Inquiring about intrinsic value was expressing a complaint about such an exclusion of nature from the philosophical arena. The quest for intrinsic value demanded that nature could be spoken of in itself, not just from the standpoint of human needs or interests. That is why environmental ethics requires a philosophy and Callicott was right to assert that such a philosophy had to be natural philosophy, not practical philosophy alone. But then we must not forget that in "natural philosophy" there is "philosophy," not just science. Philosophy, therefore, must be granted more independence, regarding science, than Callicott seems ready to concede.

NOTES

1. Richard Routley, "Is There a Need for a New, an Environmental Ethic?" in *Proceedings of the XVth World Congress of Philosophy*, ed. Bulgarian Organizing Committee (Sophia, Bulgaria: Sophia Press, 1973), 1:205–210.

2. J. Baird Callicott, "Intrinsic Value in Nature: A Metaethical Analysis," in *Beyond the Land Ethic: More Essays in Environmental Philosophy*, ed. J. Baird Callicott (Albany N.Y.: SUNY Press, 1999), 239–261.

3. J. Baird Callicott, "Intrinsic Value, Quantum Theory, and Environmental Ethics" and "The Metaphysical Implications of Ecology," in *In Defense of the Land Ethic: Essays in Environmental Philosophy*, ed. J. Baird Callicott. (Albany N.Y.: SUNY Press, 1989), 157–174 and 101–116, respectively.

4. Aldo Leopold, "Engineering and Conservation," in *The River of the Mother of God, and Other Essays*, ed. Susan L. Flader and J. Baird Callicott (Madison: University of Wisconsin Press, 1991), 7–8.

5. "A Postmodern Evolutionary-Ecological Environmental Ethic," chap. 9, in *Earth's Insights: A Multicultural Survey of Ecological Ethics from the Mediterranean Basin to the Australian Outback*, ed. J. Baird Callicott. (Berkeley: University of

California Press, 1994), 185–210; and J. B. Callicott, "Après le paradigme industriel," in *La crise environnementale*, ed. C. et R. Larrère. (Paris: INRA éditions), 1997), 205–220.

6. Callicott, *Earth's Insights*, 188.

7. Gilles Deleuze, *Pourparlers* (Paris: Éditions de Minuit, 1990), 212; Ilya Prigogine and Isabelle Stengers, *La nouvelle alliance*, 2d ed. (Paris: Gallimard, 1986); René Thom, *Apologie du logos* (Paris: Hachette, 1990); Jean Largeault, *Philosophie de la nature* (Créteil: Université de Paris, XII, 1984).

8. Maurice Merleau-Ponty, *La Nature, Notes, Cours du Collège de France*, ed. Dominique Selgard (Paris: Éditions du Seuil, 1995).

9. A nonidealistic philosophy of subject, of the "body-subject," for Merleau-Ponty, for example.

10. René Thom stresses that his "catastrophe theory" is definitely nonexperimental and does not lead to heavy technical applications.

11. In *Ecopolitics: The Environment in Poststructuralist Thought* (London: Routledge, 1995), Verena Andermatt Conley studies the "environmental awareness" of French structuralism (Lévi-Strauss) and part of poststructuralist philosophy.

12. Gilles Deleuze and Félix Guattari, *L'Anti-Œdipe* (Paris: Éditions de Minuit, 1972) and *Mille Plateaux* (Paris: Éditions de Minuit, 1980).

13. Michel Serres, *Le contrat naturel* (Paris: Françoise Bourin, 1990); Félix Guattari, *Les trois écologies*, (Paris: Galilee, 1989).

14. Emmanuel Renault, "Les philosophies de la nature d'aujourd'hui et la *Naturphilosophie* d'hier," in *Hegel passé, Hegel à venir* (Paris: L'Harmattan, 1995), 29–54, 32.

15. In the monumental, and referential, late-nineteenth-century dictionary of the French language by Littré, *philosophie naturelle* means "the whole set of astronomical, physical and biological sciences."

16. Callicott, "The Metaphysical Implications," 110.

17. Callicott, "Metaphysical Implications," 110.

18. "Entities intrinsically constituted by their irreversible interaction with the world," Prigogine and Stengers, *La nouvelle alliance*, 160.

19. "We still live in a century or a universe of concepts, beings, objects, archaic statues, or even operators, while we continually produce an environment of fluctuating interferences, which in return produce us," Michel Serres, *Éclaircissements, entretiens avec Bruno Latour* (Paris: Flammarion, 1994), 168; English translation: Michel Serres with Bruno Latour, *Conversations on Science, Culture and Time* (Ann Arbor: University of Michigan Press, 1995). He also makes a striking comparison with rugby: "The clumsy person plays with the ball and makes it gravitate around himself; the mean player imagines himself to be a subject by imagining the ball to be an object—the sign of a bad philosopher. On the contrary, the skilled player knows that the ball plays with him or plays off him, in such a way that he gravitates around it and fluidly follows the positions it takes, but especially the relation that it spawns." Serres, *Éclaircissements*, 108.

20. Holmes Rolston III, "Are Values in Nature Subjective or Objective?" *Environmental Ethics* 4 (1982): 127.

21. When he writes that what is true at macroscopic (relativity) or microscopic (quanta) level is not relevant at middle level, common experience level, the level where we see boulders, spruce trees, and hawks: "Agnosticism and relativism about the ulti-

mate structure of matter does not prevent objective knowing in a middle-level sense." Rolston, 128.

22. Karl R. Popper, Foreword to Franco Selleri, *Le grand débat de la théorie quantique*, (Paris: Flammarion, 1986), 5.

23. Prigogine and Stengers, *La nouvelle alliance*, 312–313.

24. J. Baird Callicott, "Intrinsic Value, Quantum Theory," 168. From the other extremity of the social, or human, sciences side, Lévi-Strauss comes to the same conclusion that "the mind can comprehend the world only because it is a product and a part of this world." *Le regard éloigné* (Paris: Plon, 1983), 164, quoted by Conley, *Ecopolitics*, 46.

25. Prigogine and Stengers, *La nouvelle alliance*, 299, 372.

26. Merleau-Ponty, *La nature*, 134.

27. René Thom, *Prédire n'est pas expliquer* (Paris: Flammarion, 1993).

28. Gilles Deleuze, "Qu'est-ce qu'un évènement," in *Le Pli, Leibniz et le baroque*, (Paris: Éditions de Minuit, 1988), chap. VI, 80–103.

29. J. Baird Callicott, "Metaphysical Implications," 110. He quotes Hegel, Fichte, Bradley, Royce, and Bosanquet.

30. Merleau-Ponty, *La nature*, 138.

31. From quite different philosophical grounds—Wittgensteinian ones—Vincent Descombes comes to similar conclusions. "Physics, and any science alike, implies a metaphysics of its object. The conceptual framework and the principles which are being used by physics are a metaphysics of nature, insofar it is up to philosophical analysis rather than to physical research properly spoken to make such an examination," *Philosophie par gros temps* (Paris: Éditions de Minuit, 1989), 98. He then partakes of the French consensus about a metaphysics of "event and process" succeeding a metaphysics of "substance."

32. Cf. Jean Largeault, foreword to Thom, *Apologie du logos.*

33. Thom, *Apologie du logos*, 522.

34. Thom, "La boite de Pandore des concepts flous," *Apologie du logos.*

35. According to Emmanuel Renault, Prigogine and Stenger's book "does not aim so much at providing a genuinely philosophical knowledge of nature, as to find in science the answers to fundamental philosophical questions." "Les philosophies de la nature," 43. Far from philosophy being submitted to science, science is submitted to philosophy.

36. Thom, *Apologie du logos*, 598.

37. For instance, the Linnean taxonomy draws a regional scientific ontology. Cf. Thom, "Le problème des ontologies régionales en science," *Apologie du logos*, 455–467.

38. He quotes Morowitz, according to whom "each living thing is a dissipative structure . . ." and who takes as an "instructive example" that of "a vortex in a stream of flowing water." "Metaphysical Implications," 108. But these are citations; when he gives his own explanation of internal relations, and vanishing individuals, he refers to relativity, as in "After the Industrial Paradigm," from the thermodynamic metaphor of "knots in the web of life," he goes to the relativity metaphor of "knots in the fabric of space-time," *La crise environnementale*, 215.

39. Callicott, "Intrinsic Value," 173. See also note 3 to "Metaphysical Implications," 283.

40. Callicott, "Intrinsic Value," 173.
41. Niels Bohr, "Natural Philosophy and Human Culture," *Nature* 143 (1939): 268–272.
42. Critics of such economics name such an individual (or its supposition) the "rational idiot." See Amartya Sen, *On Ethics and Economics* (Oxford: Blackwell, 1987).
43. "The reconstructed postmodern person may, if ecology exerts more influence on social modeling than physics, turn out to be an embedded self, rather than a realized self." "After the Industrial Paradigm," *La crise environnementale*, 215–216. (I am citing the English original.)
44. Callicott, "Metaphysical Implications," 101–112.
45. In chap. 9 of *Earth's Insights*, and in "After the Industrial Paradigm." See also "La nature est morte, vive la nature!" *Hastings Center Report* 22, no. 5 (1992): 16–23. Traditional French text, *Écologie politique* (Automne-Hiver 1993).
46. Alain Badiou, review of Deleuze's *Le Pli* in *Annuaire philosophique* 1988–1989 (Paris: Le Seuil, 1989), 179.
47. In "Gilles Deleuze: A Convergence with Ecological Theory and Politics," *Environmental Ethics* 19 (1997): 185–204, Patrick Hayden clearly stresses the importance of an anti-Platonic conception of diversity and multiplicity in Deleuze's naturalism.
48. In his *Foucault* (Paris: Éditions de Minuit, 1986), 118, Deleuze acknowledges that the concept of *fold*, to mean a form of subjectivization, comes from Merleau-Ponty—where it is related to what Merleau-Ponty names the "envelope phenomenon," Nature as "unwrapping" or "enfolding." So the two philosophers share a close connection: their philosophical inquiry about nature is designed to promote a new philosophy of the subject, not for nature's sake. And the example of Foucault (whose idea of subject relies on a similar concept of "folding") bears evidence that it can go without any idea of nature.
49. Deleuze, *Pourparlers*, 212.
50. Guattari, *Les trois écologies*, 69.
51. All the more because, as Conley quite rightly notes, they belong to a European tradition "that believes it has succeeded in liquidating nature centuries ago." Conley, *Ecopolitics*, 96.
52. Callicott, "After the Industrial Paradigm," 207.
53. See Catherine Larrère and Raphael Larrère, *Du bon usage de la nature, Pour une philosophie de l'environnement* (Paris: Aubier, 1997), chap. 3.
54. That is, according to Callicott, (1) "Of what is the world composed?" (2) "What is the logos, the law of nature?" and (3) "What fundamental forces drive natural processes?" "After the Industrial Paradigm," 207; *Earth's Insights*, 195–196.
55. Although he quotes Thomas Kuhn, Callicott's wide use of the concept of paradigm seems closer to Bohr's idea of a "conceptual framework" *(Begriffsgebaüde)*, which he elaborates in his 1955 article "The Unity of Knowledge."
56. Callicott, "After the Industrial Paradigm," 208.
57. See his *Conversations with Bruno Latour*, "A Synthesis Based on Relation" and "Hermes as Dispersion and Synthesis" 107–116. "Éclaircissements," the French title (i.e., "explanations") could be translated without being too much distorted by "Enlightenment," signaling Serres's ambition to be the harbinger of a new Enlightenment.

58. Actually for Serres there always has been, from Lucretius to the present day, one true physics, that of flow, turbulences, and so on. The alleged modern physics, classical mechanism, is but a mathematics, not a true physics, that is, a dynamics.
59. Serres, *Conversations*, 31.
60. Callicott, *Earth's Insights*, 189.
61. Serres, *Conversations*, 166. He already expressed such an idea in *La naissance de la physique dans le texte de Lucrèce: Fleuves et turbulences* (Paris: Éditions de Minuit, 1977). Such a statement opposes also Heidegger's idea, which distinguishes between humanity and animality by the way they inhabit the world (animals are "poorer" than human beings in world), that is, by their ethics.
62. Serres, *Conversations*, 199.
63. Which is not given, but constructed. Scientific rationality requires such a construction even though the subject-object distinction is not originary. Hence Serres speaks of "quasi-object."
64. Callicott, *Earth's Insights*, 189.
65. Callicott begins his "Metaphysical Implications" (p. 101) article by stating that such implications are not of the logical sort.
66. Callicott, *Earth's Insights*, 188.
67. Edmund Husserl, "La philosophie dans la crise de l'humanité européenne" (1935), in Edmund Husserl, *La crise des sciences européennes et la phénomènologie transcendantale* (Paris: Gallimard, 1976), 346–383.
68. Serres, *Conversations*, 180.
69. Serres, *Conversations*, 86.
70. This is one of the main themes of *Le contrat naturel*. Cf. Catherine Larrère, "Ethics, Politics, Sciences and the Environment," in *Earth Summit Ethics: Toward a Reconstructive Postmodern Philosophy of Environmental Education*, ed. J. Baird Callicott and Fernando da Rocha (Albany, N.Y.: SUNY Press, 1996), 115–138.
71. J. Baird Callicott, "Do Deconstructive Ecology and Sociobiology Undermine Leopold's Land Ethic?" *Environmental Ethics* 18 (1996): 353–372.
72. Thom, "La philosophie naturelle en quête de l'intelligible," *Apologie du logos*, 495–504.
73. See for example, *Earth's Insights*, 198: "Nature will be pictured . . . more as a vast organism than as a vast mechanism" and "nature is systemically unified by a hierarchy of internal relations."
74. One can surmise that it is not a very federalist word. It is strongly related to thermodynamics ecology, leaving aside biological ecology. So far from suggesting further extensions, it does not even express the whole of ecology.
75. Leopold, "Introduction, in *The River of the Mother of God*, 6.
76. François Jacob, *La logique du vivant. Une histoire de l'hérédité* (Paris: Gallimard, 1970).
77. As in the famous Latin fable of the limbs and the stomach.
78. Leopold, "Introduction," 6.
79. This is what Emmanuel Renault suggests in his article, "Les philosophies de la nature," 37.
80. Callicott, "Elements of an Environmental Ethics: Moral Considerability and the Biotic Community," in *In Defense*, 71–72.

81. Callicott, “The Conceptual Foundations of the Land Ethic,” in *In Defense*, 89.
82. On God’s model. Man is a cause, in the moral realm, as God can be conceived of as being cause of the world.
83. Hence the famous definition of life, as being the set of the processes that resist death.
84. Stanley Cavell, *The Claim of Reason*: (Oxford University Press, 1979). I say “American” because Cavell reads Thoreau’s *Walden* as a book about the philosophical invention of America, seeing in Thoreau and Emerson the founders of an America that was subsequently obscured by academic philosophy, and whose heir and successor he claims to be. Cf. *In Quest of the Ordinary* (Chicago: University of Chicago Press, 1988), and *This New Yet Unapproachable America* (Albuquerque: Living Batch Press, 1988).
85. “No imperative conclusion can be validly drawn from a set of premises which does not contain at least one imperative.” R. M. Hare, quoted by Cavell, *The Claim of Reason*, (Oxford: Oxford University Press, 1982), 454.
86. According to Kant, no one will ever be able to ascertain if a genuinely moral act has been ever achieved.
87. Callicott, “Intrinsic Value,” 170.

10

Quantum Physics, "Postmodern Scientific Worldview," and Callicott's Environmental Ethics

CLARE PALMER

Since the early 1980s, J. Baird Callicott has made a number of sorties into different branches of scientific theory (in particular post-Einsteinian physics and ecology) and related these to his work in environmental ethics. Most prominent amongst these sorties are his 1985 essay "Intrinsic Value, Quantum Theory and Environmental Ethics," his 1986 essay "The Metaphysical Implications of Ecology," and more recently his chapter "Towards a Postmodern Ecological-Evolutionary Environmental Ethic," which forms part of his book *Earth's Insights.* In this essay I explore the uses to which Callicott puts such scientific theories, in particular relating to quantum physics, and consider his discussion of a putative "postmodern scientific worldview." I concentrate primarily on Callicott's 1994 work in *Earth's Insights* (to which all page references herein refer) because this is most likely to reflect his current views on the subject (Callicott was still prepared to defend most of the arguments of *Earth's Insights* in *Worldviews* 1, August 1997). I pass over detailed consideration of Callicott's interpretations of ecology because others in this volume have addressed these questions far more competently than I could do; but because Callicott often brackets physics and ecology together as the vanguard of the "postmodern scientific worldview," some reference to ecology, nonetheless, is unavoidable.

CALLICOTT'S CLAIMS

In the introduction to *Earth's Insights* and in his chapter "A Postmodern Evolutionary-Ecological Environmental Ethic" Callicott makes a number of claims about changes in science, the implications and wider effects of such changes, and how they relate to his ethical proposals. These claims are founded on his belief that talking about what he calls "scientific worldviews," an expression that incorporates a number of elements, makes sense. First, the adoption of particular hypothetico-deductive methods along with constant self-conscious critical scrutiny is necessary—but not sufficient—for a worldview to be scientific; all scientific worldviews accept basically the same investigative methods.[1] In addition, this self-conscious critical scrutiny lends scientific worldviews "epistemological privilege" over other worldviews that do not exhibit such a self-critical nature. Second, by *worldview*, Callicott seems to mean "way of understanding what the world is like" or "way of describing and interpreting the world" that is manifested by the widespread use of particular metaphors (such as nature-as-mechanism) and perhaps by the acceptance of a particular kind of discourse with its accompanying underlying assumptions and structures (some of which may be unconsciously adopted by scientific practitioners). Third, he uses the term *scientific worldview* to describe the way of interpreting the world that is dominant within the scientific community at any particular time.

Building on this understanding of scientific worldviews, Callicott argues that a hugely successful, international "modern" scientific worldview exists, which one might call Cartesian-Newtonian, guided by the model of nature-as-mechanism, and characterized as *mechanico-industrial* (p. 187). This modern scientific worldview, he maintains, is now universally endorsed, differs very little from country to country, and constitutes a "cognitive lingua franca" throughout the world.

However, this modern scientific worldview is now theoretically speaking obsolete, and we should rather talk of a postmodern scientific worldview. Exactly what state this worldview is in, however, is not clear. Callicott says in different places that it is "emerging," "taking shape" (p. 188), "very much in the gestation stage and can't be definitively characterized" and that "it will sooner or later replace the waning mechanical worldview" and yet also that it has overturned the modern worldview (p. 185) out of which it grew. This postmodern scientific worldview is based on post-Einsteinian physics and recent work in evolutionary biology and ecology. As a scientific worldview, postmodern science differs from the modern scientific worldview in its paradigm, "not in the questions regarded as worth pursuing or the methods used to pursue them" (p. 197). Although it cannot be definitively characterized, Callicott thinks that the postmodern scientific worldview will have a more organic than mechanic image of human nature and indeed of nature as a whole. It will suggest that people are essentially connected to their environment, that they are viewed as "knots in webs" rather than as sharply differentiated objects. In general, organisms will be regarded as primarily relational; indeed relationships may be regarded as "more real than things." From such a perspective, the transmogrification of nature by industry will be regarded as a "dangerous outrage" (p. 188).

This has a variety of implications for Callicott's postmodern environmental ethic. Primarily because of its relation with universally endorsed scientific foundations, it can "make a claim to universality" (i.e., it can be accepted cross-culturally; p. 189). Second, his postmodern environmental ethic can "stake a coattail claim" to epistemological privilege as it is "grounded in the epistemologically privileged reconstructive postmodern scientific worldview." Alongside its universality and its epistemological status, the content of such an ethic is also determined by its relations to postmodern science. It may mean both that the "effects of human actions on individual non-human natural entities and on nature as a whole are directly accountable" and that the inability to draw sharp boundaries between people and the rest of nature means that the ethics is "a form of enlightened—or better, embedded collective human self interest."

Questions Raised by Callicott's Claims

Callicott's claims in *Earth's Insights*, as summarized previously, suggest a variety of questions on a variety of levels about his project. First, questions are raised concerning his interpretation of "the new physics" (and also about his interpretation of ecology, which I do not explore herein).[2] Second, both the general idea of scientific worldviews and the particular idea that a postmodern scientific worldview may exist are questionable. Third, and relatedly, one might ask how legitimate it is to "read up" from scientific work into philosophy/ethics as Callicott does—in particular, whether it is legitimate to read up from the microworld of quantum physics into the macroworld of individuals and things. And finally, if we grant Callicott his interpretation of quantum physics, his postmodern scientific worldview, and his reading of that into philosophy/ethics, what kind of ethics does this leave us with and what difficulties are associated with it?

My arguments on these points are not intended to provide a watertight, knockdown attack on Callicott's position. They do not develop a single, linear oppositional case. Indeed, there may be tension between the arguments (for instance, if one accepts, as I suggest, that a general scientific worldview does not exist, then the subsequent argument that no *particular* postmodern scientific worldview exists is redundant). Rather, I intend only to outline a series of individual difficulties that may lead one to feel uncomfortable with the scientific underpinning of Callicott's environmental ethics—and perhaps, as a result, with his ethics in general.

Callicott's Interpretation of the New Physics

Although Callicott does not mention the Copenhagen interpretation of quantum physics by name in *Earth's Insights*, his understanding of the new physics (by which he means relativity and quantum physics, but he does not discuss relativity) appears to draw on a version of it.[3] Exactly which version is not entirely clear either in *Earth's Insights* or in his more detailed 1985 essay.[4] This lack of clarity is in itself problematic because different versions of the interpretation may have rather different implications[5] (although Callicott's spin on the Copenhagen interpretation is contentious whichever version one adopts). However, the simple but central point I want to make here is that the Copenhagen interpretation, although still the most widely accepted in quantum theory, is just one of an increasing number of competing interpretations of the new physics. The most important of these are several versions of the many worlds theory[6]; David Bohm's ideas of the implicate order,[7] and Michael Lockwood's many minds theory.[8] Alongside these alternative theories, other physicists claim that the Copenhagen interpretation is mistaken because it is based on the belief that quantum theory is a complete, final theory, although it is, rather, incomplete and provisional. As the physics popularizer Paul Davies maintains, "no unanimous agreement [exists] amongst physicists on the approach to adopt."[9]

This is clearly problematic for Callicott because much of his postmodern scientific paradigm is dependent on accepting a very particular interpretation of quantum theory (some of the previous interpretations will do part but not all of the work for him; others will not support his position at all). That Callicott does not mention or consider any of these competing theories raises the suspicion that he has used the theory most suitable for his own preexisting purposes. His revealing recent response to a criticism of this kind is that "a good deal of mechanistic recidivism [exists] amongst contemporary quantum physicists."[10] This suggestion that, in discussing other interpretations of quantum theory, physicists are reverting to the original crime of mechanism (although this could hardly be true of, for instance, many worlds theories) indicates the ideological prejudice with which Callicott may be making his theory choice. That his interpretation of quantum physics is just one of many, and that the interpretation he chooses to underpin his philosophical position is one that is already congenial to his view, raises a question about the use of quantum physics in his philosophy.

The Idea of Scientific Worldviews and the Existence of a Postmodern Scientific Worldview

Callicott's argument in *Earth's Insights* is dependent on accepting that we can talk about a worldview that dominates the scientific community at any particular time. This definition does not, however, give us much assistance about what a scientific worldview might look like on the ground. What proportion of scientists would have to share the same discourse, assumptions, interpretations, and so forth before they could be described as constituting a scientific worldview? How many scientists would have to disagree before the idea of a dominant scientific worldview were undermined?[11] This raises a further question about the possibility of scientific worldviews. Could science be unified enough to have a governing worldview comprising a commonly agreed discourse, assumptions, structures, descriptions, and understandings? This question has been widely disputed amongst philosophers of science. Some maintain that it is impossible in principle for a unified understanding of science to exist. Most prominent among these is John Dupre who makes the claim that because of the "extreme diversity of the contents of the world" science is of necessity disunified; science could "not ever come to constitute a single, unified project."[12] But one need not accept Dupre's (nonetheless plausible) strong metaphysical claim about the disorder of things to maintain that it is, at the very least, extremely unlikely that science could ever be unified enough to have a governing or even widely accepted worldview. If one considers, for instance, the different scales at which scientists work (from microphysics and genetics to ecology, physical geography, biology); the fragmentation and specialization within many scientific disciplines and the range of competing political, personal, cultural, and financial influences on scientific work, the likelihood of convergence seems very slight.

But discounting this argument for now, we might try to seek empirical evidence for Callicott's idea of a postmodern scientific worldview. However, the idea of a dominant scientific worldview (postmodern or otherwise) is so nebulous that where one would go to find support for (or to falsify) such a claim is difficult to ascertain. Is there any evidence for the existence of the kind of dominant postmodern scientific worldview characterized by Callicott? It is certainly not obvious from a study of major scientific journals such as *Nature* or *New Scientist*, which both in general outlook and in the vast majority of articles published offer what Callicott regards as mechanistic recidivism. Although some popular science books (such as those of Fritzjof Capra or Rupert Sheldrake) adopt a line broadly compatible with Callicott's, most popular science books in biology, physics, and ecology[13] do not provide support for the existence of a postmodern scientific worldview (indeed writers such as Dawkins are famed for their reductionism, rather than their holism). Looking at physics in particular, numerous popular physics book interpret the universe in sharply contrasting ways to Callicott. For instance, myriad physicists have written about the Cosmic Anthropic Principle, which Midgley defines as "the notion that the physical universe can in some way be explained by assuming that it must be such as to contain people."[14] Whereas this view could be drawn out of more idealistic readings of the Copenhagen interpretation of quantum physics, it is surely antipathetic to Callicott's ecological-evolutionary worldview. Certainly, places exist where a worldview focusing on relationship, holism, interconnectedness, and organic images of human beings can be found. Some (although by no means all) of the so-called New Age movement is concerned with just these things. However, this could hardly be described as a scientific worldview in Callicott's sense, both because it is not widely enough held and because the rejection of (or the failure to adopt) the necessary but-not-sufficient hypothetico-deductive self-critical scientific method also characterizes the New Age movement. This means that it is not scientific in the Callicottian sense.

So what are we to make of Callicott's idea that we can speak of a postmodern scientific worldview that (in his stronger claims) has overturned modern science and (in his weaker claims) is emerging or taking shape? Callicott himself, in a comment that is both explanatory and apparently destructive of his own project, accepts that his postmodern scientific worldview is "crypto-utopian" and that he is "trying to make it happen by writing as if it is happening."[15] This suggests that actually it is not happening at the moment, which would seem to be supported by my (admittedly far from comprehensive) survey of the literature.

But is this not fundamentally damaging to Callicott's ethical project in *Earth's Insights* at least? He claims that his ethical position can make a "coat-tail claim to epistemological privilege because it is grounded in the epistemologically privileged reconstructive postmodern worldview" and that it can make a universal claim because of the universally accepted nature of science. But the basis of the general claim to epistemological privilege is the constant self-conscious critical scrutiny of the scientific community. An established dominant scientific worldview could (on Callicott's terms) claim this privilege, having been established through use of the epistemologically privileging scientific method. However, a scientific worldview that is not yet happening cannot yet have been subject to constant self-conscious critical scrutiny and hence cannot make the claim to epistemological privilege that an established scientific worldview (such as the modern one) might make. Ethics based on this not-yet worldview, then, equally cannot claim epistemological privilege; and if such ethics is not epistemologically privileged, universality cannot be claimed for it (on this basis, anyway).

The Relationship of Scientific Theories to Philosophy

This linking of ethics to a particular account of science raises questions about the relation between science and ethics. In *Earth's Insights*, for example, Callicott makes several claims about the implications of his interpretation of quantum physics. These include that the observer, as physical being, invariably affects and is affected by the physical object of observation and that the new physics may be taken to imply not merely that "the relationships of things are just as real as the things, but that the relationships are *more* real than things—that is that things are just the focus of a complex of relationships, however abstract they may seem . . . organisms are in their entire structure . . . relational entities" (p. 207). This repeats the kinds of claims that Callicott made in 1985 and 1986 and that are commonly made by deep ecologists about the holistic, interconnected relational implications of quantum physics. Such claims (dependent on the Copenhagen interpretation of quantum physics) have been countered elsewhere, and I will not repeat these criticisms here.[16] But one might also ask the more general question why scientific work at the quantum level should be presumed to have any effect on how we understand the world of everyday experience. What is strange about events at the quantum level seems to be that they are so strikingly different from events in the everyday world (where cats cannot be both dead and alive at the same time as the tale of Schrodinger's cat illustrates). Even if we were to concede some sort of sense in saying that, at the quantum level, "everything is constituted by its relationships," what meaning does this have at the macrolevel? In everyday senses, as Sylvan points out, things can still be separated from one another and have different degrees of relationship with one another (forests can be felled leaving a single tree standing that survives perfectly well without the rest of the forest, and so on).[17] Taking such scientific concepts out of their particular, limited context and applying them more generally to the world at large may result in both a distortion of the concept and in the making of a far greater claim than can possibly be justified on the evidence.[18]

Indeed, an even more general question is raised here about the relation of scientific theories to metaphysical and ethical ones. Quantum physics and scientific ecology are not, after all, the first scientific theories to have been "metaphysicalized" and normatized in the way Callicott does here. Some scientific theories have been used to support vastly different and even opposing metaphysical-ethical systems.[19] The theory of evolution is, of course, a classic case of this, providing the basis for such different philosophical-normative interpretations as social Darwinism and the progressive, onward-and-upward theological system of Teilhard de Chardin.[20] Both claimed support from evolutionary theory; both imported into it preexisting metaphysical, social, and ethical ideas. Surely the likelihood here is that preexisting metaphysical, social, and ethical ideas are read into the natural world via the vehicle of a conveniently interpretable scientific theory; then they become facts describing what the world is like and can be read back out of the world to provide the authority that underpins the same metaphysical, social, and ethical ideas.[21] This process, of course, need not be consciously manipulative, but rather reflects the widespread tendency both to accept and seek out what supports our preexisting beliefs.[22] However, it should surely make us extremely cautious when scientific theories are used to support metaphysical and ethical positions; science carries no special moral authority.[23] Such positions must surely be argued in their own right, rather than relying for special support from scientific theory.

A Postmodern Environmental Ethic

Although the questions raised previously may cast doubt on Callicott's putative postmodern scientific worldview and on his construction of a postmodern evolutionary-ecological environmental ethic, let us for the moment grant him the foundations of his position: his version of the Copenhagen interpretation of quantum physics and its contribution to a postmodern scientific worldview. What then are the ethical implications of this?

Callicott is not entirely clear here. He seems on occasions to think that the postmodern scientific worldview in itself has built-in environmental valuations; for instance he claims that "from a postmodern scientific point of view, the mechanico-industrial transmogrification of nature appears to be a grotesque and dangerous outrage, requiring us to develop an environmental ethic to temper its effects." Yet it is not obvious why such a value position would be inevitably part of a postmodern scientific worldview. Elsewhere, Callicott (p. 10) makes the claim that:

> There is a . . . more direct approach to environmental ethics which is more resonant with this emerging scientific worldview. . . . This

> approach would make the effects of human actions on individual nonhuman natural entities and nature as a whole directly accountable, regardless of their effect on other people.

This, of course, raises in acute form the questions about the relation of science to ethics discussed in the last section. But the specific claim here is also interesting: that such an environmental ethic would accept that elements of the nonhuman natural world were "directly accountable" ethically. This claim has been constantly reiterated in Callicott's work, customarily in his Hume-Darwin-Leopold construction, a construction that he repeats elsewhere in *Earth's Insights.* However problematic one might consider the Hume-Darwin-Leopold approach to ethics to be in other ways, if one accepts its premises, it does do the work Callicott wants in providing a basis for making "nonhuman natural entities and nature as whole directly accountable." However, how the same conclusions might be drawn from the postmodern scientific worldview is not at all clear.[24] Callicott's solution is to turn to the idea of the extended self, beloved of (some) deep ecologists, to do the work for him. He argues, in the time-honored formula:

1. The postmodern scientific worldview means that no clear boundaries exist between self and world; all entities are primarily relational.
2. Value is subjectively generated by humans.
3. Humans value themselves.
4. But no distinct boundaries exist between self and world.
5. Therefore, valuing self is valuing world.
6. Therefore, "we ought to care about the environment for reasons of extended self interest."
7. Therefore, "a nonanthropocentric environmental ethic, fully ecologized so to speak, turns out to be a form of enlightened—or better, embedded—collective human self interest after all" (p. 208).

Even if we put to one side the problems in the steps along the way here, how this supports his earlier claim that the postmodern scientific worldview provides a way of making elements of the natural environment directly ethically accountable is difficult to understand. In one sense, obviously, if these elements of the natural world are part of us, and we are what is ethically important, then they are directly ethically important. But this is surely a strange way of looking at things. Suppose I were cycling along the road and a car knocked me over. You came rushing to my assistance, helped me up, checked me over for broken bones. I began to thank you for your help, and you said

"No, that's all right, you're part of me, I did it for myself." I would feel that there was something missing here, that if that were really what you thought, then I was merely an instrument for your well-being and that you had not responded to my need as another individual, someone different from and other to yourself. Although this is a subtle form of instrumentalism, understanding the extended self Callicott proposes here as offering other than an indirect form of moral consideration to the natural world is difficult.

This is not the only problem raised by Callicott's idea of the extended self in ethics; indeed, extended self theories more or less similar to Callicott's have been widely criticized especially by ecofeminists.[25] In addition, the very general nature of the concept (offering, for instance, no guidance on resolving ethical conflict, or indeed, for discriminating between different elements of the environment because if boundaries have been broken down and all is part of one's self, on what grounds would anything be privileged?) makes using it in more practical terms difficult. All these factors suggest that even if one accepts the controversial underpinning of Callicott's postmodern environmental ethics, the resulting ethic itself is equally controversial.

Conclusions

Building on his earlier work involving metaphysical interpretations of quantum theory and ecology, in his 1994 book *Earth's Insights* Callicott advances an epistemologically privileged, scientifically based postmodern evolutionary-ecological environmental ethic. Herein I have attempted to show why, at a variety of levels, Callicott's claims are problematic. None of these arguments in themselves completely undermine Callicott's case. However, cumulatively I think that they raise a question both about the idea that a postmodern scientific worldview could exist and that such a thing could provide an epistemologically privileged ground for environmental ethics.

Notes

1. It is worth noting that, although I do not have space to discuss this here, I am by no means accepting either that scientists invariably practice the hypothetico-deductive method nor that science is necessarily "self-critical" and therefore epistemologically privileged as Callicott suggests. Enough work has been carried out in sociology of science, especially by feminist writers, to cast doubt on such claims. See for example, Fox and Keller (1996) and Dupre (1993, pp. 229–233).
2. See for example, Cheney & Warren (1993).
3. This is certainly what he depends on in 1985 and his interpretation of quantum physics here gives no reason to suggest a change of mind.

4. As Folse (1995) points out, many versions of the Copenhagen interpretation exist: "Although Bohr, Heisenberg, Pauli and Born can all be arrayed under the banner of Copenhagen, they in fact held quite different views."
5. A point that Sylvan (1990: 87) makes with cogency about the work of Warwick Fox.
6. The idea that there are parallel entire universes that exist at the same time, in some senses in the same space, and that usually do not communicate with one another. See, for instance, Deutsch (1998).
7. Bohm (1981).
8. See for example, Lockwood (1996).
9. Davies & Brown (1986: 39).
10. Callicott (1997: 180).
11. The issue raised here is somewhat similar to that discussed by Proctor (1997, p. 135) when he discusses what proportion of those adhering to particular indigenous worldviews must hold beliefs compatible with Callicott's ethic before an indigenous worldview could be declared resonant with Callicott's ethic.
12. Dupre (1993: 1).
13. Capra (1983), Sheldrake, (1990). Cf. Davies (1986), Dawkins (1989), Gould (1990).
14. Midgley (1992: 27).
15. Callicott (1997: 180).
16. Brennan (1988); Palmer, (1998); Sylvan (1985, 1990).
17. See also Rolston (1989).
18. Kirkman (1997, p. 206) makes a similar argument.
19. As, for example, explored in Midgley (1992).
20. See Midgley (1985) and Rachels (1991).
21. A version, therefore, of the process of externalization-objectification-internalization as outlined by, for example, Berger (1969).
22. A process outlined by Stevenson (1992).
23. I am grateful to Andrew Brennan (1998) for discussion and clarification of this point.
24. In 1985 Callicott makes the logical argument that a metaphysics based on quantum physics entails a breakdown of the Hume-Dawin-Leopold subjectivist ethic because the subject-object distinction cannot be maintained. From within this framework, value in nature is actualized on interaction with consciousness. Although Callicott later relates this conclusion to ideas of the extended self, the former does not necessarily seem to lead to the latter. However, this interpretation, although not in itself unproblematic (see Norton, 1991) is not pursued in *Earth's Insights.*
25. See for example, Cuomo, (1994); Lorentzen, (1997); & Plumwood, (1993).

Works Cited

Berger, P. (1969) *The sacred canopy: Elements of a sociological theory of religion.* New York: Anchor.

Bohm, D. (1981). *Wholeness and the implicate order.* London: Routledge.

Brennan, A. (1988). *Thinking about nature*. London: Routledge.

Callicott, J. B. (1985). Intrinsic value, quantum theory and environmental ethics. *Environmental Ethics,* 7, 257–275.

Callicott, J. B. (1986). The metaphysical implications of ecology. *Environmental Ethics, 8,* 301–316.

Callicott, J. B. (1994). *Earth's insights.* Berkeley: University of California Press.

Callicott, J. B. (1997). In defense of *Earth's Insights. Worldviews: Environment, culture, religion,* 1, 167–183.

Capra, F. (1983). *The turning point*. London: Fontana.

Cheney, J. & Warren, K. (1993, Summer). Ecosystem ecology and metaphysical ecology: A case study. *Environmental Ethics, 15,* 99–116.

Cuomo, C. (1994). Ecofeminism, deep ecology and human population. In K. Warren (Ed.), *Ecological feminism* (pp. 88–105). London: Routledge.

Davies, P. & Brown, J. (1986). *The ghost in the atom.* Cambridge, England: Cambridge University Press.

Dawkins, R. (1989). *The selfish gene*. Oxford: Oxford University Press.

Deutsche, David (1998) *The Fabric of Reality.* Harmondsworth: Penguin.

Dupre, J. (1993). *The disorder of things: Metaphysical foundations of the disunity of science.* Cambridge, MA: Harvard University Press.

Eaton, H. (1997). *Earth's Insights*—and inadequacies. *Worldviews: Environment, Culture, Religion, 1,* 113–121.

Folse, H. (1995, April 1). Niels Bohr and the construction of a new philosophy. *Studies in History and Philosophy of Modern Physics, 26B.* 107–116.

Keller E. F. & Longino, H. (Eds.). (1996). *Feminism and science* Oxford: Oxford University Press.

Gould, S. J. (1990) *Wonderful life.* London: Hutchinson Radius

Kirkman, R. (1997). The problem of knowledge in environmental thought. In R. Gottleib (Ed.), *The ecological community* (pp. 193–207). London: Routledge.

Lockwood, M. (1996). Symposium: "Many minds" interpretations of quantum mechanics. *British Journal of the Philosophy of Science, 47,* 159–188.

Lorenzen, Lois (1997) What's postmodern about Earth's Insights? *Worldviews: Environment, Culture, Religion,* 1, 123–132.

Midgley, M. (1985). *Evolution as a religion.* London: Meuthen.

Midgley, M. (1992). *Science as salvation.* London: Routledge.

Norton, B. (1991). Review of J. Baird Callicott, *In defense of the land ethic. Environmental Ethics, 13*, 181–186.

Palmer, C. (1998). *Environmental ethics and process thinking*. Oxford: Oxford University Press.

Plumwood, V. (1993). *Feminism and the mastery of nature*. London: Routledge.

Proctor, James (1997) *Earth's insights*: A geographer's perspective. *Worldviews: Environment, Culture, Religion*, 1, 131–138.

Rachels, J. (1991). *Created from animals: The moral implications of Darwinism*. Oxford: Oxford University Press.

Rolston, H. (1989). Review of Andrew Brennan, *Thinking about nature. Environmental Ethics, 11*, 260.

Sheldrake, Rupert (1990) The Rebirth of Nature. London: Century.

Stevenson, R. (1992). Thinking, believing and persuading: Some issues for environmental activists. In D. Cooper & J. Palmer (Eds.), *The environment in question* (pp. 194–223). London: Routledge.

Sylvan, R. (1985). A critique of deep ecology. *Radical Philosophy*, 40, 2–12, and 41, 10–22.

Sylvan, R. (1990). A critique of (wild) western deep ecology, in *In defense of deep environmental ethics. Environmental ethics preprint series in environmental philosophy, 18*, (pp. 37–93). Canberra: Australian National University.

11

Minimal, Moderate, and Extreme Moral Pluralism

PETER S. WENZ

INTRODUCTION

It is no surprise that moral pluralism is discussed in environmental ethics because such ethics concerns human beings, animals of various sorts, species, ecosystems, wilderness areas, and evolutionary and ecological processes. Some philosophers have maintained that no single ethic can encompass this variety, and have, therefore, advocated moral pluralism. However, both advocates and critics of pluralism have conflated three views identified as moral pluralism. To dispel the resulting confusion, I distinguish between minimal, moderate, and extreme forms of moral pluralism and defend moderate moral pluralism against its chief critic, J. Baird Callicott.[1]

I examine minimal moral pluralism using as its exemplar a general characterization of moral pluralism given by Christopher Stone in *Earth and Other Ethics: The Case for Moral Pluralism.*[2] I reject this characterization as unhelpful because, contrary to its author's probable intentions, it makes all moral theories pluralistic. Then, I draw from Stone's work a definition of what I call *extreme moral pluralism*, and reject it on grounds Callicott gives.[3] I go on to argue that the source of Stone's error is a faulty analogy among compartmentalization in the sciences, social sciences, and arts on one hand, and one's moral life on the other. I then discuss moderate pluralism, which I have endorsed in *Environmental Justice.*[4] Such pluralism, I argue, is reasonable and free of the defects that justify rejecting extreme pluralism. Finally, I show that Callicott, who believes himself to be rejecting all moral pluralism, actually incorporates moderate pluralism in his version of the land ethic. In doing so, I distinguish my position concerning Callicott's pluralism from Gary E. Varner's as yet unsubstantiated contention that the foundations of Callicott's theory are pluralistic.[5]

Minimal Moral Pluralism

Stone defines pluralistic theories by contrasting them with monistic theories. At one point he writes that a monistic theory, unlike pluralistic theories, contains "a single coherent set of principles capable of governing all moral quandaries" and yielding "for each quandary one right answer."[6] Pluralistic theories, by contrast, do not provide a single right answer to every moral question. Callicott seems initially to interpret Stone as holding that, as a practical matter, monistic theories provide ordinary moral agents with unique and specific prescriptions in every situation involving moral choice, whereas pluralistic theories do not. In other words, monistic theories, unlike pluralistic theories, provide algorithms that people can as a practical matter use to determine their duty in any situation. Although I am not particularly comfortable with this use of terms, I follow this usage here because I am joining a discussion in which the terms are already so defined. To avoid confusion, however, I call this form of pluralism *minimal moral pluralism* because the requirements for being pluralistic are minimal. For a theory to be pluralistic in this sense, it is sufficient that the theory merely lack a universal algorithmic decision procedure.

Herein I discuss Callicott's two objections to minimal pluralism and then show that these objections are misplaced, arguing that all known moral theories are pluralistic in this minimal sense. As a result, if such pluralism is objectionable, the state of human moral knowledge in general is to blame, not Stone's or any other particular moral theory. In short, when pluralism is defined in this way, it cannot be contrasted with monistic theories at all, much less invidiously.

Callicott objects, first, that in many situations a pluralistic theory does not provide practical guidance to ordinary people. This objection arises directly out of Stone's characterization of what I call *minimal pluralism.* The objection is that pluralistic theories fail to supply a single correct answer to complex moral queries. The implication is that monistic theories do not suffer from this defect, that, presumably, they do not leave ordinary people in doubt about what they ought to do.

Callicott's second objection is a corollary of the first. Because pluralistic theories do not tell people unambiguously what to do, such theories "might provide a sophisticated scoundrel with a bag of tricks to rationalize his or her convenience or self interest."[7] Presumably, again, monistic theories are free of this defect. However, they can be in this favored position only if, unlike pluralistic theories, they supply such unambiguous answers and so leave scoundrels no room to rationalize immoral behavior.

Note that the problem Callicott identifies persists in a theory that provides unity at a *theoretical* level, but not at the *practical* level. If the failure to provide unambiguous prescriptions risks self-serving rationalizations, as Callicott maintains, a theory can overcome this defect only by supplying to ordinary people unambiguous, practical prescriptions in all situations where moral choices must be made.

A moment's reflection suggests, however, that no theory is monistic in this sense. It seems that all moral theories are pluralistic in the minimal sense that they fail always to inform ordinary people through the application of a practically available algorithmic procedure of a uniquely correct action. Although we need to recognize this point, doing so gives us no reason to prefer any moral theory over its rival and certainly not monistic over pluralistic theories because no monistic theories exist to prefer to those that are minimally pluralistic. Curiously, after predicating two arguments on the assumption of a contrast that has no application, Callicott himself seems to recognize the contrast's inapplicability.[8]

The Rejection of Extreme Moral Pluralism

Stone also employs a more robust notion of pluralism that does not render all theories pluralistic. I call this kind *extreme pluralism*, which is characterized by alternations among several ethical theories. Although each such theory is accepted in its entirety, the range of application of each is limited. The extreme pluralist adopts different ethical theories for different contexts, and/or for different general subjects, of application. Stone writes that:

> . . . a senator, for example, might rightly embrace utilitarianism when it comes to legislating a general rule for social conduct (say, in deciding what sort of toxic waste program to establish). Yet this same representative need not be principally utilitarian, nor even a consequentialist of any style, in arranging his personal affairs among kin or friends, or deciding whether it is right to poke out the eyes of pigeons.[9]

Stone suggests here that a senator might be a thorough utilitarian when choosing among legislative proposals, but a Kantian in his or her relationships with family and friends, and a Leopoldian land ethicist when wilderness areas are at issue.

Callicott rejects the idea that a sane individual could reasonably alternate among three such different ethical theories. "Utilitarianism assumes a radical individualism," according to which "the community is a fictitious body,"[10] whereas the land ethic takes the community to be the fundamental unit within which individuals are organically and internally related. "To adopt Kant's moral theory is to buy into a . . . philosophy of human nature in which Reason (with a capital R) constitutes the essence of 'man,' " and all nonhumans are mere things.[11] The Kantian rejects the utilitarian concern for pleasures, pains, and preference satisfactions among human beings and other animals, and the land ethic's concern for biotic communities, wilderness areas, and endangered species. Callicott finds unreasonable an individual's alternations among such different views about the nature of human life and its place in the cosmos: "Moral pluralism, in short, implies metaphysical musical chairs. I think, however, that we human beings deeply need and mightily strive for consistency, coherence, and closure in our personal and shared outlook on the world and on ourselves in relation to the world and to one another."[12]

I do not know if Stone meant to endorse "metaphysical musical chairs," but his statements do lend themselves to Callicott's interpretation. Whatever Stone's particular views, however, Callicott argues powerfully against any version of moral pluralism that requires alternations among radically different metaphysical perspectives. I join Callicott in his rejection of it.

The Source of Stone's Error

Although Callicott accurately identifies the weakness of extreme pluralism, he does not explain the source of Stone's error. I do so here. Stone's view results from combining a keen observation with a mistaken analogy. The keen observation is that philosophers often seek more unity than is reasonable. The mistaken analogy is between scientific and social scientific discourse on the one hand, and moral judgments on the other.

Stone assumes that compartmentalization in morality is justified by analogy with compartmentalization in science and social science. In a chapter entitled "Foundations for Moral Pluralism," he writes, "Pluralism conceives the realm of morals to be partitioned into several planes. The planes are intellectual frameworks. . . ."[13] He compares these to the distinct planes of geometry and arithmetic,[14] of mathematics and poetry,[15] and of chemistry and art criticism.[16] He is explicit in a later article that his moral pluralism is inspired by the inability of investigators to "unify . . . the laws that govern the movement of subatomic particles with those that govern social conduct."[17] In his book, he reasons that "just as the rules of solid geometry are not the rules of arithmetic, so the rules that govern our relations with animals are not the same rules that govern the relations among corporate bodies.[18] Similarly, different rules govern our friend dilemmas and our stranger dilemmas,[19] as well as the personal and professional responsibilities of senators.[20]

Stone's analogy is fatally flawed. He compares differences among the sciences, social sciences, and arts on the one hand, to differences among domains of ethical application on the other. The crucial disanalogy rests on the fact that moral judgments about conduct are ideally made only after all relevant matters have been considered. Doing so restricts the propriety of compartmentalization because one is expected to take ethical considerations into account wherever, and in whatever combinations, they appear.

Moral behavior requires a disposition to be responsive simultaneously to all of one's roles and commitments. This moral requirement follows from the fact that moral judgments are properly made *all things considered*. Whenever and wherever a factor of moral relevance arises, it must be considered, even though it may not appear prima facie to be the dominant factor. Unlike moral judgments, however, judgments within certain disciplines can be isolated from judgments within other disciplines.

Stone seems to have been misled by his failure to perceive that disciplinary compartmentalization in the sciences, social sciences, and arts does not imply similar compartmentalization in morality. The difference is between acceptable specialization and unacceptable incoherence. In sum, Callicott's criticisms of what I call *extreme moral pluralism* are justified. Such pluralism is built on a faulty analogy and invites inconsistency and incoherence where consistency and reason are needed most—in our moral lives.

Moderate Pluralism

Moderate moral pluralism, the view endorsed in this essay, avoids the defects of extreme moral pluralism. First, moderate pluralism does not involve alternations among different ethical theories because it is itself a single ethical theory. It is pluralistic only in the sense that it "contains a variety of *independent* principles, principles that cannot all be reduced to or derived from a single master principle."[21] Whereas extreme pluralism involves a plurality of theories, moderate pluralism includes only a plurality of principles (in a single theory).

Second, moderate pluralism does not compartmentalize the moral life. Because the moderate pluralist confronts all situations with the same ethical theory, he or she can entertain in any situation the full range of relevant principles. For example, a senator can, and should, give some weight to the principle that promises should be kept (all other things being equal) whenever and wherever he or she has the opportunity to keep or break a promise. Subject matter, forum, and role do not affect this general principle. All other things being equal, promises should be kept concerning presents for children, pledges to the Environmental Defense Fund, and agreements to vote for minimum wage legislation. They should be kept on the floor of the U.S. Senate, in the living room, and on camping trips. They should be kept by parents, teachers, and senators. The same is true of other principles, that is, that unnecessary pain is to be avoided, that persuasion is preferable to coercion, and that wilderness areas should be preserved. Each is to be considered whenever it is applicable. Because the individual meets all situations with the same set of moral principles, his or her ethical life is coherent.

The weight accorded a principle sometimes varies with context. When I am asked to join a toast to salute the beauty of a bride, for example, the principle of honesty is outweighed by the principle of consideration for people's feelings. However, when evaluating a student's work, honesty takes precedence. Some hurt feelings are acceptable if they unavoidably accompany honest communication. Similarly, a defense lawyer may place a higher priority on promoting general human welfare when soliciting for United Way than when defending a client. Having agreed to defend the client, he or she must do so even when believing that a conviction would better serve general human welfare.

Adjustment of priorities to specific context does not require renouncing in some contexts principles that are considered of singular importance in others. Nor does it involve "metaphysical musical chairs." The individual remains prepared in all contexts to apply the full range of relevant moral principles. Thus, the lawyer defending an unsavory client may withdraw from the case if defending that client conflicts too strongly with his or her principle that human welfare should be fostered.

Coherent thinking is not jeopardized by responsiveness to more than one consideration at a time. Rather, such responsiveness is a common aspect of human life. For example, while making dinner I can choose ingredients, in part, for their nutritional value and, in part, because they are perishable and getting spoiled. I can choose to cook the potatoes (one of those ingredients) in the microwave to save electricity, at some sacrifice to my pleasure (crisp potato crusts). I can at the same time listen for my wife's call on the telephone (telling me when she will arrive home) and monitor the play of neighbor children on our old swing set. My thinking is not made incoherent because I am sensitive simultaneously to several considerations, to considerations of several types, and to considerations that suggest incompatible courses of action (saving electricity versus enjoying crisp potato crusts). Such thinking is simply normal.

Moderately pluralistic moral theories are not made incoherent by their simultaneous application of several moral principles to a given moral quandary. The failure to provide algorithmic decision procedures in these situations merely means that moderately pluralistic theories are also minimally pluralistic. This fact is not surprising because, as I argued earlier, all moral theories are at least minimally pluralistic. Such pluralism can be equated with incoherence only by imposing an unusually demanding standard for coherence. Because this standard is new and would label "incoherent" much of what we currently consider rational, it is incumbent on those who want to impose it to argue for its adoption.

The Land Ethic as Moderately Pluralistic

The theory of environmental ethics that Callicott explains in "The Conceptual Foundations of the Land Ethic"[22] and "The Search for an Environmental Ethic"[23] is pluralistic in the same way and to the same degree as the concentric circle theory that I offer in *Environmental Justice*. Callicott maintains that the land ethic adds environmental duties to our other obligations. It does not displace or override those other obligations. We retain all familiar obligations toward family, friends, country, and humanity in general. What is more, "Family obligations in general come before nationalistic duties and humanitarian obligations in general come before environmental duties," because, "as a general rule, the duties correlative to the inner social circles to which we belong eclipse those correlative to the rings farther from the heartwood."[24]

Because environmental obligations are farthest from the heartwood, one may think that their demands are always overridden by whatever demands may exist in a more interior social-ethical circle. However, this approach is not Callicott's position at all because it would constitute abandoning the land ethic altogether in favor of the culture's predominant anthropocentrism. Callicott maintains, instead: "The land ethic may . . . as with any new accretion, demand choices which affect, in turn, the demands of the more interior social-ethical circles. Taxes and the military draft may conflict with family-level obligations. While the land ethic certainly does not cancel human morality, neither does it leave it unaffected."[25] This remark accords with Callicott's statement elsewhere that: "Just as it is not unreasonable for one to suppose that he or she has *some* obligation and should make *some* sacrifice for the 'wretched of the earth,' so it is not unreasonable to suppose that the human community should assume *some* obligation and make *some* sacrifice for the beleaguered and abused biotic community."[26]

I agree with this position completely. My obligations toward my daughters are among the strongest that I have. It would certainly be wrong to sacrifice a daughter's life to save a bear. Nevertheless, it makes perfect sense to curtail her snowmobiling to protect a wilderness area. The strength of an obligation varies not only with one's relationship to its object, but also with the nature of the object's claim. The claim to protection of one's life is much stronger than the claim to enjoy a preferred pattern of recreation. For this reason, for example, restrictions often are placed on fireworks displays. Although some people prefer fireworks displays in their backyards, such displays endanger the lives of others, and human lives are (usually) more important than preferred enjoyments. The importance of protecting wilderness areas, which is of moral concern in the land ethic, may generally be placed in between protecting individual human lives, which is more important, and accommodating people's preferred patterns of recreation, which is less important. Thus, reverting to the example of my daughter, preserving her life can override environmental considerations that, in turn, override her increased satisfaction in winter recreation.

I assume that the foregoing fairly reflects Callicott's general approach to these matters. If so, he is clearly a moderate pluralist. He has many moral principles, and they are not all derived from a single, master principle. Some moral principles concern which relationships are more important than others (e.g., parenthood more than friendship, friendship more than citizenship). Other principles concern the relative values attached to different kinds of outcomes (e.g., death is generally worse than dissatisfaction, at least where human beings are concerned). Still other principles are used to identify when the normal priorities do not apply (e.g., when citizen obligations override familial duties and the value of life to justify going to war). Callicott neither presents, nor claims to possess, any master rule or principle from which one can deduce uniquely correct moral conduct in situations of moral conflict. Indeed, he writes, "How obligations . . . may be weighed and compared is admittedly uncertain. . . ."[27] Nor does he claim that the many principles he employs can be derived from any *action-guiding*, universal master rule or master principle. Thus, *at the level of action-guiding norms*, Callicott's version of the land ethic is both minimally and moderately pluralistic. It "contains a variety of *independent* principles . . . that cannot all be reduced to or derived from a single master principle."[28]

To avoid confusion, this demonstration of the pluralism in Callicott's version of the land ethic should be distinguished from the argument Varner offers. Varner defines "a pluralistic theory" as "one which acknowledges distinct, theoretically incommensurable bases for direct moral consideration."[29] He argues that Callicott's theory must be pluralistic because Callicott considers both people and biotic communities to be directly morally considerable. However, if biotic communities are directly morally considerable, "it must be for very different reasons than is usually given for saying that individual human beings are directly morally considerable. . . ."[30]

I have three comments about Varner's contribution. First, Varner's principal argument is at best inconclusive as it stands. Because he does not tell us what basis Callicott gives for maintaining that people are directly morally considerable, he cannot claim to have shown that Callicott's basis for maintaining that people are directly morally considerable is incommensurable with Callicott's basis for according direct moral consideration to biotic communities. As a result, in terms of Varner's definition of moral pluralism, Callicott may or may not be a pluralist.

Second, Varner's argument addresses the *foundations* of Callicott's version of the land ethic. In this respect, it differs from the subject matter of the current essay, which concerns pluralism at the level of normative principles.

Finally, Varner may be suggesting that normative pluralism inevitably involves pluralism at the foundational, or metaphysical, level. Such a position would involve the claim that moderate pluralism presupposes extreme pluralism, and so shares the defects of extreme pluralism. This claim is a generalization of Varner's claim that Callicott's moderate pluralism lacks a unitary foundation. Nevertheless, there is no a priori reason to accept this view. W. D. Ross's position in *The Right and the Good* is moderately, but not extremely (i.e., metaphysically), pluralistic.[31] Several independent moral principles coexist within a single metaphysical vision. A controversial, and in my view ultimately unacceptable, foundationalist epistemology ties it all together. One cannot rule out a priori a more acceptable epistemology attaching the land ethic's moderate pluralism to a unitary metaphysical vision.

Conclusion

I conclude, first of all, that all known moral theories are at least minimally pluralistic because none provides algorithms for the solution of each and every moral quandary that may arise. When Stone defines his pluralism in this minimal way, Callicott's arguments against Stone's position founder because the contrast with monistic theories that Callicott relies upon is vacuous.

A question worth exploring is whether those minimally pluralistic theories that claim to be monistic at a deeper (theoretical) level, such as some forms of utilitarianism and Kantianism, are monistic in any meaningful sense at all. It may be that any single, master principle featured in such a theory is more symbolic than operative, such as a corporate chairman of the board who is merely a figurehead. This situation would exist if and when no reasonably reliable method of using the master principle were available either (1) to generate the midlevel principles that the theory employs, or (2) to justify reliance on, or preference for, one such midlevel principle over others in cases of practical conflicts among them. In such cases, the putatively monistic theories in question would, in effect, be at least moderately, as well as minimally, pluralistic. Because the master principle would not meaningfully authorize midlevel principles, nor meaningfully adjudicate among them, the midlevel principles would be independent of one another as in a moderately pluralistic theory. I leave to another occasion arguments that putatively monistic (minimally pluralistic) theories are actually moderately pluralistic.

I conclude also that Callicott properly criticizes extremely pluralistic theories, including Stone's, as lacking the unity and integrity necessary for a coherent moral life.

I conclude, finally, that a significant conceptual distinction exists between moderate and extreme pluralism that allows the former to avoid charges of incoherence that are properly leveled at the latter. I find no a priori reason why moderate pluralism cannot coexist with unity at the metaphysical, foundational level. Thus, in the absence of additional argument, I find no reason why the moderate pluralism of Callicott's land ethic should be considered incoherent as if it were an extremely pluralistic theory such as Stone's.

Notes

1. Outside of the contexts of environmental philosophy, the term *moral pluralism* and similar terms have been associated with several topics and views. In Michael Walzer, *Spheres of Justice: A Defense of Pluralism and Equality* (New York: Basic Books, 1983), it stands for the view that different principles of distributive justice are appropriate in different contexts. To the extent that this view resembles the extreme pluralism of Christopher Stone, discussed and criticized later herein, it is susceptible of the same critique. However, because Walzer's view is more subtle than Stone's, criticisms of the latter cannot be transferred automatically to the former. The term *pluralist lexical consequentialist* is used in Samuel Scheffler, *The Rejection of Consequentialism* (Oxford: Clarendon Press, 1982), 27–28, to refer to the idea that a plurality of basic goods exists that is not commensurable. This position somewhat resembles the moderate pluralism that I endorse. A similar view is also found in Thomas Nagel, "The

Fragmentation of Value," in *Mortal Questions* (New York: Cambridge University Press, 1979), 128–141. Finally, cultural relativism in ethics, individual relativism, and subjectivism have all been referred to from time to time as pluralist views because they allow for more than one moral view to be considered correct. I trust that neither my moderate pluralism nor Stone's extreme pluralism will be confused with any of these.

2. Christopher D. Stone, *Earth and Other Ethics: The Case for Moral Pluralism* (New York: Harper and Row, 1987).

3. J. Baird Callicott, "The Case against Moral Pluralism," *Environmental Ethics* 12 (1990): 99–112.

4. Peter S. Wenz, *Environmental Justice* (Albany, N.Y.: SUNY Press, 1988).

5. Gary E. Varner, "No Holism without Pluralism," *Environmental Ethics* 13 (1991): 175–179.

6. Stone, *Earth,* 116. He repeats this point in "Moral Pluralism and the Course of Environmental Ethics," *Environmental Ethics* 10 (1988): 145.

7. Callicott, "The Case against Moral Pluralism," 111.

8. Callicott, "The Case against Moral Pluralism," 120.

9. Stone, *Earth*, 118.

10. Callicott, "The Case against Moral Pluralism," 114, where he quotes Jeremy Bentham, *An Introduction to the Principles of Morals and Legislation* (Oxford: Clarendon Press, 1823), chap. 1, sec. 4.

11. Callicott, "The Case against Moral Pluralism,' 114–115.

12. Callicott, "The Case against Moral Pluralism," 115.

13. Stone, *Earth*, 133.

14. Stone, *Earth*, 134.

15. Stone, *Earth*, 150.

16. Stone, *Earth*, 151.

17. Stone, "Moral Pluralism," 147.

18. Stone, *Earth*, 134.

19. Stone, *Earth*, 142.

20. Stone, *Earth*, 118.

21. Wenz, *Environmental Justice,* 313 (emphasis in the original).

22. J. Baird Callicott, "The Conceptual Foundations of the Land Ethic," in *A Companion to a Sand County Almanac,* ed. J. Baird Callicott (Madison: University of Wisconsin Press, 1987), 186–217.

23. J. Baird Callicott. "The Search for an Environmental Ethic," in *Matters of Life and Death*, ed. Tom Regan, 2d ed. (New York: Random House, 1986), 381–424.

24. Callicott, "Conceptual Foundations," 208.

25. Callicott, "Conceptual Foundations," 208.

26. Callicott, "The Search for an Environmental Ethic," 412.

27. Callicott, "The Search for an Environmental Ethic," 412.

28. Wenz, *Environmental Justice*, 313.

29. Varner, "No Holism without Pluralism," 177.

30. Varner, "No Holism without Pluralism," 179.

31. W. D. Ross, *The Right and the Good* (New York: Oxford University Press, 1930).

12

Callicott and Naess on Pluralism

ANDREW LIGHT

As environmental ethics approaches its third decade it finds itself at a curious crossroads. On the one hand it has produced a plurality of positions and theories attempting to extend moral consideration to nonhumans and sometimes ecosystems.[1] On the other hand the discipline is also marked by a tendency to suppose that certain approaches in the field are necessarily more promising than others, and even that these approaches are the only ones that could possibly result in an adequate environmental ethic. We are told by some theorists that we must assume that a coherent environmental ethics must embrace a restricted set of properties: nonanthropocentrism, holism, moral monism, and, perhaps, a commitment to some form of intrinsic value.

But given the relatively young age of environmental ethics as a recognizable subdiscipline of philosophy, how do we account for the presupposition that we have settled into agreement as a community of scholars on the right theoretical path to take in grounding such an ethic? Are we ready to accept the claim by J. Baird Callicott that until Christopher Stone's book on pluralism in environmental ethics came along in 1988 (*Earth and Other Ethics*), he was prepared for the discipline to "begin to work toward the creation of an intellectual federation and try to put an end to the balkanization of nonanthropocentric moral philosophy?"[2] In this article Callicott appeared to presume that a workable environmental ethics would have to be monistic and nonanthropocentric, and he was thus prepared for the field to move ahead unencumbered by the serious consideration of alternatives such as pluralism and more enlightened versions of anthropocentrism. Surely it is not odd for philosophers to think they have the right answers, but it should give us pause that in a field so young it is already the considered opinion of some of our most influential theorists, and arguably a critical mass of scholars, that anyone who is still questioning which sides they will take on the anthropocentrism-nonanthropocentrism, holism-individualism, intrinsic-instrumental value, and monism-pluralism debates is just being unnecessarily obfuscatory.

In this essay, using Callicott's work as a template, I revisit one of the presumed correct paths for contemporary environmental ethics, moral monism, and take up the debate currently in progress about the alternative, moral pluralism. Monists in environmental ethics generally argue that a single moral philosophy or ethical theory is required to ground our correct duties and obligations toward the environment. Pluralists, again generally speaking, argue that this cannot be the case either because the sources of value in nature are too diverse to account for in any single theory, or because the multitude of contexts in which we find ourselves in different kinds of ethical relationships with both humans and nature demand a plurality of approaches for fulfilling our moral obligations.

On Callicott's account, a commitment to monism is one of the strongest stances that must be maintained in environmental ethics. In one of his most important contributions to the monism-pluralism debate, Callicott, in offering a brief refutation of the suggestion that the "dean" of North American environmental ethics, Holmes Rolston III, is not a monist, remarks: "Given that even Rolston is not really a pluralist after all, one begins to wonder why *our best, most systematic, and thoroughgoing* environmental philosophers cling to moral monism."[3] Many well-respected theorists in this field may disagree with such a claim. In fact, aside from Stone, Andrew Brennan, Gary Varner, Peter Wenz, Anthony Weston, Eugene Hargrove, and many other "systematic" and "thoroughgoing" theorists have come out on the pluralist side of the debate.[4] Of the several debates in the field previously mentioned the monism-pluralism debate appears very much unsettled. Given the several responses to Callicott just cited, perhaps he would agree.

In investigating this issue further, I proceed by taking up the monism-pluralism debate as it has been more systematically advanced by Callicott in a 1994 article in the *Journal of Philosophical Research.* From there however I take a different turn from the discussion of this issue as it has played out so far. Specifically, I address the question of where Arne Naess's environmental philosophy fits into the monist-pluralist debate. In the discussions of monism and pluralism in environmental ethics so far we find little or no mention of Naess's work, even though he may have worked out some of the most interesting defenses of pluralism in environmental thought so far. Unfortunately, Naess's work has generally been relegated to the historical background to debates such as this, rather than being read as an active voice in the field. But Naess has consistently argued that the premises of the eight points of the deep ecology platform are derivable from a plurality of competing traditions.[5] Is it the case, then, that Naess's work already embodies some of the virtues of pluralism that have been defended by various Anglo-American environmental ethicists against Callicott?

My claim here will be that Naess does indeed articulate, or at least suggest, a pluralist form of environmental philosophy that avoids the foibles of pluralism that Callicott fears: the descent—or headlong rush, rather—from pluralism to moral relativism and deconstructive postmodernism. I also argue that the best frame within which to evaluate arguments for moral pluralism is to pay attention to what advocating such a position does for bridging the gap between environmental theory and practice, rather than to what such a turn to pluralism means in a more abstract sense.

Callicott's Theoretical Monism

Calls for the reassessment of the direction of environmental ethics ought to carry some weight given the object of our study: the troubling state of the environment and the complicity of humans in the creation of these hazardous conditions—hazardous to humans, nonhumans, future generations of both, and the biosphere itself.[6] Environmental philosophy, broadly speaking, and environmental ethics itself, is the attempt to bring the traditions, history, and skills of philosophy to bear on the questions of how to maintain the long-term sustainability of life on this planet. Not to be presumptuous, but if philosophers can contribute anything to the reconciliation of more stable human-nature relationships, then clarifying the direction of the discipline in relation to our contribution to the actual resolution of environmental problems is one of our most important theoretical enterprises. I say this as a theorist very hesitant to reduce environmental ethics to a set of "metaethical" debates—arguments, largely between ethicists, about how to do ethics. Nonetheless, some metaethics is required to open up the field to contributions that are more practical and immediate in relation to the problems we are aimed at helping to resolve.

Anthony Weston argues that given the early, or "originary" stages of our field of inquiry, we should assume that our field should be less settled on the right avenue for proceeding, rather than more settled. We should, according to Weston, assume pluralism. Weston suggests, partly in response to the claims cited earlier by Callicott: "At the originary stage we should . . . expect a variety of fairly incompatible outlines coupled with a wide range of proto-practices, even social experiments of various sorts, all contributing to a kind of cultural working-through of a new set of possibilities. . . . The necessary period of ferment, cultural experimentation, and thus *multi*-vocality is only *beginning*".[7] Thus, for Weston, the burden of proof at the early stages of developing a new field of ethics is on those who would wish to restrict the field to one right path for the construction of normative theories of our treatment of nature, rather than on those who resist the trends to narrow environmental ethics to a particular metaethical approach. Still, one could ask: given the environmental crisis we face, how could we afford the sorts of delays seemingly implicit in such talk of "social experiments"? Moral pluralism, the specific target of Callicott's worries over the pollution of theoretical approaches in environmental ethics, sounds to him dangerously close to the abject relativism associated with deconstructive postmodernism.[8] And relativism is something that we may agree ought to be avoided when the point of our inquiry is to respond to the deepening environmental crisis. Why? Because relativism could entail abandoning the view that there is a moral stance better than others that could guide our ethical claims about how we should treat nature. If we admit relativism then, one could argue, little by way of a moral claim stands in the way of justifying the continued abuse of nature on the ground that this is simply the way we humans customarily treat nonhuman entities.

Assuming for the moment that we can answer the relativist charge against pluralism (a point I return to later), given Weston's concerns, what would motivate us to follow Callicott down the path of moral monism? In "Moral Monism in Environmental Ethics Defended," Callicott negotiates the criticisms of various pluralists (Wenz, Varner, Brennan, Weston, and Hargrove) and advances the cause of monism by describing his approach to environmental ethics as a form of "communitarianism," where "all our duties—to people, to animals, to nature—are expressible in a common vocabulary of community," and so "may be weighed and compared in commensurable terms."[9] Before criticizing this view, I will unpack Callicott's justification for this position and outline how he thinks it is a good alternative to pluralism.

Callicott begins the article with a summary of his now-familiar objections to the pluralism of Christopher Stone. The argument is that Stone's form of pluralism—termed by Callicott "intrapersonal pluralism"—which supposes that a valuing agent can adopt different moral principles for determining either forms of value or grounds for action depending on the situation at hand (becoming, for example, a utilitarian for one purpose and a deontologist for another) is ultimately "facile" and incoherent. For Callicott, our many moral concerns as environmentalists—for fellow humans, other animals, and the larger environment—although difficult to negotiate under the terms of one moral theory, cannot be resolved simply by accepting the apparent necessity to move from one moral theory to another, depending on the sort of problem that needs to be solved. According to Callicott: "A mature moral agent, I submit, wants a coherent outlook—the one that seems true. He or she cannot comfortably live in a state of self-contradiction or as the philosophical equivalent of an individual with a multiple personality disorder."[10] Monism avoids such problems by providing a secure philosophical foundation to right action in relation to humans, other animals, and nature, grounded in one measure of value and one system of valuing.

Callicott's communitarian alternative works like this: Because we are all members of different communities—families, regions, nations, and so on—it follows that each membership generates "peculiar" duties and obligations. Therefore says Callicott, "we can hold a single moral philosophy and a univocal ethical theory, but one that provides for a multiplicity of community memberships, each with its peculiar ethic."[11] So, when faced with the infamous choice between spotted owls or the livelihood of Pacific Northwest logging communities, we have to realize that we are connected to both of these communities with each membership resulting in different obligations. Still, realizing that our membership in the "larger biotic community generates duties to preserve the old-growth forest ecosystem and the endangered species . . . that depend on it," the moral choice is between "temporarily preserving a human life-system that is doomed in any event and reserving in perpetuity an ecosystem and the species that depend upon it."[12] On Callicott's account, our duties to the ecosystem outweigh our duties to the loggers even though we are members of communities that encompass both as sources of value. Balancing our obligations to the various communities in which we belong is therefore a key to Callicott's solution to the problem of how we balance moral concerns to different subjects without resorting to pluralism. Looking at controversies such as this one though the lens of community membership thus solves these ethical dilemmas.

But if Callicott's communitarianism is enough to get us an answer to the problem of valuing all types of things in an environment under one ethical system, then why did pluralists such as Stone advocate intrapersonal pluralism in the first place? The answer may be that the original concern of other pluralists was more over the relationship between our theory-making in environmental ethics and our practice, rather than over the issues with which Callicott is concerned. Andrew Brennan, defending a form of pluralism even more extreme than Stone's, puts it this way:

> By adopting the pluralist stance, we not only start to do justice to the complexity of real situations, but we also can start to look for ways by

> which environmental ethics can be linked up with other modes of valuing and ways of responding to our surroundings. Utilitarianism and its rivals need not be abandoned, but can be considered as partial accounts of the moral life. There is scope, for example, for developing notions such as attention, humility and selflessness in our dealings with nature as part of the story of what makes a worthwhile human life. These notions should not be thought of as the truth about morality—any more than utilitarianism is. Rather they provide greater depth in characterizing our situation. Abandoning reductive monism about values and valuing makes even more sense once the force of moral pluralism in this latest form is recognized.[13]

For Brennan, the impetus for pluralism is more to build stronger connections to a larger array of human moral practices and ethical beliefs thus widening the appeal of the ethical dimensions of environmental concerns.

Whether his communitarianism actually gets the results sought for by Brennan is not clear from Callicott's account. We can easily imagine, and even expect, for example, that the ethical dilemmas the case of the owls versus the loggers raises will remain for those agents closer to the logging communities (through parental ties, etc.). Brennan's goal would be to find a way of describing the importance of the old-growth forests in ways that will appeal to the commitments of those already strongly attached to the logging communities. Pluralism opens up the possibility of finding such an appeal by making it the work of environmental philosophy to create links to existing moral priorities in specific human communities. This does not mean that we should jump from one ethical system to another depending on the situation, but rather that we should consistently look for multiple ways of describing the value of nature that appeals to a range of interests.

I think this notion of pluralism as a kind of compatibilism is the most interesting issue in the literature on pluralism in environmental philosophy. And, to his credit, Callicott does acknowledge that at least in one sense he is something like a pluralist (although, what he calls an "interpersonal pluralist") in that he upholds "everyone else's right to explore or to adopt a moral philosophy and ethical theory that seems persuasive to them."[14] Along with this suggestion is a commitment to reasoned persuasion: "Intelligent people of good will should eventually reach agreement if they take the time to thrash out their initial differences."[15] But surely toleration of the act of theorizing of different views does not get us the kind of cooperation among theorists that Brennan seems to have in mind. Brennan's pluralist would not only acknowledge other forms of theory making, but also go on to set aside possible prejudices for now of the falsity of those other forms of valuing for understanding our situation, as Brennan puts it, in greater "depth." To me, the depth of our situation ultimately will be realized in part through an acknowledgment of how the ends of moral discourse must be given priority in environmental philosophy. Normally our end is to better the environment, and so we must be concerned at least as much with creating agreement on those policies we all support as with finding the truth of how to value nature. This is not to advocate, as Callicott puts it, "deconstructive postmodern *différance*," but instead to acknowledge that the environmental situation we face requires us to adopt certain compatibilist rules for theory making, even if we are at heart moral realists.

Because Callicott's original target was Stone's argument that one can adopt different schemes of valuing for different situations and kinds of things in nature, maybe the issue of how pluralism is to be encouraged among theorists and practitioners is not really a concern for Callicott? Callicott's account does not appear to contain anything explicit that would prohibit him from consenting to something such as Bryan Norton's convergence hypothesis. The convergence hypothesis identifies and encourages the apparent agreement on ends of environmental policy among, for example, anthropocentrists and nonanthropocentrists, or those Naess calls "shallow" (or liberal) ecologists and deep ecologists, even if they do not agree on the philosophical foundations for those ends.[16] In this sense, there seems to be no objection in Callicott's argument to the motivation behind the pluralism Brennan embraces, only to the way in which theorists such as Stone endorse pluralism intrapersonally. Let us then say that Callicott appears to reject Stone's theory, which I will call a kind of "theoretical pluralism"—advocating swings from one kind of moral system to another depending on the situation—although he may assent to the "metatheoretical pluralism" of Brennan or Norton. Metatheoretical pluralism then is simply an acknowledgment of the need for divergent ethical theories to work together in a single moral enterprise despite their theoretical differences. Unlike the theoretical pluralist, the metatheoretical pluralist does not advocate shifting from one moral theory to another willy-nilly, but rather encourages a diversity of moral arguments for the same end without requiring theorists to critique each others' claims as long as they are aimed at the same end.

In fact, even though Callicott does not take up the issue of Brennan's metatheoretical pluralism in this argument, he does acknowledge the accusation of Anthony Weston that he (Callicott) is trying "prematurely . . . [to] shut off further discussion and development of the field."[17] Callicott objects, stating that Weston has confused his rejection of intrapersonal pluralism with his position on interpersonal pluralism. Callicott emphasizes again that he does not want to close off discussion in environmental ethics but rather "keep the interpersonal debate going." One might suppose then that this is enough of an answer to the question of how Callicott falls out on the metatheoretical pluralism issue. If Callicott tolerates the existence of competing forms of valuation, then surely he must tolerate the idea of divergent theories working together in a common enterprise, be they monist, pluralist, or some other variation. If metatheoretical pluralism really is the interesting issue the pluralists raise, then given what we can determine is Callicott's tacit agreement to it, perhaps his communitarianism (combined with his "interpersonalism") is sufficient as a substitute for pluralism after all.

But here I think we must be cautious. We have actually no reason to believe that Callicott would make the turn from toleration of different theory-formation (which is the heart of his interpersonal stance) to embracing compatible cooperation among theories of the sort advocated by Brennan. More precisely, even if Callicott himself did endorse metatheoretical pluralism, nothing in his position requires it. An openness to the formation of other theories does not necessarily entail an acceptance of them in any unified project. For example, one community of theorists could simply not tolerate another community even if both communities were trying to achieve the same ends. Setting aside obvious counterexamples of not wanting to endorse the positions of fascists, toleration clearly could stop at the level of theory formation for more mundane reasons.

So, what alternatives exist? In previous work I argued for what I now call *methodological environmental pragmatism* (my earlier term was *metaphilosophical pragmatism*), intended in part to provide a more coherent framework for the sort of metatheoretical pluralism Brennan embraces.[18] I will not review this argument in full here, but only reiterate that methodological environmental pragmatism is compatible with any form of environmental philosophy; it requires only new "rules for the game" of how to do environmental philosophy, rules that will promote greater compatibility in the relationship between good theorizing and effective environmental practices. Methodological pragmatism provides a litmus test against which competing ethical systems can be weighed. So the "extreme pluralism," which Callicott criticizes throughout this essay (i.e., the idea that one would adopt Aristotle's theory on one occasion, Kant's on another, etc.) is not what I have in mind as metatheoretical pluralism or methodological pragmatism. My pragmatist pluralist (and, I think, Weston, Norton, and Brennan, too), in acknowledging distinct bases for value, would at least be consistent in the application of what she thought was the best (all things considered) moral theory to a particular type of object of valuation and not change theories with the evaluation of each situation. In other words, my metatheoretical pluralist is not an ethical situationalist, and presumably, because he or she is a competent philosopher, would not apply theories in a self-defeating manner (as Callicott, at the end of his essay, hints an extreme pluralist would). But most important, and hence distinct from Callicott's view, my methodological pragmatist would not simply tolerate interpersonal pluralism, but would actively advocate multiple arguments for environmental valuation when endorsing environmental policies to generate as broad a basis of support as possible for the proposed action. Some kinds of arguments will appeal to some people, given their intuitions about other moral problems, and other arguments will appear to other people. As long as we are not contradictory about our ends, and as long as we are not advocating morally suspicious schemes of value, environmental ethicists should be making as many arguments as possible to appeal to as wide an audience as possible: both anthropocentric and nonanthropocentric claims, both moral claims about our environmental obligations to human future generations and claims about the noninstrumental value of nature. Even though my theorist's scheme of valuation would be consistent, such normative consistency would not be required across the board for all those embracing any given policy. Such a theorist would spend less time critiquing other theories of the value of nature and more time making as diverse a set of arguments as possible to as wide an audience as possible for some given environmental end.

I still think, along with Callicott, that what we should be doing is "systematic environmental philosophy." But systematic environmental philosophy cannot be systematic philosophy in a vacuum. Not to sound too vulgar (especially because I have a great admiration for people who do modal logic), but our object of concern is not the question of the spatiotemporal existence of other possible worlds but the future of the one world we are confident we do inhabit. Surely our method of interacting as philosophers must push the envelope of Callicott's interpersonal pluralism beyond what we would expect it to be for any well-trained philosopher working on any topic. Any philosopher who thought censuring the work of his colleagues appropriate only because it is different from his own, or failed to give it a fair hearing, would simply be a bad philosopher. We need not theorize about varieties of pluralism to get to that conclusion. The most interesting question concerning pluralism in environmental ethics is perhaps a question different from the one Callicott has taken up because he never really addresses the issue of what I am calling methodological pluralism. Again, Andrew Brennan provides a valuable insight: "If we accept moral pluralism as a philosophical position, the project of environmental ethics can be seen in a new light. . . . Environmental ethics is less a competitor for a certain moral position, but an investigation of a more sophisticated turn that moral philosophy has taken. Embarking upon it is a partial recognition of the complexity of our moral situation. Note, once more, that the complexity in question is intrinsic to the business of being moral. Moral pluralism is a philosophical, not a moral, thesis."[19] If we follow Brennan's line of reasoning, an argument that I think is both crucial to and lost in discussions of pluralism in this literature, then most pluralists can easily grant Callicott's arguments against Stone for now. But we should not stop there. We must also argue that Callicott, and many others who consider these questions, rethink what is at stake in the pluralism-monism debate in environmental ethics.[20] As I said at the beginning, environmental ethics does seem to be marked with certain prejudices in the field, not the least of which is how we have constructed some of the defining debates in the discipline. We need to ask whether the way we approach the practice of doing philosophy best meets the needs of the object of our concern and coheres with what is arguably the most effective contribution we can make as philosophers to the struggle to help resolve environmental problems.

Naess's Metatheoretical Pluralism

I now want to turn this discussion toward a more positive evaluation of the pluralism of another theorist. If we are going to reshape the debate between monism and pluralism along metatheoretical lines, then we also need to reassess what theories have normally been placed on which side of the dividing line between the two sides of this disagreement. In particular, I argue here that this turn to a new terrain for the monism-pluralism debate demands that we bring into this conversation the work of Arne Naess.

Many monists would probably consider Naess's formal philosophical system (his "Ecosophy T") to be a paradigm case of moral monism because it seems to emanate from a single source: the argument for the intuitive and innate ontological connection between humans and the world around them. Because other theorists such as Warwick Fox have argued for some time that the point of deep ecology is that there is no ontological divide that can be made between humans and nonhumans, that fundamental principle of deep ecology could be argued as the monistic foundation from which our duties to each other and the environment could be derived .[21]

But Naess's formal system is best interpreted as theoretically pluralist, and his comments on how to do environmental philosophy point out that he is also a metatheoretical pluralist as well. I think that the metatheoretical argument is the more interesting ground for the monism-pluralism debate, so I spend most of my time on that part of Naess's view.[22] Naess, perhaps more than any other deep ecologist, has worked to foster a form of metatheoretical pluralism both inside and outside deep ecology circles. And because Naess never really got involved in the old monism-pluralism debate as its ground was defined by Callicott and others, we can safely say that for Naess the issue of pluralism in environmental philosophy has always been the question of how to do philosophy, rather than an argument over the specific content of our ethical evaluations of different things in the world and different situations involving ethical conflict.

Naess has pointed out that behind the eight points (or principles) of deep ecology are the nonexclusive overlapping sources of deep ecological thought that constitute the multiple foundations for the theory.[23] The foundations are: Christian (some people will argue there is strong ground for the intrinsic value of nature even in the stewardship view from Genesis); Buddhist (primarily according to Naess as found in the work of Dogen);[24] and philosophical (specifically from Whitehead's process philosophy and from Naess's readings of Spinoza on the connectedness of mind and matter). We need not go into the details of how the diverse foundations inform the principles of deep ecology here, but only note that Naess is careful to argue that one may get to the principles of deep ecology from any of the foundations. As far as theoretical pluralism goes, this should be enough to get us at least the "moderate moral pluralism" introduced by Peter Wenz, where a single ethical theory is pluralistic in so far as it "contains a variety of independent principles, principles that cannot all be reduced to or derived from a single master principle."[25] On Naess's account, the pluralistic relation is between foundations and principles, rather than principles and principles, but the argument is still the same: the principles of deep ecology are derivable from distinct but overlapping foundations (Christian, Buddhist, philosophical), and one need not assent to even the coherence of any one of the foundations to derive the principles out of another of the foundations.

We can test this approach as a form of pluralism by comparing it with the previously established example of moral monism. Callicott argues persuasively that his form of monism is at the level of theory but not of principle. But, regardless of this argument, it is still the case that the idea of the importance of belonging to different communities, which he derives from Leopold, is still the foundation for the different principles that we can assume would emerge from our different obligations to different communities in which we participate. So, all principles are monistically derived. Naess, on the other hand, does not find the rationale for his multiple principles in one source or even a specific collection of sources. That one can come to the principles of deep ecology from a variety of foundations is crucial for Naess. Different people will assent to the principles of deep ecology for different reasons, hopefully building a broader movement around the policies endorsed by deep ecologists.

Perhaps, the most important article that Naess has written on the issue of pluralism is "The Encouraging Richness and Diversity of Ultimate Premises in Environmental Philosophy."[26] In this largely unknown piece (outside deep ecology circles), Naess acknowledges early on that the style of his approach to the issue of pluralism belongs to "a somewhat different tradition of metaethical discussion and methodology than the chief participants of the ethical monism/pluralism debate."[27] His first claim is that one of the central tasks of environmental philosophers is "to study different positions but not to try to reduce the ultimate differences" between those positions.[28] Such a view certainly is in line with the metatheoretical pluralism I am advocating and may even go beyond it. What Naess has in mind here is a principle of respect for different cultural approaches to environmental problems. In fact, he embraces an explicitly cultural form of pluralism (another point I return to later). But the argument Naess works out in this essay is not restricted to issues of cultural pluralism and is consistent with the positions Brennan and Norton previously outlined.

Naess gives two reasons for embracing a metatheoretical form of pluralism. His first reason is found in his intuition that a uniformity in views on valuing nature would indicate a stagnation rather than a strength of environmental ethics. Going beyond Callicott's embrace of interpersonal pluralism, Naess sees not only a potential for fruitful philosophical argument in a diversity of views, but also the basis for a claim to the strength of environmental thought grounded in that diversity. Like Weston's remarks about how we ought to find any discipline at its originary stages engaged in a plurality of theory formations, Naess suggests that the cultural richness and diversity that we may think are part of a good environmental philosophy cannot be sustained "under conditions of increasing similarity of ultimate views."[29] According to Naess, even if one could show that we could have a unified theory of all of reality, "it does not follow that adequate, verbal accounts of this oneness should or must converge or be practically translatable into each other."[30]

Such a claim may appear to conflict with my earlier endorsement of methodological pragmatism because it seems to value diversity as a primitive rather than as a means to an end, namely the end of coming up with a rich and robust environmental philosophy that helps in the formulation of better environmental policies. But for Naess, because we can expect that diversity of forms of valuing is proof of the complexity of the subject of our valuations (nature), it follows that a diversity of views should be the norm for environmental philosophy. If we are worried about the limits to our toleration of other views then, Naess argues (in a passage very similar to Norton's argument concerning the compatibility of different views for purposes of policy formation), we can assume certain restrictions: "The only reason to attack a religious or philosophical ultimate premise seems to be the assumption that a particular environmentally unacceptable position follows with necessity from it."[31]

In some respects we can say that Naess is really addressing the variety of positions among different members, or even factions, of what is loosely thought of as the deep ecology movement. In a passage similar to the ones cited at the beginning of this discussion concerning how we must not imperil different ultimate premises, Naess remarks that, "Most deep ecologists have fundamental differences from each other, and speak in a variety of terminologies. Questioning one's motives leads inevitably to philosophical positions and from there back to practice."[32] Although some may consider this concern with one group of activists and philosophers a limitation in Naess's metatheoretical pluralism, I think instead that it is a sign of why we environmental ethicists should heed his advice more keenly than that of others. Naess is one of the few environmental philosophers in the world who can claim to have directly and determinably affected a movement of practitioners, and possibly even some policymakers. For example, through the early part of its history, many leaders of the U.S. environmental group Earth First! claimed to be deep ecologists. Naess's desire to forge some sort of ground for agreement among them is a practical, rather than an applied or even theoretical, problem. We can be assured, then, that his motivations are pragmatic and consistent with Brennan's concern about how to rethink the task of philosophy given the challenge of environmental problems.

A second reason for metatheoretical pluralism in the "Encouraging Richness" article is found in Naess's distinction between what we implicitly assume in the way we theorize our moral views, and what we explicitly say in ethical conflicts. The two need not be the same. Naess is in principle skeptical that there can be any systematically articulated total view of environmental ethics, consisting in a coherent set of premises and conclusions, which would be logically complete.[33] But setting aside the objections that those of us who embrace some form of moral realism might have to this claim, we can turn our attention to the part of this argument concerning our responsibility for what we say in ethical conflicts.

For Naess, a good environmental philosopher needs to be something of a practical anthropologist, with a good understanding of moral customs such that cooperation among environmentalists is not only considered possible, but also is made the goal of the theoretical enterprise. Naess thus outlines two goals for environmental philosophy that may appear incompatible, but that are entirely consistent given the context in which he places environmental theory: on the one hand we should aim toward perfecting our theories—monist, theoretical pluralist, or whatever—and on the other we are obliged to work with the assumptions of those with whom we must work to see environmental reform even if we disagree with some of their principles (within reason). For Naess, the subject of environmental philosophy, and the need to respond to the crisis of the environment, drives these twin goals rather than something intrinsic to the generic practice of philosophy. Naess sees this strategy as called for by the particular situation we are in with respect to environmental philosophy. Again, Naess uses an intramural example from deep ecology to make his point: "The supporters of the deep ecology movement cooperate in the fight to implement decisions on the level of concrete situations with everybody who sincerely supports a decision ."[34]

Most who have followed the course of Naess's work generally regard that the limits to such pluralism and compatibilism would fall at drawing the line at cooperation with "shallow" or reform-oriented ecologists and environmentalists. After all, Naess's creation of deep ecology is often assumed to have stemmed from his distinction between deep and shallow views of the origins of environmental problems. Shallow ecologists are said to lack the deeper analysis of the origins of environmental problems in fundamental human ontological divides with nature or overarching political and economic systems such as capitalism. Like the relation between Callicott's interpersonal pluralism and some form of metatheoretical pluralism, we could assume that there is no reason why the structure of Naess's comments regarding the need for cooperation between different schools of deep ecologists would necessarily translate into reasons for broader forms of tolerance and pluralism with shallow ecologists. To overcome this assumption we would need Naess to extend his claims about pluralism explicitly to the shallow ecologists because we can assume, given some of Naess's previous work, that he would be skeptical about such claims for convergence. But in the very next sentence after the one quoted at the end of the previous paragraph, arguing for an explicit form of metatheoretical pluralism among deep ecologists, Naess sets these worries to rest. Naess asks the rhetorical question: "What could the supporters [of the deep ecology movement] achieve *without* cooperation with people whose general argumentation pattern for instance in terms of premise/conclusion relations, is shallow or merely concerned with reforms?"[35] Presumably, very little. The striking thing about this passage is that it brings Naess much closer to Norton's convergence thesis than might initially be thought. Perhaps Naess still thinks a sharper divide exists between deep and shallow ecologists than Norton is ready to admit. But surely those differences dissolve, given that the apparent point of both Naess's and Norton's metatheoretical pluralism is convergence on policies to achieve environmental reform. Certainly other deep ecologists might disagree with how far Naess is willing to take his pluralism, but we may be encouraged here that at least one deep ecologist, and a very important one at that, does not in principle see any formal reason why deep ecologists cannot find encouraging something such as Norton's convergence hypothesis.

Naess ends this essay with perhaps one of the most succinct and elegant statements for the need for metatheoretical pluralism, which again points to the need to revisit Naess's work in the context of a recast monism-pluralism debate: "The richness and diversity of philosophical and religious ultimate premises suitable for action in the ecological crisis may be in itself considered part of the richness and diversity of life forms on Earth."[36] Metatheoretical pluralism is thus the activity of environmental philosophers who wish to emulate the patterns of the objects of their concern.

Conclusions

Before closing, I want to address two issues that could be important in the reintroduction of Naess's work into a reformed monism-pluralism debate, fulfilling two promissory notes given before. First is the question of the role of cultural pluralism in this debate, which at times is the issue on which Naess hangs his comments on pluralism. Second is the remaining issue of whether Callicott's worries concerning the relation between pluralism and relativism still holds for the form of pluralism I am attributing to Brennan, Norton, and now Naess.

Today environmental philosophers often discuss cultural pluralism. Still smarting a bit from the critiques of theorists such as Ramachandra Guha and Vandana Shiva, that much of environmental philosophy is almost exclusively First World in its orientation and expectations, environmental ethicists of almost all schools of thought have begun thinking seriously about the issue of cultural pluralism.[37] Callicott has been no exception and has demonstrated a clear commitment to cultural pluralism in his book *Earth's Insights*.[38] In many ways this book is more thorough than anything Naess has written to bring about an understanding of how a variety of global cultural traditions can inform a robust environmental ethic.

In this book Callicott provides a very helpful survey of the environmental philosophies at work in a variety of intellectual traditions from around the world. The details of this account are not important here, only the unique structure of Callicott's argument.[39] Callicott not only surveys these various traditions but also uses his understanding of the very different environmental theories derivable from different philosophical and religious traditions to reinforce his commitment to a Leopoldian, nonanthropocentric, holistic monism. The demonstration of this last commitment shows how Naess's metatheoretical pluralism causes his approach to cultural pluralism to differ from the view Callicott embraces.

Callicott's move toward cultural pluralism is not metatheoretically pluralist. All of the competing world systems in Callicott's work are read through the lens of his nonanthropocentric holistic version of Leopold's land ethic. Leopold's view becomes for Callicott what he calls the "Rosetta stone of environmental philosophy," which is needed to "translate one indigenous environmental ethic into another, if we are to avoid balkanizing environmental philosophy."[40] Accordingly, the veracity of one or another cultural environmental view is judged by how closely it approximates Callicott's interpretation of Leopold's ethic. But using the version of nonanthropocentric holism that Callicott distills from Leopold's work as a yardstick for all environmental philosophy (not to mention all indigenous environmental traditions) draws more lines toward balkanization than does a metatheoretical tolerance of a multiplicity of approaches. If we may call into question the singular vision of this kind of ethic, then we may also call into question some of Callicott's appraisals of these other systems. Callicott claims for example that, "Africa looms as a big blank spot on the world map of indigenous environmental ethics."[41] Why would Callicott make such a claim? Because African environmental thought tends to be anthropocentric in contrast to Callicott's nonanthropocentric, Leopoldian ethic.

In contrast, Naess, unlike Callicott, does not evaluate forms of thought on the basis of their amenability to the foundations or even specific principles of deep ecology. Naess only points out that these schools of thought can be used to derive some common principles, even if not all of those that he endorses. Naess also suggests that although we can "infer traits of an environmental ethic" from world literatures, we cannot "pretend to be able to compare in a methodologically neutral and adequate way meanings and validity of the ultimate premises of total views."[42] Although he limits this comment to the problems of translating simple expressions into complex theoretical positions, I think the point holds in general to his approach to divergent cultural-environmental positions. Therefore, *pace* the exception of views from which we can derive antienvironmental conclusions, or morally offensive conclusions (fascism, etc.), the goal of cultural pluralism in environmental ethics is not necessarily evaluation of different worldviews, but a finding of means for convergence of environmental policies, activities, and theories.

To move to the second issue, part of resolving the issue of which approach to cultural pluralism is right will necessarily involve some answer to the original question Callicott posed, cited at the beginning of this essay: Does pluralism lead to a deconstructive, or more simply destructive, relativism? A theoretical or metatheoretical pluralism is at least neither necessarily relativist, nor incommensurable with a workable, robust, and critical environmental philosophy. Callicott is right in suggesting that there could be a tendency to move from theoretical pluralism to relativism, but without an argument for their necessary connection it seems that each version of pluralism has to be individually evaluated. Callicott's strongest case so far is therefore against Stone.[43] But I am not too concerned with theoretical pluralism here, and so finally my answer to Callicott will have to be that his question in regard to relativism needs to be rephrased against the metatheoretical pluralism that I have identified. Metatheoretical pluralism, especially given the comments on its practical importance found in Brennan, Norton, and Naess, may provide the foundations, or at least guidelines, for the types of theory development needed at this stage in the growth of environmental philosophy in relation to forming better environmental polices and practices. Because quite plausibly we may not need relativism in environmental philosophy right now (because the prejudice of many cultures is simply to deny that we have any ethical obligations to nonhumans in particular or nature in general), the development of metatheoretical pluralism can help to mitigate environmental problems by providing a nonrelative array of answers to the general skepticism toward the moral dimensions of environmental problems. If some framework is provided to prevent pluralism from lapsing into an indecisive form of relativism, and if pluralism can be argued to be important for the health of environmental ethics, then the next question becomes how we go about doing environmental philosophy so that we get the right sort of pluralism.

I have argued elsewhere that environmental pragmatism provides us with just the sort of framework we need to temper pluralism toward these goals.[44] But the question of whether this strategy, or an embrace of Naess's form of deep ecology, or something else, is best for this kind of theory development is something of an empirical question. If theorists and practitioners followed our views to their logical conclusions, and were able both to cooperate on ends and to avoid relativism (or even the unproductive debates about relativism that seems to paralyze much of contemporary philosophy), then we could actually see which form of pluralism best serves the broader goals of environmental philosophy. Of course, of all the views examined here, only Naess's pluralism seems likely to be testable in this manner anytime soon because his views have the broadest following in the world of activists and practitioners. In this respect, a careful analysis of the reception of Naess's pluralism in the deep ecology movement is one of the best next steps in advancing a reformed monism-pluralism debate.

ACKNOWLEDGEMENTS

I am grateful to Avner de-Shalit, Alastair Hannay, Eric Katz, Zev Trachtenberg, Allan Carlson, and Arne Naess for helpful comments on the original version of this article. This essay was shortened and revised from its original, appearing in *Inquiry*, vol. 39, no. 2, June 1996, pp. 273-294. Thanks to Wayne Ouderkirk for valuable suggestions for this revision. A fully revised version of the entire article will appear in *Environmental Ethics: An Anthology*, eds. Andrew Light and Holmes Rolston III (Cambridge: Blackwell Publishers, forthcoming 2001).

NOTES

1. J. Baird Callicott marks three types of nonanthropocentric theories: neo-Kantian (e.g., Paul Taylor, Robin Attfield, Holmes Rolston), Leopoldian (e.g., Callicott, William Godfrey-Smith, Richard Sylvan, and Val Plumwood), and self-realized (deep ecologists). See Callicott's "The Case against Moral Pluralism," *Environmental Ethics* 12, no. 2 (1990): 101–102. In this essay Callicott gives an excellent genealogy of the development of these areas. One may also include nonanthropocentric theories such as those of animal liberationists Peter Singer and Tom Regan, as well as anthropocentric holists such as Bryan Norton, and Gary Varner's biocentric individualism, just to name a few representative theorists.
2. Callicott, "Case," 102.
3. Callicott, "Case," 109. (emphasis added).
4. Some of the most frequently cited essays on the pluralist side are Christopher D. Stone, "Moral Pluralism and the Course of Environmental Ethics," *Environmental Ethics* 10, no. 2 (1988): 139–154; Gary E. Varner, "No Holism without Pluralism," *Environmental Ethics* 13, no. 2 (1991): 175–179; Andrew Brennan, "Moral Pluralism and the Environment," *Environmental Values* 1, no. 1 (1992): 15–33; Peter Wenz "Minimal, Moderate, and Extreme Moral Pluralism," *Environmental Ethics* 15, no. 1 (1993): 61–74; and two unpublished essays, Anthony Weston, "What Are We Arguing About?" and Eugene Hargrove, "Callicott and Moral Pluralism," both presented at the Central Division meeting of the American Philosophical Association, April 24, 1993.

5. For reasons that may become apparent later, I count Naess's "environmental philosophy" as both his attempt to characterize the deep ecology movement and his argument for a definite "total view," his Ecosophy T.

6. I do not recount the massive evidence for environmental problems here. A sufficient amount of literature exists on the disastrous effects of humans on the environment to fill any reading list and concern over the environment is one of the primary reasons for the development of the field of environmental ethics in the first place.

7. Anthony Weston, "Before Environmental Ethics," *Environmental Ethics* 14, no. 4 (1992): 333.

8. Callicott, "Case," 116–120.

9. J. Baird Callicott, "Moral Monism in Environmental Ethics Defended," *Journal of Philosophical Research* 19 (1994): 53. I find problems with Callicott's designation of his view as "communitarian," problems that unnecessarily confuse this argument with the communitarianism of Walzer, Sandel, Taylor, de-Shalit, and others. I do not think that Callicott's work shares much with these other views, especially in light of the fact that most communitarians include a substantial role for political and moral pluralism in their work. For a specific example of a more robust "environmental communitarianism," see Avner de-Shalit, *Why Posterity Matters: Environmental Policies and Future Generations* (London: Routledge Press, 1995). For de-Shalit, pluralism comes into play in communitarianism in the way that communities open themselves to the values of other communities, regarding these other valuation schemes "as potential truths rather than as something inimical" (p. 62). Ultimately, this problem is not philosophically very serious for Callicott, only a bit inelegant of him.

10. de-Shalit, *Why Posterity Matters*, 52.

11. de-Shalit, *Why Posterity Matters*, 53.

12. de-Shalit, *Why Posterity Matters*, p. 53.

13. Brennan, "Moral Pluralism and the Environment," 30. Brennan's thesis is that "no single theoretical lens [exists] which provides a privileged set of concepts, principles and structure in terms of which a situation is to be viewed" (p. 29).

14. Callicott, "Moral Monism," 54.

15. Callicott, "Moral Monism," 54.

16. See Bryan Norton, *Toward Unity among Environmentalists* (Oxford: Oxford University Press, 1991), esp. chap. 10.

17. Callicott, "Moral Monism," 56. Callicott is referring to Weston's remark that: "J. Baird Callicott . . . insists that we attempt to formulate, right now, a complete unified even "closed" (his term) environmental ethics." Weston, "Before Environmental Ethics," 333.

18. Andrew Light, "Materialists, Ontologists, and Environmental Pragmatists," *Social Theory and Practice*, 21, no. 2 (1995): 315–333, and my contributions to *Environmental Pragmatism*, ed. A. Light and E. Katz (London: Routledge, 1996).

19. Brennan, "Moral Pluralism," 30.

20. I should point out however that I am not yet ready to completely grant Callicott's arguments against Stone, Varner, Hargrove, and Brennan on theoretical moral pluralism, or what Callicott calls "pluralism at the level of theory." Strategically, however, I am ready to let the issue rest. Callicott seems to have made an interesting argument here that needs to be responded to by those original proponents of theoretical pluralism. But I still have some worries about what exactly Callicott is embracing.

Even though he claims to be a pluralist with respect to principles, I am not sure that this really individuates his theory at all. What realist in environmental ethics would argue that one single principle is sufficient to generate all duties toward the environment? Who holds such a view? If no one, then we still have a monism-pluralism debate at the level of ethics, which is at least philosophically interesting.

21. See Warwick Fox, "Approaching Deep Ecology: A Response to Richard Sylvan's Critique of Deep Ecology" (Hobart, Australia: University of Tasmania Environmental Studies Occasional Paper 20, 1986).

22. I feel obliged to point out that because I am not a deep ecologist I really do not have anything at stake in proving some form of inherent pluralism in deep ecology.

23. See Naess's diagram outlining this relationship in his contribution to *Environmental Philosophy*, ed. Michael Zimmerman et al. (Englewood Cliffs, N.J.: Prentice-Hall, 1993), 206.

24. The validity of this claim has been called into question most notably by Deane Curtin in "A State of Mind Like Water: Ecosophy T and the Buddhist Traditions," *Inquiry*, 39, no. 2 (1996): 239–284.

25. Peter Wenz, "Minimal, Moderate, and Extreme Moral Pluralism," 69, cited by Wenz from his *Environmental Justice* (Albany, N.Y.: SUNY Press, 1988), 313.

26. Arne Naess, "The Encouraging Richness and Diversity of Ultimate Premises in Environmental Philosophy," *The Trumpeter* 9, no. 2 (1992): 53–90. My thanks to Harold Glasser for calling my attention to this article.

27. Naess, "Encouraging Richness," 54.

28. Naess, "Encouraging Richness," 53.

29. Naess, "Encouraging Richness," 55.

30. Naess, "Encouraging Richness," 55.

31. Naess, "Encouraging Richness," 55. I completely agree with this limitation by Naess on the compatibilism of *environmental* metatheoretical pluralism. The point of this pluralism is to strengthen "pro-environmental" claims. If this pluralism began covering "antienvironmental" views then it would violate its environmental predicate. I will leave for another essay a discussion of the ramifications of this caveat.

32. David Rothenberg, *Is It Painful to Think? Conversations with Arne Naess* (Minneapolis: University of Minnesota Press, 1993), 136.

33. Naess could find support in advancing this position from Wim J. van der Steen, who has embraced a kind of default theoretical pluralism. Says van der Steen: "Although I opt for pluralism, I do so in a qualified way. The issue is whether we can elaborate a single, overarching theory for environmental ethics. In a sense we can if we use the term *theory* for highly abstract, general guidelines that are far removed from practical applications. However if we use the term *theory* for the mundane entities that we come across in most disciplines, pluralism is our only option. Modesty is my ultimate defense of pluralism. We should recognize our limitations, and we should be aware of fundamental limitations of science and philosophy. Plain methodology alone suffices to show that the search for grand theories that satisfy all the goals we may cherish is misguided." "The Demise of Monism and Pluralism in Environmental Ethics," *Environmental Ethics* 17, no. 2 (1995): 218. Naess too bases some of his arguments for pluralism on an analogy with science. Callicott argues explicitly in "Moral Monism in Environmental Ethics Defended" that this analogy does not follow. For Callicott, theoretical pluralism in science may be a necessity, even a virtue, but in moral reason-

ing it may only be a sign of fuzzy headedness. Although I like van der Steen's argument, I must admit that I share Callicott's worry that the analogy really does not do much philosophical work. I do not, however, agree that pluralism in moral reasoning is fuzzy headed. See for example, Nicholas Rescher's very unfuzzy *Pluralism: Against the Demand for Consensus* (Oxford: Oxford University Press, 1993).

34. Naess, "Encouraging Richness," 58.

35. Naess, "Encouraging Richness," 58 (emphasis added).

36. Naess, "Encouraging Richness," 60.

37. See Ramachandra Guha, "Radical American Environmentalism and Wilderness Preservation: A Third World Critique," *Environmental Ethics* 11 (1989): 71–83; and Maria Mies and Vandana Shiva, *Ecofeminism* (London: Zed Books, 1993).

38. J. Baird Callicott, *Earth's Insights: A Multicultural Survey of Ecological Ethics from the Mediterranean Basin to the Australian Outback* (Berkeley: University of California Press, 1994).

39. I am in no position to evaluate the rigor of Callicott's comparative project, although from what little I do know about world environmental traditions, his account seems good to me. Anyone interested in comparative environmental philosophy or theology surely will benefit from Callicott's work.

40. Callicott, *Earth's Insights*, 186.

41. Callicott, *Earth's Insights*, 158.

42. Naess, "Encouraging Richness," 60.

43. Although here I must reserve judgment until Callicott answers some of the defenses of Stone's view. See Weston's remarks in this regard in his "What Are We Arguing About?" 4.

44. See Light, "Materialists, Ontologists, and Environmental Pragmatists."

13

Beyond Exclusion: The Importance of Context in Ecofeminist Theory

LORI GRUEN

Ecofeminist theory has been developing as a remedy for the exclusionary shortcomings that plague environmental ethics. What led people to ecofeminism was the perception that struggles for social justice were linked and that emancipatory efforts would fail if they did not recognize these links. Unfortunately, most liberatory struggles have tended to focus on single issues. This narrow focus can obscure the complex network of connections that these issues exist within and often the intense focus on one particular issue—say the environment, or nonhuman animals, or women—actually undermines the strength of the arguments for recognition that are being raised. For example, many women within the environmental movement, particularly its activist, deep ecological wing, began worrying about the exclusion of women from decision-making positions and about the continuation and promotion of particularly offensive patriarchal attitudes on the part of many prominent environmental activists. When this exclusion became known it caused many feminists to reject environmentalism and many feminist environmentalists to leave the organizations with which they were previously associated. Ecofeminist frameworks emerged as a way to criticize these patriarchal premises or assumptions within environmentalism without having to abandon concern for the environment. In the area of environmental philosophy, and environmental ethics more specifically, ecofeminist ethics has developed as an attempt to provide an alternative to mainstream environmental ethics.

This attempt to provide an alternative has been misconstrued by many, including J. Baird Callicott. He has claimed ecofeminism rejects "the need for a *theory* of environmental ethics"[1] and maintains that: "There is no specific ecofeminist moral philosophy grounding a specific ecofeminist environmental ethic, identifiable as such through its particular theory of intrinsic value in or rights for nature, criterion of moral considerability, golden rule, set of commandments, or any of the other elements that we usually associate with ethics. . . . This is because such elements are allegedly masculinist, not feminist, in essence."[2] He describes an ecofeminist view as one that suggests ". . . men typically construct theories, women typically tell stories. . . ."[3] Insofar as he believes that ecofeminists do not want to engage in activities that are "essentially masculinist," such as theory construction, he thinks ecofeminism is doomed to be a cacophony of many different, often inconsistent, voices. As such, he believes ecofeminism does not deserve a place on the environmental theorists' playing field.[4]

Establishing that nature has intrinsic value has indeed been the primary goal of environmental philosophers. As Callicott has suggested "the central theoretical quest of environmental philosophy [is to establish] the intrinsic value of nonhuman natural entities and nature as a whole."[5] Yet although the notion of "intrinsic value" may come into discussions within ecofeminist ethics, in general, systematic articulations of what the concept means and the role it is meant to play in ethical theorizing are largely absent.

Ecofeminist ethics, indeed feminist ethics in general, have not focused on the articulation of intrinsic value, but rather on providing an analysis of what ethical thinking and ethical practice are and can be. Ecofeminist ethics provide a theoretical perspective through which the forces that contribute to the oppression of women, animals, and nature can be analyzed to undermine their impact and end oppression. These analyses all reject the assumed inferiority of women and by extension reject the assumption of the inferiority of all that has been associated with women. Ecofeminist ethics have an explicit normative commitment and are necessarily tied to the practice that informs them and that they inform. An ecofeminist ethic has an explicit commitment to uncover and reject assumptions, attitudes, practices and institutions that allow or promote oppression.[6] To this end, ecofeminist theorists have begun to explore those questions that many moral philosophers, feminist and nonfeminist, have "put to one side," particularly those that examine our moral relationships to the nonhuman world.

One of the central issues that too often has been "put to one side" is that of power, oppression, or social domination. Ecofeminist ethics have been concerned with the assumptions that underlie and attempt to justify social domination and are committed to examining the ethical and unethical exercise of power. This is what it means to suggest that an ecofeminist is fundamentally opposed to the "logic of domination." Environmental philosophers have tended to ignore the way various relations of power (beyond anthropocentrism) inform ethical thinking and ethical practice and this is one of the reasons why some have failed to recognize ecofeminist ethics as ethics. In addition, their failure to recognize how oppressive conceptual frameworks may operate in their own theorizing has led to the development of environmental ethics that promotes common exclusions of those whose interests and concerns are traditionally discounted—women, individual nonhuman animals, indigenous peoples, people of color, and the poor.[7]

Because ecofeminist ethics are explicitly concerned with examining the ways that supposedly emancipatory theories can advance traditional oppressive practices and institutions or can create new ones, ecofeminist ethics take seriously how values are in large part dependent on context. As Karen Warren has suggested, "at least three interrelated reasons [exist] why attention to context is of importance to ecofeminist ethics."[8] The first is conceptual; understanding what a thing is (a forest, a community, a species, etc.) depends in part on the context in which the conceptual question is being asked and the context in which the thing in question exists. Second, paying attention to context allows us to take responsibility for the construction of our own moral agency and our role in creating ethical systems. Third, understanding context is important in assessing the putatively universal claims of reason and ethical deliberation and, as I suggest later, helps us to generate standards by which we can make judgments.

The recognition of the importance of context in moral theorizing and moral practice has led many ecofeminists to embrace value pluralism—a view that recognizes the importance of specific contexts and ways of life as the raw material, as it were, for the development of full moral agents. Some environmental philosophers predictably have argued that pluralist theories of value are unsuitable for environmental ethics. Callicott, for example, suggested that: "Moral pluralism . . . implies metaphysical musical chairs. I think, however, that we human beings deeply need and mightily strive for consistency, coherency, and closure in our personal and shared outlooks on the world and on ourselves in relations to the world and to one another." He goes on:

> . . . We feel (or at least I feel), that we must maintain a coherent sense of self and world, a unified moral world view. Such unity enables us rationally to select among or balance out the contradictory or inconsistent demands made upon us when the multiple social circles in which we operate overlap and come into conflict. More importantly, a unified world view gives our lives purpose, direction, coherency, and sanity.[9]

This dispute over pluralism captures one of the central conflicts between ecofeminist ethics and mainstream environmental ethics. Ecofeminists believe that in positing a uniform and unified worldview environmental philosophers will at best be ignoring the values, interests, and concerns of those who have traditionally been excluded. Is it obvious that all human needs are the same? Is there, in fact, a singular coherent sense of self and world to maintain? Has there ever been one worldview that will grant meaning and sanity to everyone's lives?

Ecofeminists and many others believe that ethics should not attempt to reduce or flatten our rich and diverse moral experiences. Many conceptual frameworks and worldviews inform our ethical projects and deliberations. And although Callicott is undoubtedly right that consistency is an important, perhaps crucial, component of our ethical activity, one can maintain consistency within a variety of worldviews. The values and beliefs that develop within the worldview of native peoples may be very different from those that develop in a Western advanced capitalist worldview. But both may be internally coherent and consistent, and both may provide purpose and direction, maybe even sanity, to the lives of those who accept them. Callicott's rejection of pluralism and search for "the Holy Grail of environmental ethics—the . . . supertheory,"[10] highlights one of the central reasons that ecofeminist ethics emerged: to provide a critical, self-reflective, and pluralistic alternative.

In his last written work, Isaiah Berlin describes this quest for the Holy Grail as one of philosophy's most enduring projects. Underlying this quest, he suggests, lies a wider monistic thesis:

> That to all true questions there must be one true answer and one only, all the other answers being false, for otherwise the questions cannot be genuine questions. There must exist a path which leads clear thinkers to the correct answers to these questions, as much in the moral, social, political worlds as in that of the natural sciences, whether it is the same method or not; and once all the correct answers to the deepest moral, social, and political questions that occupy mankind are put together the result will represent the final solution to all the problems of existence. Of course, we may never attain these answers: human beings may be too confused by their emotions, or too stupid, or too unlucky, to be able to arrive at them; the answers may be too difficult, the means may be lacking, the techniques too complicated to discover; but however this may be, provided the questions are genuine, the answers must exist . . . the answers must be there.[11]

Such a view has much appeal. We want to be able to say that the enslavement of children is wrong; that the rape of women is morally reprehensible; that the wanton destruction of sentient creatures is abhorrent; that torture, mutilation, and murder are evil, and we want to be right when we say such things. To avoid the vertigo, apathy, and despair that our chaotic and disintegrating world engenders, the possibility that one right answer may exist that can help us get a grip on what seem overwhelming problems is indeed attractive. And given that suffering and destruction are not unique to our time, that monism has been the preferred view of moral and political philosophers throughout the ages is no wonder. Monistic mainstream environmental ethicists are in good company.

But voices of criticism have always existed, and as Berlin noted, these voices become louder during certain periods of time. We are in a time when the critics are quite vocal, for better or worse. Criticisms about what is more commonly called *universalism* contain a number of important insights—that our view of the right principles or standards is shaped by the context in which we formulate them and thus may be useful and maybe even true for one time and place but cannot be adequately extended beyond that context; that one of the things we seek to do in developing moral and political theories is organize the range of our experiences but when we look carefully at these experiences we see that they are so diverse that no overarching categorizing scheme is available. Indeed, when we look carefully at human moral experiences we see not only diversity, but also real disagreement, which renders any attempt at universalism futile. One of the most frequently voiced criticisms is that monists or universalists are imposing their own standards of rightness on others who do not share those standards and in the process either ignore or destroy other standards that have served important roles in the lives and cultures of different people. This imposition of one standard and the disrespect that accompanies it are consequences that ecofeminist ethics attempt to avoid.

Currently, significant debate has occurred among ecofeminists about what I loosely call *universalism*. On the one side of the debate are those who are opposed to what they perceive to be cultural arrogance masquerading as universalism on the part of some ecofeminists. Some of the criticisms that are raised by those opposed to ecofeminist universalism are similar to those raised about universalism generally. Other criticisms are more specific to ecofeminist positions, particularly those having to do with judgments about practices that involve the objectification, immobilization, and imposition of pain and often death on nonhuman animals, practices such as meat eating and hunting, for example.

Some have argued that an ecofeminist cannot and must not condemn meat consumption or hunting because some of these practices are central to certain cultural identities. Ecofeminist criticisms of using animals assume the point of view of white Western women from the United States and as such cannot be applied beyond that particular context. To do otherwise, critics argue, is to engage in a type of ethnocentrism that ecofeminism was supposed to move us past. If an ecofeminist ethic is meant to be pluralistic and mindful of context, then it cannot condemn the cultural practices of others.

But this sort of rejection of universalized judgments in favor of context and pluralism is open to the fairly obvious concern that when differences in values, beliefs, and attitudes emerge, adjudicating among them will be difficult. Does a commitment to context and pluralism always mean that one story is as good as the next? That if indigenous peoples decide to adapt their culture to pursue intensive factory farming or if they decide to use high-powered weapons and motorized boats to kill whales, that ecofeminists and others cannot raise criticisms? Surely an ethical theory, ecofeminist or not, must be able to provide a way of determining which among competing claims is right. If an ecofeminist ethic is to be acceptable, we must have a way of adjudicating between conflicting claims. In cases in which one worldview or set of priorities conflicts with other worldviews or sets of priorities, what can a pluralist provide by way of methods of resolution? What does ecofeminist ethics propose for resolving such disputes in light of its commitment to pluralism?

Some ecofeminists have said that the judgments of the participants in the culture are the only ones with any normative legitimacy. Ecofeminists making judgments about the practices of other cultures reveals their acceptance of what has been described as a "form of cultural hegemony which renders invisible the cultural difference of ethnically subordinated others and universalises [*sic*] the cultural ideals or experience of dominant culture or ethnic groups."[12] Condemning the eating practices of others because these practices involve killing animals, for example, would wrongfully impose the norms and values of one's own culture onto the people of another culture.

Care must be taken here. Criticizing the cultural practices of another culture, even if the critic happens to be located in a historically dominant, even traditionally oppressive culture, race, or ethnicity is not in itself imposing one's own cultural standards on another. One would have to assume that critics are unable to step back from the norms and values of their culture or that such norms directly and completely shape critics' values for the ethnocentrism charge to stick. But this does not seem a particularly plausible assumption in general and it seems false in this particular context (criticizing another culture's use of animals from within a culture that engages in the same or worse practices). Cultural practices, like any other practices, should not be immune to careful, respectful criticism—preferably from inside that culture, when possible, as well as from outside it. Historically, what gets defended as a cultural practice has often proven to be obviously morally objectionable (e.g., slavery, Chinese foot-binding, arranged marriages). Embracing pluralism does not commit one to accepting normative relativism. But how are competing cultural claims, informed by different values, to be judged? How can an ecofeminist ethic be both responsive to context (respectful of different cultures) and action-guiding in the face of real conflicts?

Two related avenues of thought should be pursued in answering these questions, although space constraints allow me only to mention them: The first attempts to develop a method of contextual justification for normative claims; the second attempts to provide a way of assessing power and social domination, "oppression" in short, that takes into account both the cultural understanding of certain practices and their larger meaning and significance. By developing both a contextual method of justification and a more thorough analysis of what oppression is and how it operates, we will be able to determine a set of norms or standards that set the conditions for and thus support our cross-cultural judgments. We will be able to determine when an ethical judgment from outside a particular culture is an expression of ethnocentrism or an instance of faulty universalization and when it is not.

Conclusion

I have suggested that ecofeminist ethics, unlike environmental ethics generally, are committed to pluralism and are concerned with moral judgments in context. By not narrowly focusing on establishing the intrinsic value of nature, a project that I believe may actually hinder finding workable solutions to our environmental problems, ecofeminist ethics can overcome some of the problems that environmental ethics currently face. An ecofeminist ethic allows us to explore the ways that we value: how we come to value what we do and how we might shift and refine our values to ultimately become better valuers. But much work needs to be done for an ecofeminist ethic to provide actual guidance in making some of the tough choices we currently face. Those engaged in both ecofeminist ethics and environmental ethics would do well, in my view, to move beyond identifying their differences and begin to focus their attention on methods for solving real, and urgent, environmental problems.

Notes

1. Revised version of J. Baird Callicott, "The Search for an Environmental Ethic," in *Matters of Life and Death* ed. Tom Regan, 3d ed. (New York: McGraw Hill, 1993), 335.

2. Callicott, "Search," 333.
3. Callicott, "Search," 336.
4. For more discussion related to this point, see my essay, "Toward an Ecofeminist Moral Epistemology," in *Ecological Feminism* ed. Karen Warren (London: Routledge, 1994).
5. J. Baird Callicott, *In Defense of the Land Ethic* (Albany, N.Y.: SUNY Press, 1989), 4.
6. This formulation of an ethic that is so obviously informed by and committed to a certain political end may initially strike some as worrying. This worry might be based on a view that political commitments necessarily undermine the "objectivity" of ethics. Strong arguments can be made that some normative commitments are compatible with certain forms of objective ethics, but I do not pursue this line of argumentation here. Rather, let me simply point out that in the context of the present discussion, environmental ethics makes an explicit normative commitment as well, namely that the natural world has moral status or is morally considerable.
7. It is interesting to note that this oversight appears more striking within environmental ethics than within ethics generally. Most consequentialist theories as well as contractarian theories attempt to explicitly reject inappropriate appeals to power. John Rawls, for example, devises the veil of ignorance in the original position precisely to avoid deliberating about matters of justice from positions of social dominance. Utilitarian thinkers too argue that the effect of actions on everyone affected by them be taken into account and that like interests be counted equally, regardless of race, gender, class, even species membership. These constraints on partiality are meant to preclude the advancement of social domination. One might speculate that this failure to pay attention to human relations of power may stem from an overly stringent commitment to nonanthropocentrism.
8. Karen Warren and Jim Cheney, "Ecological Feminism and Ecosystem Ecology," *Hypatia* 6 (1991): 184.
9. J. Baird Callicott, "The Case against Moral Pluralism," *Environmental Ethics* 12 (1990): 115, 121.
10. Callicott, "The Case Against Moral Pluralism," 105.
11. Isaiah Berlin, "The First and the Last," *New York Review of Books*, May 14, 1998.
12. Personal correspondence from Val Plumwood.

Part IV

Challenging the Implications of the Land Ethic

14

Environmental Ethics and Respect for Animals

ANGUS TAYLOR

In 1980 J. Baird Callicott gained a measure of notoriety among animal-liberation philosophers with the publication of "Animal Liberation: A Triangular Affair," in which, giving no quarter, he argued that "intractable practical differences" exist between environmental ethics and the liberation movement. Later, Callicott tempered his views in "Animal Liberation and Environmental Ethics: Back Together Again," where he advocated making common cause against the forces ravaging the nonhuman world.[1] However, this alliance was to be "on terms . . . favorable to ecocentric environmental ethics," which meant the rejection of both the utilitarian and the rights versions of animal liberation. An ethic of the general type that Callicott wishes to advance, however, does not require rejection of a strong animal-liberation position. Specifically, the values of ecosystemic integrity and the exercise of autonomy can and should jointly form the basis of a new environmental ethic.

Callicott maintains that with the special significance they attach to the possession of consciousness, animal-liberation philosophies are merely extensions of existing egoistic views of the relation between the self and the world. Because, like ourselves, some animals can suffer and some may possess a degree of self-awareness, philosophers such as Peter Singer and Tom Regan "grudgingly" conclude that these nonhumans must be granted membership in that exclusive club, the moral community. By contrast, says Callicott, we need an environmental ethic that recognizes all organisms, human and nonhuman, sentient and nonsentient, as morally considerable links in the web of life. At the same time, drawing on Hume's idea that morality arises on the basis of natural sympathy, on Darwin's contention that concern for members of one's social group has an evolutionary basis, and on Mary Midgley's idea of the animal-human "mixed community," Callicott argues that extending to wild animals the same consideration that we rightly have for members of our community would be inappropriate—indeed, ecologically disastrous. The charge of "extensionism" is a red herring, however, distracting us from the vital role that liberation philosophy, at least in one of its versions, can play in a new environmental ethic.

Reconciling an ecosystemic environmental ethic with the utilitarian version of animal liberation appears impossible, except on the assumption that utility is maximized when animals remain at liberty, not when they are fed and cared for by humans. How might such an assumption be supported? John Stuart Mill championed liberty for (rational adult) human beings on the grounds that free expression of opinions and experiments in living promote the general happiness. A utilitarian defense of liberty for wild animals obviously could not take that route. At the same time, Mill maintained that some pleasures are qualitatively superior to others and that, consequently, being a dissatisfied human being is better than being a satisfied pig. Similarly invoking quality of pleasure, we might conclude that being a wild animal exercising her natural powers in the struggle for survival is better than being that same animal fat and pampered in captivity. A utilitarian case for wild nature is not obviously absurd, then, but may well not be persuasive.

The rights version of animal liberation, by contrast, rejects the utilitarian calculus and rests squarely on positive regard for the exercise of autonomy. Autonomy here is to be understood not in Kant's strict sense of the capacity to govern oneself according to a rational understanding of principles, but in the broader sense of the capacity to act on the basis of preferences, what Regan calls "preference autonomy." Animals with preference autonomy are held to be worthy of respect, which is to say that they are never to be treated by us as mere instruments or resources.

Regan is quite clear about the implication of his rights view for wildlife: We should give wild animals the opportunity to live their lives as they see fit. Respecting the rights of wild animals does not mean intervening to alleviate suffering: "The total amount of suffering animals cause one another in the wild is not the concern of morally enlightened wildlife management. Being neither the accountants nor managers of felicity in nature, wildlife managers should be principally concerned with letting animals be, keeping human predators out of their affairs, allowing these 'other nations' to carve out their own destiny."[2] At first glance, Regan's position seems strikingly similar to that of Callicott, who has written that in the realm of wild nature "each being should be respected and left alone to pursue its *modus vivendi*—even if its way of life causes harm to other beings, including other sentient beings."[3] I say "at first glance" because, as I indicate later, the way Callicott articulates his notion of respect elsewhere in his work does not include leaving wild animals alone.

One might expect Callicott to welcome Regan as an ally. Instead, he devotes himself to arguing that the rights view would actually commit us to intervening in nature to prevent harm befalling wild animals from nonmoral sources (such as other animals), just as we are committed to rescuing people from attacks by the criminally insane.[4] Callicott seems determined not to see any merit in the ascription of rights to animals; yet his reluctance is unwarranted. Rights to life and liberty, like the right not to be made to suffer unnecessarily, are commonly (although not always) understood simply as claims of noninterference against moral agents. The right to life, for example, is a claim against moral agents that they not kill you against your wishes unless special circumstances provide morally compelling reasons for them to do so, and, arguably, that they protect you against being wrongfully killed by other moral agents. As such, the right to life does not include a right not to be killed by lightning, or by cancer, or by a hungry cougar, or by a knife-wielding madman.

We are rightly concerned with the overall well-being of other individual members of our family and immediate community, concerned both about their suffering and about their natural capacities to conduct their lives. Such duties as we have to care for and nurture those near to us are *in addition* to our basic duty of noninterference. As Evelyn Pluhar puts it, "We have the unacquired duty not to interfere with the freedom or well-being of others, unless they pose a threat to us. We have additional, acquired duties to beings whose existence or living conditions have resulted from our choices."[5] Pluhar emphasizes that acquired duties are independent of genetic kinship; if we typically have acquired duties to many human beings, this is not *just because* they are human and these acquired duties do not allow us to disregard the basic interests of nonhuman beings.

Ascribing rights to many animals, including many wild animals, recognizes that different kinds of beings have different natures and that the intrinsic needs and drives of sentient organisms differ from those of nonsentient organisms. Having different needs and drives, different beings merit (often significantly) different treatment. In this vein, Midgley distinguishes between "ecological" claims on us and the dramatically different "social" claims of sentient beings, although she makes the crucial point about these two sorts of claims that "since habitat is so important to animals, they converge much more often than they conflict." The special importance of sentience in another being "is that it can give that being experiences sufficiently like our own to bring into play the Golden Rule—'treat others as you would wish them to treat you.'"[6] Midgley, one should note, does not restrict application of the Golden Rule to our treatment of members of our own community.

To say that sentient animals have the right not to be made to suffer unnecessarily and that many or all of them have prima facie rights to life and liberty is to give them their due on the basis of what differentiates them from the nonsentient world. It is to say that we ought to refrain from harming them unless we have no other means of preventing significant harm to ourselves. As such, our duty to allow animals to exercise their natural powers without harmful interference by us does not include a responsibility to protect them from natural hazards.[7] Importantly, however, it does include some responsibility to restore environmental conditions that allow wild creatures to exercise their natural powers effectively.

One environmental philosopher has argued that rights make sense only as claims within the power structure of human society and that, consequently, to take the idea of animal rights seriously would be to extend this power structure to the entire biosphere, thus arrogantly "humanizing," or domesticating, the whole planet.[8] But surely wild nature today is already vulnerable to the dictates and vagaries of human power. Recognizing that wild animals have noninterference rights does not bring uncontrolled creatures under human control. On the contrary, it says to human beings, "Enough is enough. Back off, and let these creatures live in their own ways, in the ways that have resulted from the long course of evolution, and in environments conducive to the exercise of their autonomy."

Although promoting the exercise of autonomy should generally take precedence over minimizing suffering where wildlife is concerned, we should not rule out all intervention to reduce suffering. Rescuing oil-soaked birds from the harm humans have caused is certainly permissible, perhaps even obligatory. And though we may not have any duty to help a beached whale, there normally will not be any pressing reason not to act on the basis of the sympathy we may well feel for a fellow creature in need. Indeed, it could be said that a whale on the beach in front of us, like an injured stray cat in our back yard, acquires a claim on us by virtue of proximity: fate has unexpectedly connected us with this individual.

Callicott has suggested that traditional Native American views on the treatment of animals provide us with the key to reconciling the holism of Aldo Leopold's land ethic with the demand that individual animals be respected.[9] Callicott claims to find support here for his own position, but I believe that he draws the wrong lesson from traditional Native American views. He does so because he fails to understand the concept of respect for animals in historical context.

North American aboriginal people, says Callicott, saw spirit everywhere in nature and viewed wild animals and plants with reverence. For the Lakota, human beings and other living beings ideally form, in effect, an extended family, whose members are mutually dependent and mutually supporting. Callicott is particularly taken with the Ojibwa worldview, which he sees as similar to the land ethic in which humans and nonhumans form an integrated economy of nature. Callicott makes much of the fact that Ojibwa culture represents animals as participating in a voluntary economic exchange with human beings. Animals allow themselves to be killed; humans, for their part, must show animals respect by refraining from using them wastefully and by burying their skeletal remains so that the individual animals who have given up their present bodies can be reborn. The lesson that Callicott draws from this is that today we show respect for individual wild animals if we kill them "humanely" and use their bodies without wasting or degrading them.

The belief that hunted animals partake in a voluntary exchange with human beings is factually wrong. However, this does not mean that in the past Native Americans were wrong to hold such a belief. Hunting was a morally legitimate activity insofar as it was a necessary part of wresting a living from the environment. Furthermore, the myth of voluntary exchange served to place limits on the exploitation of animals by emphasizing that animals are not to be reduced to objects of utility. That animals were seen as engaging in voluntary exchange implied that they were active agents in the world, entitled to be recognized as such. The ways animals were typically treated by humans were legitimated by the consent animals gave to be used to satisfy basic human needs. But for humans to exploit animals more than necessary was to violate the moral order. The rule, in effect, was, "Take what you need and no more."

Callicott's belief that today we can show respect for individual animals by appropriating their lives and bodies for our exclusive purposes without their consent is odd to say the least. His notion of respect departs radically not only from the animal-rights concept, but from the concept of respect implied in the Native American myth of voluntary exchange. Indeed, it is the animal-rights concept of respect, and not Callicott's, that in its core idea (of never treating another autonomous being as mere means) resembles the traditional view of Native Americans. Because today we understand that animals do not consent to being killed and because today we have, generally speaking, no need to kill them, we cannot show respect for them when we kill them.

Callicott deplores the factory farming of animals. Nonetheless, in the case of animals of the traditional barnyard, he maintains that we have an "unspoken social contract" with them that includes our killing them. But this is incorrect; no such contract exists. One could argue that an unspoken agreement exists to care for them in exchange for their good behavior, perhaps, or for their allowing themselves to be milked or shorn. But to imagine that they agree to suffer or to be killed is to indulge in a convenient fantasy. And it will not do to claim that because certain creatures have been bred to be abusively exploited, it is in their natures to be abusively exploited, and we therefore do no wrong when we abusively exploit them.

I have been arguing that we can and should have it both ways: ascribe rights to all sentient beings and recognize special duties to members (human or nonhuman) of our extended families. One could say, then, that two lines are to be drawn when assessing our duties: one between the sentient and the nonsentient, and another between the near and the far. But this puts things too bluntly. The nonsentient natural world is morally considerable because it is the home of sentient life. When we understand the intimate connection between the flourishing of sentient beings and the flourishing of their natural environments, we see that strong respect for sentient life entails an intensely "green" attitude to the world. Consequently, we may want to say that the nonsentient natural world has intrinsic value of some kind, although not the kind that comes from valuing one's own experiences (what, following Regan, we can designate "inherent value") and that qualifies one as a subject of rights. And when it comes to the distinction between the near and the far, there is, as in the case of beached whales or endangered rain forests, no tidy, once-and-for-all division between "here" and "there."

I see no fundamental conflict between ecosystemic integrity and the exercise of autonomy. The flourishing of individuals, human and nonhuman, depends on appropriate environmental conditions, conditions that cannot be established by human fiat but that are the product of eons of ecological evolution. Historically, the mistake of liberal ideology has been to imagine that individuals are not fundamentally social and natural beings whose welfare is tied to supportive social and natural environments. Equally, however, ignoring the intrinsic drive of all living things for self-fulfillment, including the drive of sentient creatures to exercise autonomy, would be a mistake.

The reluctance of many environmentalists to ascribe rights to wild animals seems to arise from (1) a misunderstanding of what is intended in liberation philosophy by the ascription of rights to animals, (2) a desire for continued human interference in wild nature in the form of hunting and fishing, interference that was once necessary but (certainly in the case of hunting) is seldom necessary today, and (3) a belief that preserving the integrity of ecosystems sometimes requires harming individual animals (as when deer populations must be reduced to preserve endangered plant species). This last item is the likely philosophical sticking point. Liberationists must concede that the rights of wild animals may be on occasion overridden to protect the basic interest that animals and human beings have in living in a flourishing natural environment. For their part, environmentalists must understand that a heavy burden of proof always lies with those who would override rights.

The philosophy of animal rights is consistent with an environmental ethic that calls for a radical lessening of human interference in wild nature and a new regard for the nonhuman world in general. Ecosystemic integrity and the exercise of autonomy as joint fundamental values mean:

1. A "hands-off" policy toward wildlife, except where compelling reasons dictate otherwise. This includes an end to hunting and fishing except where such activity is necessary for subsistence.
2. An end to destruction of wildlife habitats, and the restoration of habitats to provide the basis for the exercise of autonomy by wild creatures.
3. An end to factory farming of animals, both because it treats these animals without respect and because it infringes on the exercise of autonomy by wildlife through habitat destruction. This means an end to the meat industry.
4. The use of domesticated animals only with their implied consent (i.e., only when they are neither distressed nor harmed by use). This means an end even to the production of so-called organic meat, although not necessarily to all "free-range" eggs and dairy products.

None of these objectives will be achieved soon or easily. The industrial juggernaut will continue to ravage the nonhuman world. An alliance of the animal-liberation and environmental movements is not only possible but also imperative.[10]

Acknowledgment

The author thanks Thom Heyd for his comments on a draft version of this essay.

Notes

1. Both these essays are included in J. Baird Callicott, *In Defense of the Land Ethic: Essays in Environmental Philosophy* (Albany, N.Y.: SUNY Press, 1989).

2. Tom Regan, *The Case for Animal Rights* (Berkeley: University of California Press, 1983), 357.

3. Callicott, *In Defense of the Land Ethic*, 57

4. J. Baird Callicott, "The Search for an Environmental Ethic," in *Matters of Life and Death*, ed. Tom Regan, 3d ed. (New York: McGraw Hill, 1993), 352–353.

5. Evelyn B. Pluhar, *Beyond Prejudice: The Moral Significance of Human and Nonhuman Animals* (Durham, N.C.: Duke University Press, 1995), 269.

6. Mary Midgley, *Animals and Why They Matter* (Harmondsworth, England: Penguin, 1983), 91.

7. For more on this and related matters, see David DeGrazia, *Taking Animals Seriously: Mental Life and Moral Status* (Cambridge, England: Cambridge University Press, 1996).

8. John A. Livingston, *Rogue Primate: An Exploration of Human Domestication* (Toronto: Key Porter, 1994).

9. J. Baird Callicott, "The Search for an Environmental Ethic." See also "Traditional American Indian and Western European Attitudes toward Nature: An Overview," in *In Defense of the Land Ethic*, and J. Baird Callicott, *Earth's Insights: A Survey of Ecological Ethics from the Mediterranean Basin to the Australian Outback* (Berkeley: University of California Press, 1994).

15

J. Baird Callicott's Critique of Christian Stewardship and the Validity of Religious Environmental Ethics

Susan Power Bratton

The Issue

In the first edition of Tom Regan's edited volume *Matters of Life and Death,*[1] J. Baird Callicott's "The Search for an Environmental Ethic" opens pursuit for "a living, practical morality, a real world ethic" for the environment. To evaluate environmental ethical alternatives, he establishes three criteria that emerge primarily from the practice of philosophical or scientific logic: (1) "an ethic, like any other sort of rational pattern of ideas, must be self-consistent in order to be acceptable and persuasive"; (2) the ethic must be "adequate" and address the appropriate moral problems—"those problems that together constitute the 'environmental crisis'"; and (3) the ethic is practical in the sense that one can "actually live in accordance with the precepts of the ethic."[2]

In the first edition of Regan's book, Callicott makes the Judeo-Christian tradition of stewardship the centerpiece in a section entitled "How Not to Do Environmental Ethics." After repeating Lynn White Jr.'s and Ian McHarg's accusations of Christian despotism and human-centered orientation, Callicott dispatches Christian ethics on the grounds that they fail the test of consistency because the cosmology expressed in Genesis 1 and 2 (originally sacred scriptures of Judaism) conflicts with Darwin's evolutionary worldview developed almost two millennia after Christianity was founded and more than three millennia after the texts in Genesis were first edited into their present form. In the third edition of Regan's book,[3] Callicott moderates his condemnation of Christianity (and de facto of the other Abrahamic faiths), and also recognizes that not all Christians may hold the same ethical views. Ignoring historic and denominational differences in hermeneutics, Callicott distinguishes between a despotic reading and a stewardship reading of Genesis.

Adding a fourth criterion for a viable ethic, he identifies "parsimony or conceptual economy" and suggests: "If we have a choice between theories that are equally consistent, . . . adequate and practicable, then we should prefer the one that involves the fewest assumptions." He declares Christian stewardship, if constructed as "citizenship environmental practice" to be adequate and practicable, but continues to deny Christianity and Judaism "consistency." He then suggests that:

> the citizenship reading of the God-ordained relationship of man to the environment implies an atavistic return to nature, that would be, if not utterly impracticable, for most people so unappealing as to be thoroughly unacceptable. We can rescue its practicability by allying it with the deep ecological practice of Self-realization, but only at a cost in consistency with the scientific world view.[4]

Considering the long history of Christian mysticism, including nature mysticism, and the number of Christian writers who describe the experience of merging with the divine or the cosmos, it is rather odd that Callicott suggests deep ecology's self-realization or "the realization that one does not exist either in isolation from or in opposition to all other living beings and natural processes . . ."[5] would be a valuable addition to the tradition.

Callicott's Analysis

Rather than weighing the merits of Callicott's arguments specifically concerning Christianity, taking a step back and asking whether Callicott's project of finding "an" environmental ethic is valid and whether his criteria are appropriate for evaluating religious ethics in general would be wise. The first version of Callicott's critique attempts to apply metaethics to specific social phenomena. Callicott actually goes through a process of self-correction between the first and second versions of the article, and he recognizes that if he argues Christianity is not a valid basis for environmental ethics because its cosmology is not based on Darwinian evolution, then Hinduism could be rejected on the same basis. Callicott's original criteria, in fact, would find the vast majority of the world's religions "inconsistent," and thus encourage rejection of all forms of religious ethics out of hand. Although academic critique of religious ethics is perfectly appropriate, Callicott's first version of the article verges on "anti-religious discrimination" because he promulgates folklore common in academia about Christian views of Darwin, and he so strongly stereotypes Christians (and de facto the other Abrahamic faiths) as antirational.

Callicott is taking a step in the right direction when he attempts to compare the validity of ethical models. He draws his criteria, however, out of classroom exercises in philosophical logic. Callicott seems to be on the quest for general theory in a case where this may be a priori impossible. First, any ethic has a cultural context, including the language and symbol systems that convey it. Callicott's abrupt treatment of religion in the first version of the article betrays a disinterest in how real societies make day-to-day ethical judgments or establish social norms. Community values and ethical priorities differ among Amish farmers in Pennsylvania, Muslim shepherds in Jordan, and agnostic philosophers who raise journal articles instead of sheep. Second, as Shrader-Frechette and McCoy so aptly point out, the scientific field of ecology lacks general theory, so successful ethical approaches may have to rely partially on accumulation of ecosystem specific data bases and on the construction of accurate case histories.[6] The intrusion of human economics into most environmental issues further complicates the question of viable general theory.

We should ask, therefore, if we are searching for "an ethic"—the Holy Grail of the environment, or whether we should be investigating ways to facilitate environmental problem-solving and sensitivity within existing ethical systems. Is Callicott's assumption that he may, based on a single historically mismatched comparison, declare two world religions (Christianity and Judaism) ethically unfit, getting off on the wrong foot and doing far more harm than good in terms of encouraging our great variety of human cultures and societies to confront and resolve environmental problems and conflicts?

Callicott assumes the reader will accept the validity of his criteria while making no effort to prove they are critical to establishing environmental norms in actual human communities. Although he likes to end his criteria in "y," he is actually drawing from two different categories of values. Self-consistency and parsimony are the criteria of logic and are employed in constructing scientific models and explanations. Practicability (i.e., can one live with the ethic) is a social criterion. Adequacy can be employed in logic, but Callicott's use appears to be more social, particularly when he desires an ethic that addresses the entire, rather ill-defined "environmental crisis," which is a complex and continually changing phenomenon.

Religious ethics are rarely internally completely self-consistent, and religions often appear to avoid parsimony rather than cultivating it. Religions evolve through time, adding and losing rituals, myths, and norms—addressing first one ethical issue and then another. They accumulate bits and pieces of culture, much as biological species accumulate bits and pieces of DNA. As with nature they are often overflowing with information and productivity, rather than being efficient and concise. A Gothic cathedral, for example, is anything but parsimonious. During any one era, not everything a religion incorporates may be "useful," and religions, like philosophy, may not engage new challenges in a timely way. Conversely, religions may prove to be important sources of societal adjustment to change. Expressions of a religion in terms of the cosmology, coda, and community structure vary greatly even within sects and regional variants. Not all Amish "orders" have the same norms concerning wagon tires and use of motors, for example. Religions that have survived through the centuries, however, have proven their practicability (on the average, not necessarily relative to every issue or every cultural context) and have evidenced at least partial social adequacy as well. Religions continue to foster dialog over ethical concerns of all sorts, and many people are attracted to religious commitments and lifestyles for just this reason. One of the great advantages of religious ethics is they have followings, and are supported by ritual, community activities, and popular education. If you asked people worldwide if they are Christian or Muslim or Buddhist, vast numbers would answer in the affirmative. If you asked them if they are Kantians or Heiddeggerians, probably a fraction of 1 percent (primarily well-educated Westerners) would say "yes."

A major difficulty concerning Callicott's approach, therefore, is that the question he is asking can not be properly answered by traditional philosophical methods, but must be approached using the techniques of social science. Whether Christianity has an effective environmental ethic cannot be determined by classroom logic—Callicott has not even properly constructed the hypotheses for this kind of test. He is employing methods normally used to check self-consistency in exercises such as if a = b and b = c then a = c. Callicott does not bother to define *Christianity*. Who is he really looking at? How is he going to handle historic change?

To turn this exercise into academically testable hypotheses, the questions must be framed so they can be verified via historic, sociological, or anthropological information because the task is to relate schools of thought and specific cultural groups to their actions and lifestyles. For example (and this will be a general attempt at hypothesis testing), have human communities that based their ethics on Christian precepts (as defined by New Testament texts) consistently developed community ethics that ignored environmental care? This makes the question more objective because all we have to do to answer it is to screen for environmental elements in context specific Christian teaching.

The answer to the question, of course, is "no"; numerous Christian communities have developed ethics of environmental preservation or conservation. Experts often cite desert and Celtic monasticism, the Benedictines, and agriculturally oriented Anabaptist groups as positive environmental models, even in secular literature.[7]

The next step might be to ask what characterizes Christian communities with well-developed environmental ethics, or what sects or denominations fail to develop an environmental ethos? Here, Callicott and other critics of religion who wish to have the arguments on their own terms, actually may be diverting proper ethical investigation by religious scholars. Callicott, in his early work, wants to evaluate Christianity in terms of beliefs concerning the origin of the cosmos. Christian sects with well-developed values of environmental protection are just as likely to be characterized by their nature-friendly eschatology—where they believe we are all going in the end—or by a belief that saintliness brings peace with the surrounding environmental order.[8] Conservation-oriented sects often have a strong emphasis on community life and care for children or for the socially marginalized. St. Francis of Assisi knew nothing of Darwin and yet was concerned with the suffering of bees and worms—and of lepers and the displaced urban poor!

In one of the most recent permutations of his commentary on religion, *Earth's Insights: A Multicultural Survey of Ecological Ethics from the Mediterranean Basin to the Australian Outback*,[9] Callicott continues to strike a contrast between "despotic" Christian environmental ethics and "stewardship" alternatives. Using the methods of comparative study of religions, he outlines the environmental ethic of more than a dozen world religious traditions. Callicott's discussion, however, is not based on consistent criteria. He ignores variance of Christian sacramentality, and perhaps due to his upbringing in the southern United States, emphasizes Christian biblical interpretation and the problem of misplaced literalism. When analyzing Asian religions, in contrast, Callicott expands to wisdom literatures and incorporates an entire chapter comparing four branches of Buddhism including Tendai and Zen. For east Asia, Callicott quotes monastic and ascetic sources, while largely disregarding their Christian and Islamic equivalents. As Thomas Merton perceived, from the point of view of the practitioner, Zen and Christian asceticism have much in common. In *Earth's Insights*, Callicott has selected "exemplary" traditions, based primarily on practicability and on the aesthetics or charm of their myths or literature. He ends the volume by providing three case histories where religious environmental activists, including a Christian Farming Stewardship Program, are helping to improve care for farm fields and forests. Although Callicott still subtly favors non-Western religion and presents Christian alternatives in dichotomous categories, this account is very much an improvement over his cynical essay in the first edition of Tom Regan's *Matters of Life and Death.*

In *Earth's Insights,* Callicott has thankfully dropped the sweeping philosophical challenges to religious validity and argues instead that "purely secular programs—bureaucratic, technological, legal, or educational—aimed at achieving environmental conservation may remain ineffective unless the environmental ethics latent in traditional worldviews animate and reinforce them."[10] He also steps back from the pursuit of "an ethic" and peruses a series of culturally specific alternatives. Callicott has, however, in the process, diverted from his project of establishing criteria for systematically evaluating the validity of ethical approaches. *Earth's Insights* leaves many unanswered questions, such as: when is a traditional regional religion able to adapt to industrialization-induced environmental change, and when does it fail to do so? What are the common threads in environmentally friendly religious teachings? To what extent has U.S. and European Protestantism influenced contemporary environmental responses from other religions, and has Protestant asceticism and volunteerism synchronized with regional religions, creating new religious hybrids? Callicott's journey from open rejection of Christian traditions to cautious appreciation still begs the question: how can we objectively determine the role of religion in environmental ethics, and how can we fairly compare among religions, sects, and theological or philosophical schools?

Taking a Sociohistoric Approach

Over the past four years, my students and I have been investigating the development of environmental ethics among commercial fisherpersons. We are attempting to compare ethical values and norms between communities that are more or less traditional, and more or less industrialized. The ethical role of religion varies strongly among the communities we have already studied, and we assume that as we move into an increasingly cross-cultural framework, this variation will increase.

One of the questions we would like to answer is: what type of language do fishers use when they are discussing right action, virtue, or duties? Thus far, among Irish and Pacific Northwest fishers, using religious language in response to general questions about values and duties is rare. The Irish from Gaeltacht (Irish-speaking areas), however, participate in a number of common religious rituals and practices that are tied to the fishing trade, and Catholicism appears to be a continuing source of community identity and cohesion, particularly during emergencies, such as drownings. In recent interviews at Westport, Washington, in contrast, only two respondents out of twenty-one expressed strong religious views (this proportion is subject to change as we increase our sample size). Although the types of fishing are very similar, one community is far more secularized than the other. Among those who express religious values or even superstitions, the diversity of beliefs is greater at Westport, which has a far more eclectic cultural heritage than rural Ireland. In practical terms, religion is one means of addressing environmental concerns and is perhaps most helpful when it addresses other major community needs.

Social Criteria for Evaluation

If we are intent on justly and honestly judging the viability of religious ethical responses to environmental problems, I propose replacing Callicott's criteria with seven social benchmarks. The key is not adopting one form of ethic as the "philosopher's stone," but rather determining how each vehicle for conveying environmental information and establishing cultural values and norms interacts in actual applications. Rather than using dichotomous decision making—the ethical model either meets a set of criteria or it does not—we should look at these criteria in terms of relative degrees of adaptation. Callicott admits of degrees for all his criteria except self-consistency. In real case histories, social criteria are more likely to be partially rather than completely met.

The first criterion is how well does a religion or an environmental ethic *understand* nature or the nonhuman portion of the cosmos, and how well does it understand the relationship between the human and nonhuman? Does it convey information about how nature operates and how humans should react? Environmental ethicists and historians often find themselves attracted to prescientific regional religions, and thus present these as ideals for religious environmental care. Regional religions, developing over centuries, often do incorporate an astute and time-tested portrait of the surrounding environment. Hawaiian myths about Pele, the volcano goddess, for example, contain detailed descriptions of volcanic eruptions and encourage respect for the unpredictable "goddess" and her violent activities. Tales of her capricious and sometimes vengeful nature reflect the behavior of volcanoes. Another Hawaiian example is the story of a trickster god who, when pursued by a chief's warriors, placed his giant canoe at the top of large waterfall and blocked a mountain stream. The warriors foolishly walked up the dry stream bed, and when the god removed the canoe, a great flood of water drowned the entire party.[11] This myth contains important information about infrequent but deadly environmental risks, such as a debris dam temporarily blocking a creek, or the massive withdrawal of water that precedes a tsunami. The story of the god's canoe reminds the forest traveler to avoid walking down a dry river bed when it is pouring rain. Flash flooding, in fact, is a continuing problem on the sides of the steep Hawaiian *pali*, even where it is well vegetated.

A second criterion is *comprehensibility*, that is, the community must be able to understand the language or symbols that convey an ethic. Where religion is a principal vehicle for social ethics, integration of environmental values into the religious system, or accentuating and interpreting the environmental values it already fosters, is one of the best means of having an enhanced environmental ethic widely accepted. Respectful environmental and scientific dialogue with Islam alone is important because Islam forwards a deep respect for the "created order" and provides a forum for ethical dialogue and a source of common vocabulary for myriad municipalities, bioregions, and ethnic groups. Regional religions (as opposed to say, academic philosophy) are often extremely comprehensible to the indigenous general populace. In contrast, the comprehensibility of Christian ethics might be questioned at Westport, Washington, where few fishers are involved in Christian practice or ritual.

A third criterion is problem identification and analysis or what Callicott terms *adequacy*. The first question is not, however, whether an ethical model or system can solve the entire "environmental crisis," but whether the ethics of a community can recognize and process the environmental problems encountered on a day-to-day basis. Science is rather good at this (until it runs into the chaos generated by economics). Regional religions, as long as surrounding ecosystems are not undergoing massive change, develop ethics of care that specifically address excessive human exploitation of the environment. Religions have developed taboos that protect species or natural processes sensitive to human disturbance. Religions ranging from ancient Judaism, to Polynesian, to native Alaskan prohibit or limit interference with breeding or nesting animals, for example.

A question concerning the "great" or "world" religions is whether they sometimes lack environmental discourse and adequate problem definition because they have adapted to urbanized societies and have spread widely among cultures and regions. Within Christianity, sects and religious movements have differed greatly in their degree of environmental concern. Ancient Irish monasticism, which was an indigenous adaptation of Christianity without conquest or cultural displacement, includes many tales of environmental protection in its hagiographies, and it emphasizes natural imagery in its poetry. The community-conscious Amish consider care of agricultural lands to be a primary virtue. When Calvinists arrived in the region they dubbed New England, they brought their European, middle-class ideals and farming methods with them.[12] They legislated environmental protective measures relatively early, but did not, like early Celtic Christians, convey these values repeatedly via religious myth. Despite their theocratic leanings, for the Puritans, the environment was already falling within the sphere of policy.

The fourth criterion is a viable and just environmental ethos must be adapted to the "human ethical ecology" of society; that is, it must be *compatible and integrated* with other ethical concerns. Generally, ethical language must overlap and be comprehensible between spheres of community problem-solving. Callicott wades into the anthropocentric-ecocentric argument, but in functional terms, environmental ethics can not seriously conflict with social norms that concern primarily human issues, or the application of the environmental ethics will be limited by other priorities. A simple example from the fishing communities are the high priorities given to work ethics and to forwarding vessel safety. These responses to very real and constant challenges to sound decision making arise from difficulties in trying to get small groups to work together efficiently and compatibly in limited space (e.g., if a crew person is late, the vessel can not leave port), and the risks posed by the changeable physical environment of the oceans. In our first series of Irish interviews, all the respondents had lost a friend or relative at sea. One older fisher's son had died off Cornwall two months prior to the interview. The Gaeltacht community of Helvic Head had tragically lost "two boys" two years before. Furthermore, the fishers sometimes experience difficulties in supporting their families because their income and success in fishing fluctuates. They may not always have money when they need it, and at the worst will lose their vessels if they cannot pay the mortgage or taxes. The unpredictability of the marine environment is not just a problem for humans. As I write this, El Niño is warming the waters of the Washington coast. This natural phenomenon changes patterns of nutrient upwelling and the geographic locations of fish populations. Starvation is currently widespread among sea lions and sea birds.

The integration of environmental ethics with other ethical and social concerns is one component of Callicott's overriding criterion of practicability. For the fishers (and for sea lions), the environment is dangerous and chaotic. One might question whether the environmental regulations that force fishers into short harvest seasons are morally justifiable, for example, if they pressure captains to set out in bad weather and increase the risk of sinkings due to storm waves, ice, or winds. In some cases where religious ethics do not appear environmentally relevant or seem to ignore nature, religion may actually be in dialogue with an environmental concern. Fisherpersons from Helvic Head cross themselves as they sail out of port past St. Declan's holy well, and they keep the bread of St. Nicolas (saint of fisherpeople) and a crucifix on their vessels. These rituals and icons tie them to their homes and families in a periodically frightening milieu and establish a spiritual link between land and sea. In this case, a viable environmental ethic must also respect traditional Irish culture and the difficulties of navigating small, family-owned craft on the unruly North Atlantic. We should also note that Irish fishers are already actively protective of many creatures, such as sea birds, and are concerned about pollution in the waters they fish.

A fifth criterion for a viable contemporary environmental ethics is that it must encourage *dialogue and interaction* among ethics, science, religion, and economy. In the techno-industrial world, science often has the power of problem identification, but cannot by itself elicit an appropriate societal response. Thus religion is a more viable component when it can accurately incorporate rather than reject science. For many religions developed in a prescientific framework, this dialogue may not be natural or easy to initiate. Most religions are not inherently unfriendly to science, however, and some Christian groups, such as the Franciscans and the dissenting denominations in early industrial England, have historically forwarded the advance of science or of science education.

A sixth criterion is that the viable ethic must *initiate and maintain community engagement* with ethical problems. Academic publication is fleeting and may forward less robust societal leadership than one might hope for. Grassroots religious movements are often very successful contributors to long-term societal change. Consider, for example, the role of the black churches in the U.S. civil rights movement or the contributions of religious pacifists in averting unnecessary warfare and mitigating interethnic violence. One of the advantages presented by ecofeminism, in fact, is its broad audience among women activists, some of whom are not primarily interested in environmental issues, but are committed to lasting improvements in social justice.

A seventh criterion is that a viable vehicle for environmental ethics must be *adaptable* to new issues and to technological and cultural change. Religion can be conservative and may be more intent on maintaining the status quo than on righting societal wrongs. From the Reformation to "cargo cults," however, religion has been an important means of human adaptation to and conversation about social change.

Callicott first presents Christianity as if it is unable to adapt when he infers that the entire religion (including all those liberal Protestant denominations) has remained in conflict with Darwin. He then modifies his stance, and presents Christianity as divided into two teams; the unadaptable literal interpreters and the adaptable Christian stewardship crowd. Environmental historian Donald Worster, in contrast, presents the U.S. environmental movement as the legitimate (if genetically recessive) offspring of radical Protestant values. Worster's hypothesis that Western religious asceticism is living on in granola-munching protesters sitting on platforms high in rainy old-growth forest, at least acknowledges the roots of U.S. social reform movements in the revivals of evangelical righteousness during the eighteenth and nineteenth centuries.[13] Callicott's method forces ethical responses into dichotomous categories and tends to neglect questions of historic origins and the possible interrelations among several related schools of thought. Most of the major schools of social activism in the United States have religious ties or roots.

Ethics and Agendas

As Callicott has repeatedly addressed Christian stewardship through his career, he has moderated his view. In the first edition of "The Search for an Environmental Ethic," Callicott demonstrates almost no practical knowledge of late-twentieth-century Christianity. In recent publications, he has actually termed Christian stewardship worthy of further study. His changing position reflects more careful analysis of Christian literature and, presumably, exposure to at least a handful of credible Christian environmentalists.[14]

Callicott's willingness to judge, however, and the notion that it is appropriate for him to do so without detailed study of religious sources or dialogue with involved Christian stewards, are unnerving. His assumption of sweeping academic authority is reminiscent of philosophy programs that tout their friendliness to ecofeminism and offer a women's studies course or two, but have historically had few or no women on their faculties, and in the worst cases, have a man repeatedly teach ecofeminism because the department still has no established female scholars. Callicott's trek toward religious tolerance asks a difficult question: Do such philosophical approaches assume too much social authority while ignoring the necessity of real dialogue (rather than projected dialogue) among ethical perspectives? Is it the task of environmental ethicists to dictate norms, or is it their task to facilitate the greater society's development of norms and values?

Philosophical environmental ethics has been wildly struggling against an inevitable academic reality—some of its favored questions are beyond its methods. The field is repeatedly proposing "solutions" to societal problems *without* properly testing the viability of its approaches. The commonly used basic text book *Environmental Ethics: Divergence and Convergence* cites not a scholar of religion, but Eugene Hargrove (who is trained in philosophy and has little formal background in religious thought) in its introduction to Western monotheism.[15] In the early 1980s, Hargrove did the academic ethical community a favor by editing *Religion and the Environmental Crisis*,[16] thus encouraging the incorporation of a diverse selection of religious scholarship into the quickly consolidating field of environmental ethics. Botzler and Armstrong note that Hargrove "proposes evaluating how major religions, including Christianity, can best respond to the environmental crisis. Whatever eventual responsibility is assigned to Christianity for the environmental attitudes held in Western society, we believe Hargrove is correct in moving beyond the White debate. His proposal to determine if Christianity and other major religions are able to provide a strong and workable foundation for an environmental ethic is a good place to start."[17] But who, pray tell, is going to judge if Christianity is adequate? Hargrove and Callicott? To be equitable, we should judge philosophy and religion by similar standards. We in general do not even know if the ethical models presented in journals such as *Environmental Ethics* or *Ethics and the Environment* are helpful in normative ethics or if they are being absorbed by the general public.

If one is interested in environmental politics and policy, Callicott's approach in the first version of his article appears to be misguided: why summarily dismiss the vast majority of religious practitioners, the single largest constituency for regular ethical study on the entire planet? If we are looking for "an ethic," is all this religious diversity too much to deal with? Or is the tendency to dismiss religion primarily due to the fact it lies outside the realm of control by academics and their institutions? Academics can block religion out of their dialogs, but cannot completely control its impacts in the greater society. If we really care about correcting and preventing *real* environmental problems, thoughtful participation of both majority and minority religious groups can only be beneficial. Can we, for example, really discuss environmental justice for blacks and Latinos, if no Christians or Muslims are at the table? And what if we limit Christians to the "rational," mainline Protestant men? Should our discussion of the "peoples of the book" be limited to issues surrounding the exegesis and errancy or inerrancy of the Bible and the Quran? Any attempt at cultural diversity in environmental ethics must respect religious views, not just Jewish, Christian, Islamic, and Native American, but also Catholic, Protestant, and Orthodox—reformed, mainline, evangelical, and charismatic. All the great religions have nature-friendly sacred texts, mystical traditions, or sects. Most if not all regional religions have myths or coda concerning right relationship with the environment. Judging one to be better than another is less helpful than encouraging all to do their best in realistically engaging contemporary environmental concerns.

Conclusion

I would like to conclude, first, with a petition for greater historic accuracy. Over the last three years, I have been investigating the roots of environmental racism in Europe—specifically environmental anti-Judaism. Medieval images of animal-like Jews adopt pre-Christian (pagan) artistic conventions, and weave them into Christian myth. As Christian theological influence weakens in the eighteenth and nineteenth centuries, academic philosophers pick up the anti-Jewish polemic, declaring Judaism and the Jewish people to be "unnatural." To understand the ultimate terror of the mid-twentieth century, one has to look at the interactions among philosophy, art, religion, and literature. Considering pre-twentieth-century Western culture to be "Christian" and "backward" is a way of avoiding realistic assessment of the cultural origins of our most serious environmental difficulties. Even in the second version of his article, Callicott separates religion and philosophy, which diverts attention from actual historic relationships of Western ethical traditions.

Second, in our survey of Irish fisherpersons, we found widespread agreement that their fisheries were declining. When we asked who was responsible, roughly half of the first group of respondents said it was the "others" (i.e., the French and Spanish), and the other half said it was "everyone."[18] Although Christianity may be historically culpable for many environmental misdeeds (and certainly is in my study of environmental racism in Europe), for philosophers to summarily dismiss potential Christian virtue or ethical viability is the pot calling the kettle black. We should develop standards of adequacy that apply to *both* philosophy and religion and that we can test by use of sociological data or via careful historic comparisons. This is necessary to avoid turning environmental ethics into just one more form of environmental denial where we blame someone else for our own sins.

Third, we should view religious ethics in terms of "community ecology," that is, how they are expressed in real human societies defined by complex and ever-changing relationships, both among humans and between humanity and our environments. If everything in an ecosystem is related to everything else, this is true in social ethics as well. In the survey of Helvic Head, when both fishers and nonfishers were asked what the best or most noble action they could remember fishers actually performing was, many respondents pointed to the case of the recent drowning of the two boys, where the entire Helvic fleet had ceased to fish at the height of the season and had searched for the young men until they were found. This continued even after it was almost certain both had died, but only one body had been recovered, and the economic losses to the fishers were mounting. The nonfishers of Helvic, including many of the women, came down to the dock each day to pray the rosary and repeat the Fisherman's Prayer in Irish. The people of Helvic valued the continuing search by the fishers because it forwarded community integrity and care for others. Economics or immediate productivity was not more important than the injury or death of community members, or more important than the grief of their relatives and friends.

The religious response is a realistic one, considering the capricious and often life-threatening nature of the sea. Helvic fishers still express both love and respect for the ocean and for marine creatures. Christian beliefs and rituals are part of the cement that holds the community together and assist in coping with the unforgiving side of the North Atlantic. If one were to encourage the fishers of Gaeltacht to a new and strengthened environmental "ethic," pushing those Irish-speaking women, including widows of fishermen, off the dock because their prayers are not self-consistent or parsimonious would be destroying one set of community ties to supposedly build another. Helvic Head is an anomaly in modern industrial fishing because Helvic fishers consistently report strong support from their home community for their trade. The bioregionalists' ideals of connectedness to the land (or sea) and to a supportive human network are strongly expressed in this little relict of Munster Gaeltacht, where traditional religious practices still emerge to give the community a single, compassionate voice. Abstract, agnostic environmental ethics delivered by outside experts could actually do much damage in this context and could certainly be destructive to Irish traditions, where nature is often personified and religion is deeply imbedded in community identity.

Undocumented judgments of supposed religious failures exclude a diversity of voices from environmental dialogue. Even the blessing of the philosophers on religious ethics may be dangerous if pursued without careful study of the cultural context of religious expression. Lacking disciplined sociological and historic study, such affirmation may emphasize the wrong themes and variables. If environmental ethics are to result in environmental care without causing social injustice, religious responses should be nurtured and facilitated so religious practitioners in diverse human societies may find their own voice and receive a community-based blessing in a call to earth-service and earth-ministry.

Notes

1. J. Baird Callicott, "The Search for an Environmental Ethic," in *Matters of Life and Death*, ed. Tom Regan (Philadelphia, Pa.: Temple University Press, 1979).
2. Callicott, "Search," 381–382.
3. Tom Regan, ed., *Matters of Life and Death: New Introductory Essays in Moral Philosophy*, 3d ed. (Boston: McGraw Hill, 1993).

4. Callicott, "The Search for," in Regan, *Matters of Life and Death*, 3d ed., 347.
5. Callicott, "The Search for," in Regan, *Matter of Life and Death*, 3d ed., 330.
6. K. S. Shrader-Frechette and E. D. McCoy, *Method in Ecology: Strategies for Conservation* (Cambridge, England: Cambridge University Press, 1993).
7. Susan Bratton, *Christianity, Wilderness and Wildlife: The Original Desert Solitaire* (Scranton, Pa.: University of Scranton Press, 1993). See also Donald Kraybill, *The Riddle of Amish Culture* (Baltimore, Md.: Johns Hopkins University Press, 1989), and Donald Kraybill, *The Amish and the State* (Baltimore, Md.: Johns Hopkins University Press, 1993).
8. Bratton, *Christianity, Wilderness and Wildlife.*
9. J. Baird Callicott, *Earth's Insights: A Multicultural Survey of Ecological Ethics from the Mediterranean Basin to the Australian Outback* (Berkeley: University of California Press, 1994).
10. Callicott, *Earth's Insights*, 184.
11. Vivian L. Thompson, *Hawaiian Myths of Earth, Sea and Sky* (Honolulu: University of Hawaii Press, 1966). See also Tony Swain and Garry Trompf, *The Religions of Oceania* (London: Routledge, 1995); Michael Kioni Dudley, *Man, Gods and Nature* (Honolulu: Na Kane O Ka Malo Press, 1990).
12. Carolyn Merchant, *Ecological Revolutions: Nature, Gender and Science in New England* (Chapel Hill: University of North Carolina Press, 1989), has a discussion of how the view of the landscape changed historically in New England, and she includes some discussion of religion, including folk religion.
13. Donald Worcester, *The Wealth of Nature: Environmental History and the Ecological Imagination* (Oxford, England: Oxford University Press, 1993).
14. Callicott has published other articles that acknowledge the possible ethical viability of Christianity, including a chapter on ethics in Gary Meffe and C. Ronald Carol, ed., *Principles of Conservation Biology* (Sunderland, Mass.: Sinauer, 1994).
15. Richard Botzler and Susan Armstrong, *Environmental Ethics: Divergence and Convergence* (Boston: McGraw Hill, 1998).
16. Eugene Hargrove, ed., *Religion and the Environmental Crisis* (Athens: University of Georgia Press, 1989).
17. Botzler and Armstrong, *Environmental Ethics*, 200.
18. Shawn Hinz and S. P. Bratton, "Ethical Responses to Fisheries Decline in the Republic of Ireland," in *Ethics and the Environment* (forthcoming). See also S. P. Bratton and Shawn Hinz, "Religion and Fisheries Decline in Traditional Irish Fishing Communities with a Comparison to the Pacific Northwest Region USA," *Ecotheology*, 8 (2000): 114–131.

16

Callicott's Last Stand

Lee Hester, Dennis McPherson, Annie Booth, and Jim Cheney[1]

The Land Ethic Discovers—and Claims—the New World

Assimilation redux. Early in *Earth's Insights: A Survey of Ecological Ethics from the Mediterranean Basin to the Australian Outback*, J. Baird Callicott presents in summary fashion his vision of the reconciliation of his land ethic—"proffered as a universal [and univocal] environmental ethic, with globally acceptable credentials"[2] —with the multiplicity of worldviews and values found in Indigenous cultures:

> One might therefore envision a single cross-cultural environmental ethic based on ecology and the new physics and expressed in the cognitive lingua franca of contemporary science. One might also envision the revival of a multiplicity of traditional cultural environmental ethics, resonant with such an international, scientifically grounded environmental ethic and helping to articulate it. Thus we may have one worldview and one associated environmental ethic corresponding to the contemporary reality that we inhabit one planet, that we are one species, and that our deepening environmental crisis is worldwide and common. And we may also have a plurality of revived traditional worldviews and associated environmental ethics corresponding to the historical reality that we are many peoples inhabiting many diverse bioregions apprehended through many and diverse cultural lenses. But this one and these many are not at odds. Each of the many worldviews and associated environmental ethics can be a facet of an emerging global environmental consciousness, expressed in the vernacular of a particular and local cultural tradition.[3]

In this essay we critically examine Callicott's attempted intellectual coup d'état of Indigenous thought and find it deeply problematic.

The Indian problem. Although Callicott purports to celebrate the rich diversity of traditional and Indigenous cultures and their diversity of worldviews and "associated environmental ethics," he has deep misgivings concerning the ethical pluralism that seems to be implied by such a celebration. "Untempered pluralism," he says, "courts conflict rather than mutual understanding and cooperation. The endpoint of untempered 'claims of otherness and an ethic of difference' . . . is the violent ethnic conflict now plaguing the world. . . . What is needed is a Rosetta stone of environmental philosophy to translate one Indigenous environmental ethic into another, if we are to avoid balkanizing environmental philosophy."[4]

This Rosetta stone, of course, is Callicott's "postmodern evolutionary–ecological environmental ethic"—his land ethic. He explicitly invokes the land ethic "as a standard for evaluating the environmental attitudes and values associated with traditional cultural worldviews. For example, . . . the woodland Indian concept of multispecies socioeconomic exchanges [is,] abstractly speaking, identical to the ecological concept of a biotic community, which is foundational to the Leopold land ethic." Callicott likens his "Rosetta stone" to ecosystem integration: "The myriads of species that make up biological diversity do not . . . exist in isolation from one another. Each is integrated into an ecosystem. How, analogously, might we unite the environmental ethics of the world's many cultures into a systemic whole?"[5]

In his earlier "The Case against Moral Pluralism," Callicott raises an additional objection to ethical pluralism based on the premise that to buy into a particular ethic is to buy into a particular worldview; ethics is grounded in metaphysics. For example, "Utilitarianism assumes a radical individualism or rank social atomism completely at odds with the relational sense of self that is consistent with a more fully informed evolutionary and ecological understanding of terrestrial and human nature," and "To adopt Kant's moral theory is to buy into a vintage Enlightenment philosophy of human nature in which Reason (with a capital *R*) constitutes the essence of 'man.' "[6] Indeed, a hallmark of Callicott's work in environmental ethics has been his concerted attempt to defend his land ethic by defending the "metaphysics of morals" that he believes grounds that ethic.

Callicott's project in *Earth's Insights*, then, is threefold: (1) to defend a particular ethical theory, (2) to articulate and defend (or, at least, throw in his lot with) the worldview—the "metaphysics of morals"—that grounds that theory, and (3) to somehow, consistent with this, celebrate a rich diversity of traditional and indigenous cultures and their diversity of worldviews and "environmental ethics."

The missionary ameliorates the yoke of conquest. No reader of this volume needs yet another summary of Callicott's version of Aldo Leopold's land ethic. It suffices for our purposes to note only that Callicott has consistently worked to align his land ethic with the latest developments in evolutionary and ecological biology. He has also taken pains to show the ways in which Leopold has either anticipated these developments or provided the conceptual resources for their accommodation.[7] Callicott distances his land ethic from specifically modern aspects of the scientific worldview and embraces what he calls a "postmodern evolutionary–ecological environmental ethic." It is this move that allows him to claim "Indigenous non-Western systems of ideas [as] cocreators of a new master narrative for the rainbow race of the global village."[8] In claiming that the postmodern scientific worldview is substantively non-Western in its epistemology and metaphysics, Callicott takes himself to have Indigenous non-Western credentials for his land ethic while at the same time holding that the land ethic is the touchstone by which Indigenous environmental ethics are either validated or rejected, the touchstone by which their relative "greenness" can be assessed.

The relevant postmodern aspects of the land ethic for our purposes are these:

1. With Darwin, "our species becomes a part of nature, a creature among creatures":

> Darwin's evolutionary epic not only undermines the Cartesian dogma that our fellow creatures are soulless automata. More profoundly, it undermines a cornerstone of the Cartesian modernist epistemology—viz., that we human beings are essentially disembodied passive observers of nature. From a Darwinian perspective, reason is a survival tool. . . . Darwin thus set the stage for the great epistemological upheaval of postmodern physics, in which the observer, as a physical being, invariably affects and is affected by the physical object of observation, and always observes from a finite and immersed, rather than from a synoptic and privileged, point of view.

2. Adding Elton's notion of a biotic community to Darwin's notion of evolution, Leopold's "vision of nature as an integrated community and an organic whole . . . points beyond the bifurcated Cartesian model of nature" toward the view that "any entity (oneself included), from an ecological point of view, is a node in a matrix of internal relations."[9]

Science, Callicott says, is "a legacy of the Western intellectual tradition." Concerning the difference between modern and postmodern science, he says that they "differ in the substantive worldview or paradigm each presents, [but] *not in the questions regarded as worth pursuing or the method used to pursue them.*" In virtue of its *substantive differences* from the modern scientific worldview, however, the "postmodern scientific worldview need not be received in non-Western cultures as yet another Western import or imposition."[10]

Callicott is aware that many might perceive his view as "an arrogant assertion of philosophical imperialism, a bid for intellectual hegemony." Conceding that "science is Western in provenance" and that, therefore, "one cannot pretend that a scientifically grounded environmental ethic is culture-neutral," Callicott argues against the charge of imperialism that "science is now practiced internationally" and that his land ethic may "make a claim to universality simply to the extent that its scientific foundations are universally endorsed." Western science and its associated technology has "inoculated all other cultures with Western attitudes and values."[11]

Furthermore, Callicott claims that the Western postmodern scientific worldview is "epistemologically privileged," because in virtue of its methodology "it is self-consciously self-critical" and "comprehend[s] more of human experience" than any other.[12] The epistemic privilege of this worldview confers similar epistemic privilege on the land ethic itself: "the evolutionary-ecological environmental ethic . . . may stake a coattail claim to epistemological privilege, since it is grounded in the epistemologically privileged reconstructive postmodern scientific worldview."[13]

Despite the epistemic privilege of both the Western postmodern scientific worldview and the land ethic, non-Western worldviews can, Callicott says, play an important role in the development of postmodern natural philosophy. He allows that these worldviews "can provide a multiplicity of critical perspectives, bringing to light 'areas that we may have failed to see as important' and deep assumptions that might otherwise go unnoticed." This "can" is strictly hypothetical in *Earth's Insights*, however. What is given most weight is the idea that "Indigenous worldviews around the globe can contribute a fund of *symbols*, *images*, *metaphors*, *similes*, *analogies*, *stories*, and *myths* to advance the process of articulating the new postmodern worldview," and can *in this way* "be cocreators of a new master narrative." Because "interesting similarities [exist] between the ideas of the new science and non-Western traditions of thought," and because the postmodern scientific worldview is radically different from its modern predecessor, Callicott can generously conclude that, although historically and developmentally the postmodern worldview is Western, substantively it is not.[14]

The heart(lessness) of the matter. We can certainly agree when Callicott insists that his is no "starry-eyed" attempt to transplant non-Western views, attitudes, and roles into the West. The asymmetry of the relation between Callicott's land ethic and Indigenous thought is precisely expressed in the following passage: "Traditional environmental ethics can be revived and, just as important, *validated* by their affinity with the most exciting new ideas in contemporary science, while the abstract and arcane concepts of nature, human nature, and the relationship between people and nature implied in ecology and the new physics can be *expressed* in the rich vocabulary of metaphor, simile, and analogy developed in the traditional sacred and philosophical literature of the world's diverse cultures."[15] Indigenous thought is validated by the land ethic. The task of Indigenous thought is to express the abstractions of the Land Ethic in the "rich vocabulary" of Indigenous cultures. Indigenous thought provides color commentary in the local vernacular, and Indigenous practices show how to put the land ethic into practice in local bioregions.

The remainder of this essay is a critical commentary from an Indigenous perspective on the project of *Earth's Insights.*

The Greasy Grass

Callicott's moral monism and Indigenous pluralism. Callicott's view that we need a single environmental ethic to function as a standard for evaluating environmental attitudes and values on the grounds that "untempered pluralism . . . courts conflict rather than mutual understanding and cooperation" flies in the face of the historical fact that hundreds of Indigenous cultures have existed side by side on this continent "forever" without the "violent ethnic conflict now plaguing the world" about which Callicott is so concerned. Certainly, conflicts occurred between various indigenous cultural groups, but they were of an entirely different nature and of a vastly different magnitude than the Balkan conflicts. The Balkan conflicts were the result, not of the advocacy and acceptance of cultural diversity and ethical pluralism, but—quite the contrary—of pitting monistic ideologies one against the other, each claiming to be in possession of the "One Truth." It is this condition that did not prevail on the North American continent prior to European contact. It was, in fact, the acceptance (and even celebration) of a rich cultural and ethical diversity, of the differences between cultures, that made it possible for hundreds of cultures to flourish side by side "forever."

Much of the conflict between "factions" has been fostered by larger powers seeking to gain control. (This has been particularly true with respect to the incorporation of Indigenous peoples into larger states.) Only after the conquest is the new political unit's breakup considered unthinkable by those in power. Conflict is caused, ultimately, by those in control of the larger political unit. The conqueror is unwilling to allow its conquest and power to slip away. Callicott's use of the "Balkans defense" of moral monism fits right in with this imperialist conquest model: we need, Callicott says, a single ethic because pluralism courts conflict. On the contrary, Indigenous peoples (and other groups not "inoculated" by ideologies claiming to be in possession of the one truth) seeking independence would be more than willing to achieve independence peaceably.[16]

To evaluate fairly ethical pluralism we must look closely at actual exemplifications of it, not at Western instances that exemplify pluralism only on the surface—that is, instances in which various ethnic groups, each with a different monistic conception of the One Truth find themselves no longer under the top-down state control that keeps these conflicts at bay. Differing monistic cultures are kept from one another's throats by imposed political order, whereas cultures that endorse pluralism exhibit an emergent ecological order. The moral monism that Callicott proposes is, in fact, closer to state-imposed political order than it is to the emergent ecological order he claims as an analogue to the unification of "the environmental ethics of the world's many cultures" by means of his privileged Rosetta stone.

Two interconnected aspects of Indigenous moral perception set its pluralism apart from the pseudo-pluralism that prevailed in balkanized Europe: the Indigenous notion of respect and the Indigenous use of narrative.

Imagine a deep practice of universal consideration[17] for all beings (including what Euro-Americans would call "things"), a consideration (perhaps a considerateness) that is not instituted as a moral principle or rule governing behavior, but is a dimension of one's very perception of the world. Such a conception is present in the notion of "respect" for all beings that is pervasive in Indigenous cultures. To Western ears, the term *respect* may have overtones of hierarchically structured relationships or it might have a Kantian flavor of obedience to moral law. But to Indigenous ears it signals a mode of presence in the world the central feature of which is awareness, an awareness that is simultaneously a mode of knowing—an epistemology—and what might be called a "protocol" or mode of "comportment," as Carol Geddes explained in response to a question concerning the meaning of the Tlingit notion of respect: "It does not have a very precise definition in translation—the way it is used in English. It is more like awareness. It is more like knowledge and that is a very important distinction, because it is not like a moral law, it is more like something that is just a part of your whole awareness. It is not something that is abstract at all."[18]

Next, consider the way in which Indigenous peoples tend to characterize concerns related to what Euro-Americans call "environmental ethics." Geddes, in explaining the difference between Western ethics grounded in scientific knowledge of the nonhuman world and Indigenous thought on "respectful" relationship with the nonhuman world, tells the following story:

> I would like to tell you a small story about a very great lady in the Yukon. Her name is Mrs. Annie Ned. This illustrates, in a way, what bothers me about thinking about environmental ethics in the way we do today. Mrs. Annie Ned . . . was taken to a scientific conference in Kluane National Park. . . . Well, Mrs. Ned listened to all of the scientists giving their ideas about physical events in the park: what sort of things happened in the park, the geography of the park, and various other subjects. Mrs. Ned just very quietly listened to this all day. Then as they were leaving that evening . . . Julie [Cruikshank] said to Mrs. Ned, "How did you like the conference?"
>
> Mrs. Ned said to her, "They tell different stories than we do."
>
> This is very, very important, in fact, it is profoundly important that we hear that. That is what they are, different stories. . . .
>
> This is . . . a source of confusion for me: that I would be able to understand environmental ethics within the context of narrative as the way First Nations people were taught about the environment. We would never have a subject called environmental ethics; it is simply part of the story. When you are a child you first hear the animal mother story, about how animal mother gave the animals to the world, and how people have to consider this as a gift from the animal mother; and if we do not take care of the animals, then the animal mother will start to take the animals back. We see that happening now. That is the context with which we understand environmental ethics, within that narrative, within the storytelling.
>
> On the other hand there is all the scientific knowledge that we also learned in school, the different stories as Mrs. Ned said, the new paradigm. Too many people say, well let's take lessons from First Nations people, let us find out some of their rules, and let us try and adopt some of those rules. Let us try to look at it the same way that First Nations people do. But it is not something that you can understand through rules. It has got to be through the kind of consciousness that growing up understanding the narratives can bring to you. That is where it is very, very difficult, because people have become so far removed from understanding these kinds of things in a narrative kind of way.[19]

For Indigenous cultures, stories convey an attitude of respect for all beings that is not primarily theoretical and not what Euro-Americans would call *ethical* or *moral*. Respect is presence in the world, a practical awareness that is simultaneously a mode of knowing and a protocol or comportment reflecting that practicality rather than an abstract and imposed morality. Deeply practical knowledge arising out of this respect, this very practical awareness, is perhaps best conveyed in, and understood as, *stories* of engaged experience in the world—it is best kept close to the bone, rooted in the experiences out of which knowledge arises.[20] And it is the storied nature of respectful relationship to all beings in Indigenous cultures—in Geddes's example, stories of gifts given and reciprocity—that makes possible a rich pluralism untainted by subtextual claims to be in possession of the One Truth about the way the world really is or about proper ethical or moral behavior.

The notion of respect and use of narrative that underwrite Indigenous acceptance of cultural and ethical pluralism contrast strongly with the coupling of a "metaphysics of morals"—the postmodern scientific worldview—and ethics at work in Callicott's land ethic.

Callicott holds that his land ethic provides us with a universal and univocal environmental ethic that is grounded in the postmodern scientific worldview. Rather than ground ethics in a "metaphysics of morals," Indigenous peoples give primacy to the grounding practice of respect. Respect, as a practice, is a particular way of being aware in the world. Respect is what we might call a "thick" epistemological notion. That is, it is a practical epistemology—an epistemology honed to a particular way of being in the world—and, in its engagement in the world, gives rise to practical values that in turn inform the epistemology. Implicitly, Indigenous epistemology rejects the idea that an epistemology is a method of coming to understand how the world really is. An epistemology is a mode of engagement in the world.

Respect for Indigenous peoples does not propose a theory about how one ought to behave. Respect is *practical engagement and presence in the world.* To survive in this world, and to live fully and well, one must be attentive. To impose agendas on the world (e.g., ethical, political, economic, scientific) is, to some extent, to cease to pay attention, it is to organize one's perception of the world according to the dictates of the mode of control (theoretical as well as physical) one wishes to impose. Out of practical engagement and presence arise techniques and technologies for survival, values and disvalues, protocols for engaging the world, and, importantly, stories telling of this engagement, what one has witnessed and learned. In this way one does really become, in Leopold's words, "plain member and citizen" of the land community.[21] One may well ask if the project of exporting the land ethic globally is the work of one who is a plain member and citizen of the land community. It seems that we have in the Indigenous practice of respect and Callicott's project of exporting the land ethic a clear example of the difference between emergent, ecological order and imposed, ideological order.[22]

Callicott's "metaphysics of morals." Callicott's claim that the scientific worldview is privileged in part because it is self-consciously self-critical is at best puzzling. To hold that the scientific worldview is both epistemically privileged and self-consciously self-critical is on the face of it to subscribe to mutually exclusive views. A rigorously self-conscious, self-critical method, although certainly desirable, would, one would think, include a refusal to claim epistemic privilege. To hold such apparently mutually exclusive views, one must be placing very impressive limits on one's self-criticism. There must be some premise or element of one's method of generating a worldview that is held to be somehow above suspicion to merit the appellation of "epistemic privilege." Callicott, in other words, does not think of himself as simply telling another story; he is explicitly attempting to construct a new, epistemologically privileged master narrative—as he himself puts it. "Master narrative" is not far off, actually, but calling it a *narrative* suggests that he believes it is just another story, which he does not.

The claim that Western science is self-critical and therefore privileged also seems to imply that other traditions, because they are not privileged in this way, are not self-critical (or not as self-critical). This is a dubious conclusion at best. Arguably, Indigenous peoples are far more self-critical than Euro-Americans. They engage in self-criticism communally, over the long haul, and expect that dissension will always occur because there really are different stories. We might call this "unself-conscious examined interaction." It comes quite naturally; it is not forced, it is not artificial, and it does not have a rigid protocol—all to the good: it gives a flexibility of mind that Western science does not have. The way Indigenous peoples engage in self-criticism, along with the results of that criticism, just appear uncritical to the outside observer. Indigenous people swap stories and come to understand each other better and understand the various ways of examining the subject matter of the stories better. They are not uncritical. It is just another form of criticism and arguably should stand Indigenous people in better epistemic stead than any form of privilege-assuming self-conscious self-criticism if only because it does not and never would claim primacy and privilege. Indigenous peoples know that their stories differ from those of most Western science. The stories of Western science are good for some things and Indigenous stories are good too. But to suggest that Indigenous people are not self-critical in the way that supposedly stands Western science in such good stead comes off as arrogant and misguided. The claim to privilege on behalf of the science and the ethic it "grounds" can only deepen the problem, not help solve it.

Callicott's claim that modern and postmodern science differ in substantive worldview but not in questions raised and methods of addressing them is also problematic, and calls into question Callicott's claim to have produced a genuinely postmodern evolutionary-ecological environmental ethic. Surely the questions and methods of science are too intimately bound up with its substantive views to be so easily disentangled, except at a highly abstract level. The understanding Western science has of nature is partly a function of the questions, values, purposes, and methods of observation and theory construction that it brings to the world. Historically, the methods and theoretical concepts of science were in significant measure (although by no means exclusively) forged as tools for understanding nature instrumentally because so ordering science's understanding of nature that it could be effectively harnessed for exploitation. These methods and concepts constitute a method of knowing that invites (although it does not require) the manipulation of nature far beyond biological and cultural needs, far beyond the reach of available knowledge, and not only leads to destructive practices, but also entices environmentalists into managerial and stewardship models of right relationship to nature. Conservation biology—to which Callicott conceives his land ethic to be a contribution—has recently been criticized by Jack Turner on just this score:

> In the face of biodiversity loss . . . conservation biology demands that we do something now, in the only way that counts as doing something—more money, more research, more technology, more information, more acreage. . . . In short, the prescription for the malady is even more control.
>
> This mirrors the mode of crisis response familiar from Michel Foucault's studies of insanity, crime, and disease. Like psychiatry, criminology, and clinical medicine, conservation biology is a theoretical discipline that seeks control in pursuit of a morally pure mission: to end a crisis. . . .
>
> Unfortunately, instead of striking at causes, modern theoretical disciplines such as conservation biology strive to control symptoms. Their controls are directed at the Other, not at our own social pathologies. . . .[23]

The lesson, here, is that Callicott's claim that Western science "comprehend[s] more of human experience" than any other is misleading. If the "knowledge" embedded in a culture results in practices of "controlling" nature that leave it ever more out of control, can that culture be said to "comprehend more of human experience" than any other in any sense relevant to the matters at hand? Culturally unmoored knowledge is not "comprehensive" in the relevant sense.[24] To think that the "metaphysics of morals" invoked by the land ethic could provide those moorings is not to think ecologically.

In contrast to Callicott's procedure of beginning with a scientific account of the world—one historically founded on the value of control over nature—from which an ethic is derived, Indigenous peoples begin with practical respect that structures an epistemological orientation to the world not shaped a priori by values of control and manipulation. This contrast illustrates a fundamental difference between Western environmentalism and Indigenous attitudes toward nature. Western environmentalism is, in one way or another, control oriented, as Indigenous relationships to nature are not. Whether that control be exercised in seemingly benign ways—as in conservation biology and Christian stewardship—or in obviously destructive ways, it is fundamentally different from the ethos of Indigenous peoples.

John Rodman asks whether environmental ethicists are like missionaries, ameliorating the yoke of conquest, "or whether they criticize the process of conquest in the interest of liberation." In speaking to his own question, he offers another way of understanding the hubris we have been describing. Rodman views Natural Law theory as stemming from "a desperate nostalgia for a state of nature . . . in which the prohibitions now prescribed by God, Conscience, and the State would have operated 'naturally' (i.e., from inside the organism, as a matter of course)." From this perspective, he says, moral law in contemporary moral theory represents a kind of externalization of this natural "law" in which the domination of external nature is replicated in the ethical domination of the "natural" self. To invoke such a "law" on behalf of the liberation of nature is to be of two minds: it is to work for the liberation of nature by using an ethic modeled on the externally imposed order that is at the heart of the problem of the domination of nature. Rodman argues that "from the standpoint of an ecology of humanity, it is curious how little appreciation there has been of the limitations of the moral/legal stage of consciousness," and urges that we "become less fixated" at this stage of consciousness.[25]

Callicott's land ethic also seems to work both sides of this split. It purports to be an "evolutionary-ecological environmental ethic," suggesting that it is in large part a description of an evolutionarily emergent ethic, one which is simply informed, and thereby expanded, by contemporary ecological science. On the other hand, Callicott, in his worry about pluralism, argues strongly for a privileged place for the land ethic as the Rosetta stone by which the greenness of each Indigenous environmental ethic is validated (or not). He is clearly uncomfortable with a truly emergent environmental ethic. This doubleness (or ambiguity) in Callicott's work is epitomized in his claim that the uniting of "the environmental ethics of the world's many cultures into a systemic whole" by means of the land ethic—clearly an instance of *imposed* moral order because the relative greenness of the various Indigenous ethics is determined by their congruence with the land ethic—is analogous to the integration of species into an ecosystem—clearly a case of *emergent* order.[26]

Callicott's disclaimers to the contrary notwithstanding, his use of his land ethic as a Rosetta stone of environmental ethics operates clearly within a Western framework of control over nature (however seemingly benign) and represents yet another cooption of Indigenous thought. Callicott does not adopt Indigenous views; rather, he adapts distorted versions of those views to his land ethic and sells them as real Indigenous views. At the same time that Callicott touts the land ethic as the Rosetta stone of Indigenous ethics, he seeks the mantle of authority of Indigenous cultures by proclaiming that the substantive views of the postmodern scientific worldview are distinctly non-Western.

As an example of this cooption, witness Callicott's account of the Ojibwa land ethic:

> The Ojibwa land ethic . . . rests on the same general concept as Leopold's. Human beings, plants, and animals, if not soils and waters, are members of a single, tightly integrated economy of nature, or biotic community. Human beings are not properly "conquerors of the land community"; neither ought we to be stewards of it. Rather, we should assume the role, as Leopold would have it, merely of "plain members and citizens" of the land community. In the Ojibwa land ethic, as in Leopold's, human beings ought principally to respect their fellow members of the biotic community. . . . When the mythic and scientific detail is stripped away, an identical abstract structure—an essentially social structure—is revealed as the core conceptual model of both the totemic natural community of the Ojibwa and the biologists' economy of nature. In form, the Ojibwa land ethic and the land ethic of Aldo Leopold are identical.[27]

Yet Callicott's land ethic, not the Ojibwa worldview or the Ojibwa notion of respect (which Callicott transforms into a moral "ought"), is named the Rosetta stone. Why is this? Because presumably the land ethic is based on a knowledge system that is privileged by its self-consciously self-critical attitude and its comprehensiveness. Oddly, however, the scientific worldview that is the basis for the land ethic is the historical product of a methodology underwritten by the value of control over nature, which is precisely the problem, or the root of the problem, the land ethic is brought on stage to deal with. Indigenous "knowledge," on the other hand, is *not* the basis of an Indigenous "ethic." Rather, Indigenous respect is closer to the bone. It is a mode of awareness, an epistemology, honed to practical engagement in the world.

Indigenous knowledge and wisdom are based on and grow out of respect. Indigenous knowledge, that is to say, is fundamentally shaped by practical respect in relationship to all that exists. The ontologies of Indigenous worldviews are a kind of residue of Indigenous practices of respect and the modes of attaining knowledge associated with those practices. That is, ontology is a kind of picture, or metaphor, of epistemological orientation to the world and the practices of respect on which those epistemological practices are founded. For Callicott's land ethic, ethical respect for the land comes *last*: First there are the parental, filial, and other social sentiments or instincts nurtured into existence through evolutionary time; these in turn are informed by the scientific worldview, which broadens the scope of moral sentiment to encompass our true community (the nature of which is revealed by evolutionary and ecological biology). For Indigenous peoples, on the other hand, respect for all beings is primary; it shapes epistemology; and, in shaping epistemology, it shapes Indigenous worldviews. The land ethic and Indigenous respect may cross paths in a sense, but they are fundamentally different orientations to the world.

Callicott assimilates Indigenous practical respect to his land ethic. That is, he adapts that respect and transmutes it into a land ethic—he does not adopt it. Callicott attempts to bridge a gap in the only way a Western philosopher knows how—by providing an overarching theoretical construct. But there is no need to bridge a gap if only we all realized that a gap does not exist. The gap arises precisely because some people assume that there is one truth out there. There is no gap because we all live in the world and the world is made up of all our stories. We grow closer and appreciate each other by listening to the stories and allowing them to make their own sense. Indigenous people do not sense the gap Callicott is trying to bridge, but they are well aware of the widening chasm brought about by the apparent lack of respect and understanding caused by Callicott's attempt at bridging a nonexistent gap.[28]

Respect, epistemology, and ceremonial worlds. The world just is. Any theory is distortion. A story, on the other hand, makes sense or not in its own right within the world, and only has those implications beyond itself that are brought in by listeners in their own interaction with the world.

Epistemologies, we have said, shape ontologies.[29] We have been describing the very different epistemologies of Indigenous peoples and Euro-Americans. The close-to-the-bone practicality of Indigenous epistemology embedded in practical respect in engagement in the world lends itself to the practice of embedding knowledge in stories—stories of engagement with respect—leading to knowledge and wisdom. (Differing stories, we should note, do not generate genocidal war; differing conceptions of the One Truth do.)

What this suggests is that because worldviews are fundamentally shaped by epistemologies, we should not understand the accounts of the world that emerge as accounts of the way the world really is but as what we call *ceremonial worlds.* The various practices of cultures are enactments of ceremonial worlds within which they live. These worlds are not mere fantasies; they are more or less adequate to a culture's purposes in enacting them and can be evaluated and modified as such.

We can best get at this notion of a ceremonial world by turning to J. L. Austin's focus on the performative functions of language.[30] Consider the following example. Sam Gill reports (as have many others) that nonliterate people are often highly critical of writing. He says of this, however, that he does "not believe that it is actually writing that is at the core of their criticism. The concern is with certain dimensions of behavior and modes of thought that writing tends to facilitate and encourage. And these dimensions are linked to the critical, semantical, encoding aspects of language. . . . We interpret texts to discern systems of thought and belief, propositional or historical contents, messages communicated. Put more generally, we seek the information in the text. We tend to emphasize code at the expense of behavior, message at the expense of the performance and usage contexts."[31] The written word conspires with visual metaphors of knowledge to turn the world into a passive object for human knowledge and to focus our attention on language as a sign system primarily designed to encode beliefs.

In a number of articles, Sam Gill has attempted to reinstate the fundamental nature of the performative function of language, using Navajo prayer as a case study. Invariably, when he asks Navajo elders what prayers mean, they tell him "not what messages prayers carry, but what prayers *do.*" Furthermore, "the person of knowledge in Navajo tradition holds that [theology, philosophy, and doctrine] are ordinarily to be discouraged. Such concerns are commonly understood by Navajos as evidence that one totally misunderstands the nature of Navajo religious traditions."[32]

Generalizing from his analysis of prayer acts to religious practice generally, Gill asserts that "the importance of religion as it is practiced by the great body of religious persons for whom religion is a way of life [is] a way of creating, discovering, and communicating worlds of meaning largely through ordinary and common actions and behavior."[33]

We would like to generalize even further, arguing that the performative dimension of language be understood as fundamental—not just in obviously religious settings, but generally. There are alternative ways of intelligently engaging the world, alternatives to construing one's thinking in terms of belief.[34] We do things with words. Foremost among these performative functions is the creation of the ceremonial worlds within which we live. Other performative functions of language are possible only within these ceremonial worlds—promise making, for instance, is possible only within an accepted set of social conventions, as is the progress achieved within science.

Take, for example, Diamond Jenness's report of a Carrier Indian of the Bulkley River who says, "The white man writes everything down in a book so that it might not be forgotten; but our ancestors married the animals, learned their ways, and passed on the knowledge from one generation to another."[35] We suggest that this be understood as saying that Carrier Indians passed down the means of creating, or recreating, the worlds, the ceremonial worlds, within which the ancestors lived—the stories, the ceremonies, the rituals, the daily practices. They passed down modes of action, which when written down come to be understood as information. Euro-Americans want to know what beliefs are encoded in the utterances of Indigenous peoples, they want to treat these utterances as mirrors of Indigenous worlds. This may, however, be asking the wrong question. In fact, these utterances function primarily to produce these worlds. Euro-Americans are concerned with ontology, correct descriptions of Indigenous worlds. Indigenous peoples, on the other hand, are concerned with right relationship to those beings that populate their worlds; they are concerned with respect.

N. Scott Momaday in justly famous words says: "It seems to me that in a sense we are all made of words; that our most essential being consists in language. It is the element in which we think and dream and act, in which we live our daily lives. There is no way in which we can exist apart from the morality of a verbal dimension."[36] Momaday is speaking not so much of sets of beliefs by which people constitute themselves, but more fundamentally of performance, enactment, the bringing into being of cultural and personal identity by means of action and practice, primary verbal. It is the difference between, for example, the sacred as object of knowledge or belief (and, derivatively, of acts of faith and adoration) and sacramental practice—a matter of comportment, which brings into being a world, a ceremonial world, around it.

Ceremonial worlds are not fantasy worlds. We do, of course, experience the world, and given their epistemological practice of respect, no people experience the world more fully, perhaps, than Indigenous peoples. Experience is taken up into ceremonial worlds. It is part of the self-correcting feedback loop that makes it possible for the day-to-day activities of food gathering, child rearing, shelter building, and so on to take place, to succeed, not only on the terms set by the world, but within the context of richly textured ceremonial worlds. In such worlds, as Paul Shepard has observed, "everyday life [is] inextricable from spiritual significance and encounter."[37]

Moving away from epistemologies based on the values of domination and control toward an epistemology based on the Indigenous notion of respect, we move closer to an older sense of the word "knowledge": knowledge as a kind of intimacy. Contrary to the emphasis some place on the constructed nature of the worlds we live in, reflected in the catch phrase "It's words all the way down," we suggest a very different emphasis: It's *world* all the way *up*—even into the language of the ceremonial worlds we have been discussing.

Language is rooted in being, rooted in the world as are we who speak forth that world in our language. And our language is a mode of interaction with, and hence a mode of knowing, that world. Knowing can take shape as a form of domination and control—it can break faith with language rooted in and expressive of the world. It can also take shape as a way of "stepping in tune with being."[38]

The postmodern notion that "it's words all the way down" is a vestige of modern dualism in which the knowing mind is cut off from (or, rather, cuts itself off from[39]) the world it inhabits and can never bridge its imposed epistemological gap between self and world. Rather than attempting to *model* the world to control it (the epistemological orientation of an estranged, alienated consciousness), Indigenous ceremonial worlds are fundamentally expressions of the world (an epistemological orientation that reflects awareness *in* the world rather than awareness *of* the world).

Jeannette Armstrong emphasizes the ecological dimension of Momaday's thought that "we are all made of words": "The Okanagan word for 'our place on the land' and 'our language' is the same. The Okanagan language is thought of as the 'language of the land.' This means that the land has taught us our language. The way we survive is to speak the language that the land offered us as its teachings. . . . We also refer to the land and our bodies with the same root syllable. . . . We are our land/place."[40] Momaday's notion that we are "all made of words" is not the postmodern "it's words all the way down" but, rather, an articulation of the notion of ceremonial worlds, language rooted in the earth and expressive of it.

If language is performative and if we have our being and identity fundamentally within ceremonial worlds, then the coherence we should be listening for is not merely the logical coherence of one sentence with another, one belief with another, but something more like the harmonic coherence of one note with another. Practices, including linguistic practices, create ceremonial songs of the world, worlds of meaning, within ecological niches. Within these ceremonial songs of the world language is a mode of interaction with the world.

Ceremonial practice defines the worlds in which we live and work. The ontology of a culture's world is a residue from its practice in engaging the world and the modes of attaining knowledge associated with that practice. This residue is highly prized and receives intense scrutiny in Western literate culture, but the epistemology embedded in our practice is the fundamental dimension of our relationship to, and understanding of, the world. Ontology is a kind of picture, or metaphor, of epistemic practice. It is in this context that we can understand both the reality and the ceremonial nature of practices such as "running with the sun" to help it on its journey.

The comprehensiveness of Indigenous worldviews is a function of the fact that they are ceremonial worlds designed to encompass Indigenous life in the world in which it finds itself. The kind of overall coherence for which ceremonial worlds strive is a mosaic of language (in the broadest sense) that serves many purposes at once. In the life of a community it must articulate a sense of those processes that bind the community together and to the land; and it must do this in a language that functions effectively to call forth appropriate responses. Above all, in such a world "natural things are not only themselves but a speaking."[41]

As an example, consider the difference between the "search for one's roots" that is fashionable in the West these days and the relationship between Indigenous identity and place. Indigenous practice shows that the need for roots can take other shapes than that of a search for redemption in the mode of a search for the "Truth" of one's origin and identity. Choctaw legend maintains that the Choctaw people migrated long ago to Mississippi carrying the bones of their ancestors with them. When they reached Mississippi they are said to have built the mound of *Nanih Waiyah* to house these bones. Yet, Choctaw legends also state that *Nanih Waiyah* is the great "Productive Mound" from which all people emerged. From the point of view of the One (literal) Truth this seems contradictory—the new burial mound could not be the Choctaw place of origin, emergence. From the point of view of Choctaw practice, however, a different meaning of emergence and origins arises.

When forcibly removed to Oklahoma, the first Choctaw capitol in Oklahoma was called *Nanih Waiyah* and even today there is a Lake Nanih Waiyah near the Choctaw Nation of Oklahoma council building in Tuskahoma. Although they were forced to leave their ancestors behind and although many of their loved ones died on the Trail of Tears, the Choctaw people of Oklahoma are rooted in their new land. Choctaw practice has the consequence that as a people the Choctaw are always at home on this earth, never detached from tradition and tribal history—these are always present in the tangible form of the emergence mound. Practice and the social meaning embedded in that practice are central. Choctaw knowledge grows from this practice.

Understanding the worlds we construct as performances, enactments of worlds for various purposes, makes it easier to understand the Indigenous view that the fundamental issues are not ones of "Truth," but of respect and well-being—the telling of "stories" that enable a culture to live in balance. The world's Indigenous cultures represent a vast array of stories within which cultures live well and achieve balance.

Indigenous ecological ethical reflection. Having said that Indigenous peoples do not have ethics in Callicott's sense of the term and having treated the most likely candidate for a key ethical concept (namely, respect for all beings) as nonideological, practical engagement, and presence in the world (that is, as a *practical epistemology* that gives birth to *practical* values) the reader might well wonder whether ethics exist at all in Indigenous worlds.

What we have stressed in this essay is the kind of ethic that does not appear in Indigenous thought, namely, ethical theory, at least theory of the kind Callicott insists on. His reason for insisting on a universal and univocal ethical theory is to provide "a means, in principle, to assign priorities and relative weights and thus to resolve . . . conflicts in a systematic way" because "when competing moral claims cannot be articulated in the same terms, they cannot be decisively compared and resolved. Ethical eclecticism leads, it would seem inevitably, to moral incommensurability in hard cases. So we are compelled to go back to the theoretical drawing board." The alternative to resolution by means of theory, according to Callicott, is, at best, a "mere coalition of convenience."[42] For Callicott, as we have seen, such an ethic is to be grounded in an acceptable "metaphysics of morals" reflecting state-of-the-art scientific theory.

We have argued that this approach, along with Callicott's desire to validate Indigenous environmental ethics to the extent that they conform to his land ethic, is decidedly nonecological, despite his claim that the land ethic serves to "unite the environmental ethics of the world's many cultures into a systemic whole," as, analogously, "the myriads of species that make up biological diversity . . . [are] integrated into . . . ecosystem[s]."[43]

Indigenous ethical reflection, on the contrary, is "ecological." Its roots are in a very practical attentiveness and respect regarding the world in which one is engaged. This respect is not moral respect. Rather, it is a circumspection, or practical humility, one is well advised to exhibit in relation to a world full of surprises, unexpected powers, dangers, and gifts. As a practical epistemology, respect seems to be inherently holistic. That is, practical awareness and engagement focuses on relationship or interrelationship; it looks for interconnectedness between things. Holism is the most practical epistemological assumption. By assuming that interconnectedness is pervasive, such an epistemology maximally heightens practical perception. Perception that is relationship focused will monitor events with better practical effect than isolating, object-focused perception. The resulting practice will most likely be more attentive to context and will be maximally sensitive to ripple effects.[44]

The fruit of respect in an Indigenous culture is a rich ceremonial world that weaves the threads of knowledge that respect yields into a tapestry that functionally relates these threads one to another, to the practical values that knowledge suggests, and to the virtues required to engage fully in the world in a practical and fulfilling way with harmony and balance. There is little to wonder at that this tapestry reflects Indigenous epistemology, that interconnectedness is the central feature of Indigenous ceremonial worlds, and that the central value is harmony—at all levels: within and between tribes as well as between tribes and the rest of nature. The harmony that is given ritual and ceremonial status in Indigenous religion circles back and underwrites perception itself and the practical concerns of survival and well-being that are at the heart of Indigenous epistemology.

This tapestry, which is a culture's ceremonial world, must grow organically, ecologically, with attention given to the interconnections between the threads. The order this tapestry exhibits, its particular cultural beauty, is emergent and closely tied to the specific place on earth in which that particular culture is engaged. The order is not abstract, but deeply expressive of a particular place and a particular culture's engagement with that place.

This flowering of ceremonial worlds from practical respect in some measure transmutes respect itself. This respect now values the world in which it is engaged, not abstractly, but in very particular ways embedded in practice. A culture's ceremonial world is that culture's mode of engagement with the world. Effective—rather than imposed, totalitarian—moral authority exists only when it is emergent, embedded ecologically in a culture in the way we have described.

How, we may ask, could ethical theory have the kind of authority that emerges ecologically in this way?

Going Home: A Smudge

Emory Sekaquaptewa recounts a science project in ecology for Hopi children in a school operated under the Bureau of Indian Affairs:

> It called for the children to bring various living things (insects, animals, etc.) into the classroom, putting them in a cage and accepting responsibility for their care while they watched what happened. It seems that the Hopi children were not interested in taking care of the animals while they studied them. It didn't matter to them whether the animals died or survived. The teachers became very concerned about how to teach the Hopi child about ecology if he didn't show any interest. He had no feeling for the animal.

> We never resolved the problem as far as the teachers are concerned. But I would like to make the statement that perhaps ecology, or learning how to live with the environment, is not a matter of taking sides with the environment, is not a matter of taking sides with one or some other living things; rather it is acceptance of the fact that if a certain living thing cannot survive on its own, that is a fact. Must we intervene with our special powers as human beings to control and bring about ways to help this poor thing to survive outside its natural ability to survive? . . . Learning to live with the environment is not a matter of taking sides, but of accepting facts.[45]

(We might add that Hopi children would not have thought to bring these living things into the classroom and put them in cages in the first place.) The critical thing to note, here, is that the Euro-American, whether one who exploits and ravages nature or one who "takes care" of nature, is governed by a model of human control of nature, whereas the Indigenous person is not inclined in either of these directions.

Respect implies a lack of the arrogant pride or presumption (hubris) that brings one to take charge of the world, whether for ill or apparent good. This hubris, in turn, implies a lack of "at homeness" in the world tied to various forms of dualism that are at the heart of the Western worldview. Paul Shepard, who has examined these dualistic splits as well as any Euro-American, says: "In our society those who would choose the owl [over the oil well] are not more mature. . . . Fear and hatred of the organic on the one hand, the desire to merge with it on the other; the impulse to control and subordinate on the one hand, to worship the nonhuman on the other; overdifferentiation on the one hand, fears of separation on the other: all are two sides of a coin. In the shape given to a civilization by totemically inspired, technologically sophisticated, small-group, epigenetically fulfilled adults, the necessity to choose would never arise."[46]

Euro-Americans can as easily imagine "taking responsibility for the fate of the earth" as they can acknowledge being responsible for environmental destruction. But responsibility is tied to the notion of authority, which in turn implies both power and authorship. We become responsible, or take on responsibility, by having or acquiring the power and authority to do so. Indigenous peoples do not think of themselves as having either the power or the authority to become responsible for nonhuman nature. Indigenous peoples have, certainly, played a role in shaping their biotic homes, but for Indigenous peoples it is the sheerest pride to think that one is, or could be, responsible for the biotic community, that one has that sort of power and authority. If one does something, is the author of an action, then, of course, one must or should take responsibility for the fruits of that action, but this is a far cry from the hubris involved in taking responsibility for the biotic community.

Indigenous peoples understand that they play a part, but not an essential part, in maintaining balance and integrity in the world. Humans might enhance the world with joy and appreciation of the natural world with the creation of beauty and the telling of stories, but such limited powers as they have are given to them as gifts from the nonhuman world. For Indigenous peoples, knowledge, which is power, is not forged and maintained by unifying and controlling consciousness.

What we miss in the evolutionary-ecological scientific worldview that grounds Callicott's land ethic is an account of a presence in the world adequate to the task the land ethic sets itself. Callicott sees that the postmodern scientific worldview and its associated land ethic resonate with at least some Indigenous worldviews and the practical respect associated with them. Assuming for the moment that one of them can be said in some sense to validate the other (as Callicott holds), we may ask whether it is really the land ethic that is the touchstone that validates Indigenous views or whether it might not be, if anything, the other way around.

In the one case we have a worldview that—going back to its Baconian roots—developed in part from values of domination and control of the natural world in a social setting in which individualism was highly prized. As it happened, this worldview developed in such a way that many Euro-Americans now understand themselves as members of one species among others, brought into existence by the forces of evolution and kin to all living creatures, part of an ecological web of interdependence and reciprocity. Although this understanding may help mitigate the destructive uses to which the scientific worldview has been put, and although it may for some evoke a deep, inclusive, and even sacred sense of kinship with the world, one must keep in mind that that worldview (even in its postmodern dress) is still a powerful engine of destruction. Looking at the storyline of Western science, it is as though science inadvertently (and quite contrary to its intentions at the outset) stumbled on a conception of self and world that bears some resemblance to Indigenous worldviews—a conception that seems to call to task science's own initial intentions. The scientific worldview is close to the Western heart and the land ethic is validated to the extent that it is grounded in that worldview. Euro-Americans bootstrap themselves up to a better ethical place in this way, although, as we have seen, along with this comes a strong tendency to adopt a managerial stance with respect to the environment, whether that manifest itself in "Save the Earth" bumper stickers, conservation biology, or a desire to "manage" the world's Indigenous cultures on behalf of the land ethic.

The practical respect cultivated by Indigenous peoples has served the world far better than Callicott's land ethic could ever hope to. Indigenous peoples, when trained in Western science, can, of course, see that evolutionary and ecological theory resonate in some fashion with their worldviews, but they are perplexed that something so immediate as respect is treated (and transmuted) in such an abstract way in environmental ethics that it must route itself through evolutionary and ecological science and be first grounded in that theorizing and then imposed as a moral rule on behavior.

From this perspective, Indigenous worldviews validate the emerging postmodern scientific worldview of which Callicott speaks. Or, more properly, Indigenous worldviews can point the direction for the development of a Western practical respect that has become at least a conceptual possibility for the West with the development of evolutionary biology and ecology. Indigenous thought can provide the West with a model of moral pluralism in its understanding of worldviews, not as so many pretenders to the throne of the One Truth, but as multiple stories of the world built on respect.

At a recent conference on ecological resistance movements in Madison, Wisconsin (following Callicott's address on the themes of *Earth's Insights*), Walter Bresette spoke—seriously, but at the same time a bit tongue in cheek—about a conference in Estes Park, for which he had been asked, as a Native American, to give a keynote address. He observed that the people at that conference (all white) cared more about the earth than any people he had ever met. And, he said, they knew more about Native Americans than he did. He confided to a friend with some concern: "I think I'm at the wrong conference. There's nothing I can say here that I can bring anything new to." His friend responded that they just had a different problem than they thought. The education has been accomplished, but one thing was missing. "If they are not going back home, those who get it, how can we expect Exxon or any corporation to go back home? So we need to adopt them and make this their home, so that they will then accept it in an entirely different way than in the past." After telling this story, Bresette adopted everyone at the conference, saying: "When I do that, the earth becomes your mother. You have no choice. Now if you want to pimp her, if you want to sell her, if you want to treat her the way you are treating her, it becomes your choice—after tonight. You have no excuses. . . . You ain't going to become a Chippewa, but . . . you better become a Native American!"[47]

The West can have its own story; it can become native in its own way. But it ought not engage in the hubris that leads it to say that its thought is the touchstone, the very Rosetta stone, of environmental ethics. Callicott's land ethic only just barely begins to glimpse the balanced worlds of Indigenous peoples. The World Trade Organization and North American Free Trade Agreement may require that Callicott be permitted to export his land ethic to

Indigenous cultures (as he wants to do), but those cultures, we hope, will not be required to buy it (because they will not).[48]

Ayaangwaamizin.

The authors dedicate this article to the memory of Anishinabe Native American rights and environmental activist Walter Bresette.

Notes

This essay incorporates material previously published in the following articles: Lee Hester, Dennis McPherson, Annie Booth, and Jim Cheney, "Indigenous Worlds and Callicott's Land Ethic" *Environmental Ethics* 22 (2000): 273–290; Jim Cheney and Anthony Weston, "Environmental Ethics as Environmental Etiquette: Toward an Ethics-Based Epistemology," *Environmental Ethics* 21 (1999): 115–134; Jim Cheney, "The Journey Home," in *An Invitation to Environmental Ethics*, ed. Anthony Weston (New York: Oxford University Press, 1999), 141–167.

1. This essay was shaped by Western problematics in environmental ethics. Indigenous coauthors provided some Indigenous content, but the problem is Western and in most ways so is the voice of this essay. Although this essay attempts to bridge some of the gap between these two philosophical worlds, to understand the Indigenous perspective one must be willing to engage Indigenous people on their own ground.
2. J. Baird Callicott, *Earth's Insights: A Survey of Ecological Ethics from the Mediterranean Basin to the Australian Outback* (Berkeley: University of California Press, 1994), 186 (emphasis added).
3. Callicott, *Earth's Insights*, 12.
4. Callicott, *Earth's Insights*, 186. This passage might leave some readers wondering whether Callicott's concern is with ethical pluralism or with cultural pluralism. Other passages, and Callicott's work in general, make clear that his concern is with ethical or moral pluralism. The ethnic conflict (or the dimension of ethnic conflict) that concerns Callicott is that which stems from differences in ethical belief. Although this essay traces what we take to be a widespread thread of agreement in practice at the very heart of Indigenous ethical practices, we also acknowledge the ethical pluralism that is a dimension of Indigenous cultural pluralism.
5. Callicott, *Earth's Insights*, 189, 186.
6. J. Baird Callicott, "The Case against Moral Pluralism," *Environmental Ethics* 12 (1990): 114–115
7. See for example, J. Baird Callicott, "Do Deconstructive Ecology and Sociobiology Undermine Leopold's Land Ethic?" *Environmental Ethics* 18 (1996): 353–372.
8. Callicott, *Earth's Insights*, 192.
9. Callicott, *Earth's Insights*, 199, 200, 206.
10. Callicott, *Earth's Insights*, 193 (emphasis added), 197.
11. Callicott, *Earth's Insights*, 189. This argument does not defeat the charge of imperialism. Indeed, many would argue that such "inoculation" is far more effective than military or biological assault as a strategy of imperialism. Cultural vaccination versus spreading the smallpox virus with blankets: "It's for their own good," versus "The only good Indian is a dead Indian."

12. Callicott, *Earth's Insights*, 190–191.
13. Callicott, *Earth's Insights*, 190–191.
14. Callicott, *Earth's Insights*, 190–191 (emphasis added). The phrase "areas that we may have failed to see as important" is quoted from Val Plumwood.
15. Callicott, *Earth's Insights*, 12 (emphasis added).
16. A universalist or monolithic view such as Callicott's is unwise in any case because it easily ignores or minimizes relevant regional, historical, and other differences.
17. On "universal consideration" in environmental ethics, see Thomas H. Birch, "Moral Considerability and Universal Consideration," *Environmental Ethics* 15 (1993): 313–332.
18. Carol Geddes, panel discussion by Yukon First Nations people on the topic of "What is a good way to teach children and young adults to respect the land?" Transcript in Bob Jickling, ed., *A Colloquium on Environment, Ethics, and Education* (Whitehorse, Yukon Territory: Yukon College, 1996), 46.
19. Geddes, "What is a good way," 32–33.
20. For a compelling argument for the view that the origins of pragmatism are to be found, at least in part, in the philosophical perspective of Indigenous peoples of North America, see Scott L. Pratt, "Native American Thought and the Origins of Pragmatism," *Ayaangwaamizin: The International Journal of Indigenous Philosophy* 1 (1997): 55–80.
21. One reader of this essay commented: "The authors seem to imply that Callicott has misappropriated Leopold as well as Indigenous views. Arguably, Callicott imposes on Leopold a reading of him as 'environmental philosopher' when, in fact (especially in his latest writings), he was talking about 'respect as presence.' I would be sympathetic to this view." We would indeed be pleased if the referee's suggestion about Leopold is correct.
22. Having ethical, political, economic, and scientific agendas is not always and everywhere wrong—these, too, may have their place in life. And it is not wrong to do science—the stories science tells also have their place.
23. Jack Turner, "Wildness and the Defense of Nature," in *The Abstract Wild*, Jack Turner, ed. (Tucson: University of Arizona Press, 1996), 114. See also Thomas H. Birch, "The Incarceration of Wildness: Wilderness Areas as Prisons," *Environmental Ethics* 12 (1990): 3–26.
24. This has been a recurring theme in the writing of Wendell Berry. See his *The Unsettling of America* (San Francisco, Calif.: Sierra Club Books, 1977).
25. John Rodman, "The Liberation of Nature?" *Inquiry* 20 (1977): 98, 103.
26. Callicott, *Earth's Insights*, 186.
27. Callicott, *Earth's Insights*, 130.
28. Our use of the term *gap* in this paragraph may seem contradictory and, worse, seemingly contradicts the use of that same term in our disclaimer in note 1. The careful thinker will see that this is not so. In correspondence on this matter, one Indigenous coauthor wrote [in his White mode]:

> Yep, have to remember about those darn "contradictions." If only people would realize that there are different phases of the world, different parts. Though it may seem unlikely that the physical world is ever contradictory, that a chair both is and isn't on the porch, say, it turns out that for things like

human beings and their understanding there really are many "contradictions." Interestingly, because of the academic enterprise, including the need to publish, non-Indian philosophers are much more likely to say that there both is and isn't a chair while at the same time saying that their reasoning and beliefs don't contain contradictions. You see, they know they're right, but they can't get published if they say there is a chair there. Since they know that rightness consists in part of being "rational" and rationality means "non-contradictory" among other things, they know they must be non-contradictory. It follows from their rightness. An Indian would just sit in the chair, unless some fool philosopher said it wasn't there. Then he might be polite and refrain from sitting so as not to befuddle the philosopher. He might even think it was an interesting story, this no-chair story, but kinda silly after a long day. Actually, it is my western philosophy training that most helps me swallow the whoppers. There are a lot of them in the literature. The problem is when they tell whoppers about things they really shouldn't. I can swallow the chair not being there and still sit in it. Unfortunately, the western philosopher is about as likely to remain standing, not out of politeness, but out of principle. He has swallowed the biggest whopper of them all, his own "rationality." They seem to have forgotten the Delphic imperative. They'd rather believe a fictional account of themselves. Luckily, this fictional account fits in nicely with domination and its concomitant subjugation. This is why folks like Hume are not at the forefront of western philosophy. He told some of the truth, and it included some "irrationality" on the part of humans. So the gap, in my story at least, comes from Callicott and others swallowing the whopper of their own rationality. Changing the disclaimer to something that has absolutely no tension with the [paragraph to which this note is appended] would mean we really are writing one of their stories.

29. The view that ontology is relative to epistemological orientation has recently come to the fore in ecosystem ecology itself in the rejection by what is perhaps unfortunately known as "hierarchy theory," of the view that an ecosystem can be subdivided in only one way. Ecosystems are not constructed out of bottom line components. Which analysis is appropriate varies from problem to problem depending upon the questions one asks and the observation set, in particular the spatiotemporal scale, appropriate to the problem at hand. There is "no need to force the phenomena into a set of components that are observable and interesting at some other space and time scale." Contemporary biological theory naturalizes ontology by privileging nonfoundational, contextualist epistemology over foundational ontology. R. V. O'Neill, D. L. DeAngelis, J. B. Waide, and T. F. H. Allen, *A Hierarchical Concept of Ecosystems* (Princeton, N.J.: Princeton University Press, 1986), 68, 177.

30. J. L. Austin, *How to Do Things with Words* (Cambridge, Mass.: Harvard University Press, 1962).

31. Sam Gill, "Holy Book in Nonliterate Traditions: Toward the Reinvention of Religion," in *Native American Religious Action: A Performance Approach to Religion* ed. Sam Gill (Columbia: University of South Carolina Press, 1987), 139–140.

32. Sam Gill, "One, Two, Three: The Interpretation of Religious Action," in *Native American Religious Action: A Performance Approach to Religion*, ed. Sam Gill (Columbia: University of South Carolina Press, 1987), 162–163, 151.

33. Gill, "One, Two, Three," 162.
34. See Leroy N. Meyer and Tony Ramirez, " '*Wakinyan Hotan*' ['The Thunderbeings Call Out']: The Inscrutability of Lakota/Dakota Metaphysics," in *From Our Eyes: Learning from Indigenous People* ed., Sylvia O'Meara and Douglas A. West (Toronto: Garamond Press, 1996).
35. Diamond Jenness, "The Carrier Indians of the Bulkley River," *Bureau of American Ethnology Bulletin* No. 133 (Washington, D.C.: Smithstonian Institution, 1943), 540.
36. N. Scott Momaday, "The Man Made of Words," in *Native American Traditions: Sources and Interpretations* ed., Sam Gill (Belmont, Calif.: Wadsworth Publishing, 1983), 44.
37. Paul Shepard, *Nature and Madness* (Athens: University of Georgia Press, 1998), 6.
38. "Poetry is one among the many forms of knowing. . . . Knowing freed from the agenda of possession and control—knowing in the sense of stepping in tune with being, hearing and echoing the music and heartbeat of being—is what we mean by poetry." Robert Bringhurst, "Everywhere Being Is Dancing, Knowing Is Known," *Chicago Review* 39 (1993): 138.
39. See Susan Bordo, *The Flight to Objectivity: Essays on Cartesianism and Culture* (Albany, N.Y.: SUNY Press, 1987).
40. Jeannette Armstrong, "Keepers of the Earth," in *Ecopsychology: Restoring the Earth, Healing the Mind*, ed. Theodore Roszak et al. (San Francisco, Calif.: Sierra Club Books, 1995), 323.
41. Shepard, *Nature and Madness*, 9.
42. J. Baird Callicott, "Animal Liberation and Environmental Ethics: Back Together Again," in *In Defense of the Land Ethic: Essays in Environmental Philosophy* ed. J. Baird Callicott (Albany, N.Y.: SUNY Press, 1989), 50, 59.
43. Callicott, *Earth's Insights*, 186.
44. See Sean Kane, *Wisdom of the Mythtellers*, 2d ed. (Peterborough, Ontario: Broadview Press, 1998). By contrast, the epistemology of control that underwrites modern science isolates, takes apart. Its model of knowing is the controlled experiment, which limits and controls the variables—that is, isolates the experiment from a wider, ecological context as far as possible. When the results of such experiments are brought back into the wild world, the results are, as often as (or oftener than) not, disruptive of natural systems.
45. Emory Sekaquaptewa, "Hopi Indian Ceremonies," in *Seeing with a Native Eye,* ed. Walter Holden Capps (New York: Harper and Row, 1976), 42–43.
46. Paul Shepard, *Nature and Madness*, 123.
47. Walter Bresette, speaking at the Ecological Resistance Movements: Religion, Politics, and Ethics conference, University of Wisconsin–Madison, November 10–11, 1995.
48. For another critique of Callicott from an Indigenous perspective see V. F. Cordova, "EcoIndian: A Response to J. Baird Callicott," *Ayaangwaamizin: The International Journal of Indigenous Philosophy* 1 (1997): 31–44.

17

The Very Idea of Wilderness

WAYNE OUDERKIRK

Surprisingly, J. Baird Callicott, one of the world's leading environmental philosophers, has argued against the idea of wilderness.[1] Not so surprisingly, his motivation for doing so is to promote more positive relations between humans and the environment. I think Callicott is mistaken in his thinking about wilderness. An examination of that thinking will show not only that his central argument is flawed but also that, as other thinkers in this volume have argued, he has not yet extricated himself from the bonds of modernism, so his view on wilderness would have the opposite effect from his intent. Another, more viable way exists to achieve Callicott's purposes.

Actually, Callicott presents three main arguments against the "received" concept of wilderness.[2] Before summarizing them, I have to note that Callicott says his complaint is not against those areas of the earth that remain nearly free of human influence or harm. He admits that those areas are important, valuable refuges, and he wants to keep them. Rather, he is opposed to the concept of wilderness, which he thinks is deeply problematic.[3]

His three arguments are:

1. The idea of wilderness as a place without humans and their activities is dualistic; it continues and reinforces the metaphysical separation of humans and nature that many, including Callicott, see as a major conceptual component and justification of environmental destruction.
2. The idea is "woefully ethnocentric" because it views the world through Euro-American eyes, which saw, for example, North America at the time of colonization as devoid of human influence. In fact, aboriginal peoples were probably more numerous and their influence on the landscape more profound, than Westerners suspected until recently.[4]

3. The wilderness idea, with its emphasis on nature as European colonists found it, ignores the dynamism and constant change inherent in ecosystems.

I want to concentrate on the first of these arguments, but I begin with some brief comments on each of the others. By now the myths of the "pristine" New World and of the environmentally noble savage are well exploded.[5] To that extent, Callicott's ethnocentric charge is well taken. However, in responding to Callicott, Dave Foreman makes two relevant observations. First, the idea of wilderness is not unique to Euro-Americans. If one defines *wilderness* as a place where humans visit but do not remain, as in the U. S. Wilderness Act of 1964 (although there is no necessity to that particular definition), Foreman finds instances of "wilderness as reality and idea" in both contemporary and ancient non–North American cultures. It thus looks a bit less ethnocentric. Second, Foreman argues that wilderness areas "have a track record unmatched by any other land designation anywhere in the world . . . [for] protecting ecological processes and some of the most sensitive species in North America."[6] Thus, despite the concept's origins, it still has something important to contribute to our present situation.

The charge that the wilderness concept ignores the dynamism inherent in ecosystems and strives to "maintain things as they were when the 'white man' first came on the scene"[7] is a strange one. Although some supporters of wilderness might talk in such an unecological manner, preservationists usually view wilderness areas as places where nature can proceed on its own, independent of human interference. That is exactly the wilderness definition, translated into policy, that many Americans questioned when extensive areas of Yellowstone's forests were left to burn in 1988. If the wilderness concept were one that denied or attempted to prevent change, the fires would never have caused a policy debate, let alone a media event, because they would have been fought as soon as they broke out.

To me, Callicott's argument that wilderness perpetuates a destructive dualism is the most philosophically interesting one. I agree that the modernist, mechanistic worldview has at its core a concept of humans (usually, "man") separate from nature. I agree also that dualism has had "insidious" effects on our thinking about nature, indirectly justifying environmental destruction. So I agree that we must avoid dualism, but not in the way Callicott tries to do so.

The main argument he presents against dualism is based in the science on which he develops his interpretation of the land ethic; it goes something like this: Modern science, which in its beginning era embraced dualism and underwrote the destructive resource/mechanism view of nature, has given way to newer scientific paradigms that envision humanity as part of nature. Thus, he says, "Since Darwin's *Origin of Species* and *Descent of Man*, however, we have known that man is part of nature."[8] Elsewhere he says that, "mechanism and dualism are going out of style and have been for most of the twentieth century," citing new developments in a number of scientific disciplines, including physics and ecology.[9]

More explicitly, the argument is this: Evolution demonstrates that humans have evolved in the same way as have all other species; other contemporary scientific developments also point toward unity in nature. Thus, we humans are simply parts of nature. Therefore, dualism, which claims we are separate from nature, is wrong. From there, Callicott draws another conclusion: "If man is a natural, a wild, an evolving species, not essentially different in this respect from all the others, . . . then the works of man, however precocious, are as natural as those of beavers, or termites, or any of the other species that dramatically modify their habitats."[10] Note, first, that this antidualist argument does not actually refute dualism. True, the scientific evidence points toward a more naturalistic account of human nature; however, that evidence does not preclude a separation between humans and the rest of nature. In the incredible story of evolution, various phenomena can be called emergent that are not simply identical to the processes that precede them (e.g., saying that locomotion is "nothing but" an elaboration of amoebic movement is false). Callicott needs much more argumentation to show that his reductionism regarding human nature—and it is a form of reductionism—is credible. Indeed, many thinkers assert that because humans have developed language, or higher cognitive skills, or whatever, they are different from other species. For example, Holmes Rolston III agrees that humans have evolved but asserts that we have evolved out of nature into culture, which he sees as quite different from nature.[11] Although not a classical form of dualism, Rolston's view shows that dualism remains a possibility if we limit our critique of it to Callicott's kind of argument.

Next, note that our alleged naturalness defeats environmentalism. If we are as natural as any other species, and if our works are as natural as others', then on what basis can we condemn human environmental destruction? Callicott emphasizes that because our works are natural, they can be "well tuned and symbiotically integrated with other contemporaneous evolutionary phenomena, with coral reefs and tropical forests, as well as the opposite."[12] But he elsewhere admits that "Most anthropogenic change is certainly not okay. Indeed, most of what we do in and to nature is destructive."[13] But on what criteria can we base such judgments? Callicott calls for "human economic activity that does not seriously compromise ecological integrity,"[14] and he envisions applications of "biosphere reserve style conservation" and other ecotopian revisions of current practice. Those are good ideals, but why should we adopt them? Once we equate human activity with that of other species, we must think of our actions and their consequences as simply part of the biosphere. We thereby pull the rug from under the feet of any moral limits on human activities. Callicott's attempted naturalization of humanity cancels our environmental ethic.

Callicott offers ecosystem health as a standard by which we might judge human treatment of the environment because health is both normative and objective.[15] And certainly ecosystem health is something we ought to promote. However, the question remains, why ought we? If we are completely natural beings, then promoting ecosystem health actually amounts to self-interested prudence, not a moral project. Besides, given the difficulties of finding generally accepted definitions of *ecosystem integrity*, *stability*, and *community* McIntosh and Shrader-Frechette emphasize in their contributions to this volume, that health will fare any better in ecological research and theorizing seems unlikely. Given the successful public relations efforts of corporate polluters to clothe themselves in green attire, the promotion of ecosystem health by environmentalists who also maintain that our works are as natural as other species' will pose no threat to the status quo. Anything goes, and probably will.[16]

These responses to Callicott's arguments point to the logical flaw in his reasoning: In moving from the facts of evolution to his conclusion that we are natural beings, thence to the naturalness of our works, he in fact equivocates, playing on two senses of "natural." Yes, we are natural in the sense that we are part of the natural world, having evolved in the same manner as have other species; every species is natural in this sense. However, we also are human, and in that sense we are different from the nonhuman. In this second sense of the term, the natural is everything *besides* us. Callicott's argument works only if we forget that when we talk about humans per se and their works, identifying us as human, we are using the second sense of the term.[17]

The source of this mistake is, ironically, dualism. Despite his rejection of dualism, Callicott allows it to set the terms of his analysis and argument. Of the many interrelated forms of dualism created by modernism, the human-nature or culture-nature dualism is most pivotal here.[18] Callicott sees the culture-nature dichotomy as responsible for environmental destruction, so he reacts by eliminating one of the two poles, namely, culture. So dualism lurks in his solution, which tacitly accepts the original division of reality. But as I pointed out, once we are totally identified with nature, any solution to the question of wilderness, or to the environmental crisis generally, lacks a moral ground on which to stand.

A perceptive reader will by now be asking what my own view is because it looks as if I want to have things both ways: to join Callicott in rejecting dualism and to criticize him for his antidualist theory. Interpreting evolution is part of the problem and part of the solution. Evolution is a natural process, and as I admitted earlier we are natural in the sense that we evolved. However, as I also pointed out, evolution produces novelties that are not simply equivalent to their predecessors. It happens that many of us humans conceptualize the novelties produced in us by natural selection by employing an unfortunate overemphasis on difference. The result is dualism. Callicott overemphasizes our continuity with the rest of nature. For an adequate picture of the human-nonhuman relationship, we need both aspects (differences and continuities) but that there are two aspects does not imply dualism.

Rolston strongly emphasizes the differences between nature and culture. For example, he says, "There is nothing unscientific or non-Darwinian about the claim that innovations in human culture make it *radically different* from wild nature."[19] He thus sounds like an irredeemable dualist. However, I think we can develop a view of our differences from the rest of nature that is not, as Rolston says, unscientific, but that does not end in the "radical" differences he sees. To do so, we can exploit some strands in Rolston's thought and in the thinking of Val Plumwood, which I believe are pointing to a new concept of wilderness.

The first step is to repeat that evolution produces novelties, things that require different explanations or understanding from those regarding their evolutionary predecessors. Nevertheless, those novelties are rooted in evolution, in nature. Culture might be considered in the same way. Culture, Rolston says, "presupposes" nature.[20] That means, in my view, that *we cannot have or understand culture without having and understanding nature*. In other words, I agree that culture is different from other processes we currently call "natural" or "wild," but we have to acknowledge that culture is an evolutionary emergent. Its difference is that it requires additional concepts for understanding it, concepts that are not necessary for and that are inappropriately applied to understanding the rest of nature. To speak in outmoded philosophical terms, if culture emerges from and presupposes nature, then an essential connection exists between them. We cannot have the one (i.e., culture) without the other (i.e., nature). They cannot be separated.

This is the idea at the core of Plumwood's response to dualism. She combines a cogent and comprehensive critique of dualism with an account of the human-nonhuman relationship that acknowledges both the differences and continuities between us and the rest of nature. No metaphysic or ethic is complete without either aspect.[21] The denial of either is blatantly false, both because we are a species that evolved and so have strong and undeniable connections to the rest of nature *and* because we are a distinct species with characteristics different from those of other species.

Because a destructive dualism has dominated our thinking and behavior toward the rest of nature for so long and because it is embedded in our social institutions and languages, emphasizing the continuity between us and the nonhuman sometimes makes rhetorical and political sense. However, as a metaphysical proposal, Callicott's radical naturalism is as mistaken as dualism.

Accepting and emphasizing both continuities and differences is not dualism, nor is it a classical monism like mind-body identity theory, nor is it a view like Callicott's, which allows dualism to set the terms of the discussion and so limits the possibilities in advance. If anything, it is a metaphysical pluralism. In nonmetaphysical terms, it represents a community. In it, wilderness is not separate from us. Rather, it is part of our nature, where we came from. We need it to survive and to be what we are, although what we are is more than the sum of our parts or origins, just as wild ecosystems are more than the sum of their parts.[22]

In his response to Callicott's anti–wilderness-concept arguments, Rolston points out a contradiction in Callicott's presentation. Although insisting that we are as natural as any other species, Callicott also admits that "the cultural component in human behavior is so greatly developed as to have become more a difference of kind than of degree."[23] The apparent contradiction can be dissolved within a view such as the one sketchily proposed here. That our works are cultural cannot be denied, but that means only that they are more than the sum of their parts. They thus cannot be regarded as literally the same as the works of beavers and termites. They are human works. But those works do not and cannot occur independently of their essential connections to that wild nature of which beavers and termites and we are part. We cannot and ought not ignore our ecological embeddedness.

And what is the justification for the moral dimension of that "ought"? Because nonhuman nature is importantly different from us (while still connected to us), it has its own processes and "ends" that we should respect. Expressed differently, it is "Other" and we have an obligation to respect and appreciate its independent functioning.[24] To say so is not to contradict my main idea, that we are both connected to and different from nonhuman nature. Rather, it is only to acknowledge that just as we cannot be reduced to nature, nature cannot be reduced to us, as it would be if we inhabited all of it, used all of it.

Callicott has taken the next step in his critique of the wilderness concept by proposing "that we rename wilderness areas 'biodiversity reserves.' "[25] In part, he justifies his proposal by reference to his critique of the concept of wilderness, which he sees as demonstrating that that concept carries too much "baggage" from its past. His new label would also have the advantages of emphasizing that the preserved areas are primarily "for the nonhuman inhabitants of such places"; of injecting "a more universal and higher-minded conservation aim" into debates about preservation; of allowing for carefully selected and planned human economic activity in such areas; of enabling us to preserve additional areas that were "overlooked by the historic wilderness preservation movement" and that will more likely contribute to the preservation of biodiversity and especially of endangered species.[26]

No one interested in promoting improved treatment of the nonhuman world could argue against those advantages. However, it is unclear to me that they cannot be secured with our traditional term, *wilderness.* Jettisoning that term might cause more confusion than would working with it, and the new name is not the guarantee of protection Callicott thinks it is.

First, a conceptual or methodological point: Although Callicott does not connect his proposal for a new name for wilderness areas with his call for the development of a postmodern worldview, we certainly should see it as part of that call if only because of his rejection of the dualism he finds in the modernist worldview and in the concept of wilderness.[27] However, I believe that we cannot develop a new worldview or a new concept of wilderness or of nature or of human nature ex nihilo. We have to work with the concepts we have, responding to them, chipping away at their unhappy rough edges, refining them as we go. In their own time, modernist thinkers were rejecting the medieval worldview, but they continued to use its concepts, such as substance and cause, to develop their new perspective. The mistake of many who realize the limitations of modernist concepts is to rush to create a completely new postmodernism. That is simply impossible. Swimming against the still robust currents of modernism, completing the needed changes will take time, and we do not know exactly what the resulting concepts will be. So my own proposal uses the concepts we have, modifying them, I hope, in positive directions.

But even on the level of the practical advantages Callicott claims for the new label for wilderness areas, making the change would be a mistake. First, we are able to communicate and enforce the idea that some areas are for the good of their indigenous species without the suggested change. For example, hunting endangered species in wilderness areas is prohibited. In the Adirondack Mountains, hikers have adapted easily and well to the idea that they ought not walk on the few acres of alpine vegetation found on a few of the highest summits there. In addition, the concept of wilderness has been a significant, integrated part of U.S. environmentalism for so long that any concerted, explicit effort to change terminology will inevitably confuse sympathetic citizens as well as some activists. Finally, we already have areas designated by a name similar to that proposed by Callicott, namely, wildlife refuge areas. That name makes clear the purpose of the areas, but that has not stopped those who discover something besides wildlife in them from attempting to compromise their status to get what they want. The controversy about the oil reserves under the Arctic National Wildlife Refuge are a case in point. The effort to protect such areas does not end with the conferral of a name.

I think that Callicott's critique of the wilderness concept is valuable because it teaches us that important ideas can carry with them problematic histories, and we have to work deliberately to rid our ideas and our practices of those problematic features. So my own view is that yes, we should continue to preserve wilderness, even while refining and reforming the concept and its importance for us. Although it might still carry some negative connotations from its origins, the concept of wilderness remains a useful and significant ideal for environmentalists.

Notes

1. J. Baird Callicott, "The Wilderness Idea Revisited: The Sustainable Development Alternative," in *The Great New Wilderness Debate: An Expansive Collection of Writings Defining Wilderness from John Muir to Gary Snyder*, ed. J. Baird Callicott and Michael P. Nelson, (Athens: University of Georgia Press, 1998), 337–366. Originally published in *Environmental Professional* 13 (1991): 235–247. See also his "Should Wilderness Areas become Biodiversity Reserves?" in Callicott and Nelson, 586–587.

2. In "A Critique of and an Alternative to the Wilderness Idea," *Wild Earth* 4 (1994): 54–59, Callicott discusses six "realizations [that] are subverting this simple philosophy of nature conservation through wilderness conservation."

3. Callicott, "A Critique of," 54; Callicott, "The Wilderness Idea Revisited," 339; and Callicott, "Should Wilderness Areas," 587. One wonders, along with Dave Foreman, about the legitimacy of that distinction. Foreman, "Wilderness Areas Are Vital: A Response to Callicott," *Wild Earth* 4 (1994): 65. William Cronon, "The Trouble with Wilderness, or, Getting Back to the Wrong Nature," in *Uncommon Ground: Toward Reinventing Nature,* William Cronon, ed. (New York: Norton, 1995), uses the same distinction in his own critique of the concept of wilderness, which is similar in some respects to Callicott's, including the charge of dualism.

4. See William Denevan, "The Pristine Myth: The Landscape of the Americas in 1492," in *The Great New Wilderness Debate: An Expansive Collection of Writings Defining Wilderness from John Muir to Gary Snyder*, J. Baird Callicott and Michael D. Nelson, eds. (Athens: University of Georgia Press, 1998), 414–442; and Steven R. Simms, "Wilderness as a Human Landscape," in *Wilderness Tapestry: An Eclectic Approach to Preservation*, Samuel I. Zeveloff, L. Mikel Vause, and William H. McVaugh, eds.

(Reno: University of Nevada Press, 1992), 183–202. Perhaps the best-known work documenting the influence of Native Americans on North America's landscape is William Cronon, *Changes in the Land: Indians, Colonists, and the Ecology of New England* (New York: Hill and Wang, 1983).

5. Simms, "Wilderness as a Human Landscape," and Cronon, *Changes in the Land.*
6. Foreman, "Wilderness Areas Are Vital," 67.
7. Callicott, "A Critique of," 56.
8. Callicott, "The Wilderness Idea Revisited," 350.
9. Callicott, "La Nature es morte, vive la nature!" *Hastings Center Report* 22 (1992): 22.
10. Callicott, "The Wilderness Idea Revisited," 350.
11. Rolston, "The Wilderness Idea Reaffirmed," in *The Great New Wilderness Debate: An Expansive Collection of Writings Defining Wilderness from John Muir to Gary Snyder*, ed. J. Baird Callicott and Michael P. Nelson (Athens: University of Georgia Press, 1998), 368 and passim.
12. Callicott, "The Wilderness Idea Revisited," 351.
13. Callicott, "A Critique of," 56.
14. Callicott, "The Wilderness Idea Revisited," 355.
15. See for example, his essays, "Environmental Wellness," and "Aldo Leopold's Concept of Ecosystem Health," in his recent collection, *Beyond the Land Ethic: More Essays in Environmental Philosophy* (Albany, N.Y.: SUNY Press, 1999).
16. Foreman, "Wilderness Areas Are Vital," 65.
17. For discussions of the several meanings of *nature* and *natural*, see Kate Soper, *What Is Nature? Culture, Politics and the Non-Human* (Oxford: Blackwell, 1995), and Neil Evernden, *The Social Construction of Nature* (Baltimore, Md.: Johns Hopkins University Press, 1992).
18. For an extended analysis and critique of dualism, see Val Plumwood, *Feminism and the Mastery of Nature* (London: Routledge, 1993).
19. Rolston, "The Wilderness Idea Reaffirmed," 368 (emphasis added).
20. Rolston, "The Wilderness Idea Reaffirmed," 368.
21. Val Plumwood, *Feminism and the Mastery of Nature*, esp. 136 ff. See also her "Wilderness Skepticism and Wilderness Dualism," in *The Great New Wilderness Debate: An Expansive Collection of Writings Defining Wilderness from John Muir to Gary Snyder*, ed. J. Baird Callicott and Michael P. Nelson (Athens: University of Georgia, 1998), 652–690.
22. This argument points in the direction of, and supports, the concept of biophilia. See the essay by Partridge, this volume. For discussions of cultures that include nonhuman nature in their moral communities, see Deane Curtin's *Chinnagounder's Challenge: The Question of Ecological Citizenship* (Bloomington: Indiana University Press, 1999). Significantly, Curtin defends wilderness preservation on similar grounds to mine: that we and the nonhuman are part of an integrated community and that the nonhuman constitutes part of our identity (pp. 187 ff.)
23. Rolston, "The Wilderness Idea Reaffirmed," 369; Callicott, "The Wilderness Idea Revisited," 351.
24. See Plumwood, *Feminism and the Mastery of Nature*, 161–164; and Plumwood, "Wilderness Skepticism," passim.
25. Callicott, "Should Wilderness Areas," 590. See also his "A Critique of and an Alternative to the Wilderness Idea."

26. Callicott, "Should Wilderness Areas," 590–592.
27. Callicott calls for and envisions this postmodern worldview developing in many of his writings. See for example, *Earth's Insights: A Multicultural Survey of Ecological Ethics from the Mediterranean Basin to the Australian Outback* (Berkeley: University of California Press, 1994), esp. chap. 9, "A Postmodern Evolutionary–Ecological Environmental Ethic." Also, in his "Introduction" to *Beyond the Land Ethic*, he indicates the need for such a worldview because dualism is embedded in the modernist worldview (p. 16). And in another essay in that same collection, "After the Industrial Paradigm, What?" he envisions a "reconstructive" postmodernism as "better adapted to the natural environment" (p. 317). This theme has been part of Callicott's thinking for a long time. A well-known instance is his "Intrinsic Value, Quantum Theory, and Environmental Ethics," in his *In Defense of the Land Ethic: Essays in Environmental Philosophy* (Albany, N.Y.: SUNY Press, 1989), esp. 165 ff.

Part V

Callicott Responds

18

My Reply

J. BAIRD CALLICOTT

OF THANKS AND PROVOCATION

I begin to write this essay around the Thanksgiving holiday in the last year of the second millennium. I have much to be thankful for, both personally and professionally. For my personal blessings, I will have expressed my thanks privately to family and friends gathered at my home for the holiday. For my professional blessings, I express my thanks here publicly to my colleagues and critics, both to those represented in this volume and those who are not.

My first love in philosophy was that of the ancient Greeks. And among the many legacies that they bequeathed to the subsequent Western tradition was critical engagement. One might go so far as to say that the force driving the rapid development of Greek philosophy from Thales to Aristotle was the critical engagement of each generation of thinkers with one another and with the thought of their predecessors. I still occasionally teach the history of ancient Greek philosophy and always invite my students to marvel at how, for example, the critical philosophy of Zeno challenged Empedocles, Anaxagoras, and Democritus to think far more subtly and incisively about the nature of matter and motion than had the Milesians. Nearly one-third of a century ago, I dimly foresaw the possibility of a new environmental philosophy. From the beginning, therefore, I believed (and still believe) that if environmental philosophy was to emerge (and to persist) as a robust field of inquiry, a community of thinkers engaged in vigorous mutual criticism would be essential. Accordingly, I deliberately spoke and wrote boldly, provocatively, hoping to attract critical engagement and doing so has become a lifelong habit. This volume of critical essays is, therefore, especially welcome to me, although some of the criticisms of my work registered here are painful to read. I sincerely thank all my present critics, but I want to thank especially those—Kristin Shrader-Frechette and Holmes Rolston III (both outstanding in this regard), Ernest

Partridge, John Barkdull, Robert P. McIntosh, Bryan G. Norton, Eugene C. Hargrove, Catherine Larrère, Peter Wenz, and Andrew Light—who have the grace to express some appreciation for my work, however wrong-headed they suppose it to be, before expressing their disagreement. And finally I want to thank the editors of this volume, Wayne Ouderkirk and Jim Hill, for their effort and persistence in assembling and introducing the critical (and mostly original) essays of so distinguished and diverse a group of environmental philosophers and for seeing this book through to publication. With some exceptions, my reply to each author follows the order in which they appear in this book. To avoid fragmentation and, to the extent possible, to craft an essay that has a unity of its own, not only have I occasionally considered a criticism out of the order in which it is presented, I occasionally grouped some criticisms by themes other than those identified by the editors. All this for the sake of efficiency, the avoidance of redundancy, and the thematic integrity of this, my own contribution.

Of Partridge and Barkdull and Hume and Smith

The essays by Partridge and Barkdull remind me of one of the earliest scholarly disputes in environmental philosophy. Arne Naess (1977) had grounded his seminal version of Deep Ecology in the philosophy of Spinoza. But Spinoza himself is on record as being unequivocally, even militantly, anthropocentric. Naess's critics were quick to point this out (Lloyd, 1980). Naess (1980), however, was undaunted. He argued that Spinoza had a great insight about the unity of nature, including human nature, the full moral implications of which Spinoza himself could not have seen, because the prevailing humanism (anthropocentrism) of his times clouded his vision. Had Spinoza lived in the age of environmental crisis, he would surely have been a Deep Ecologist.

About the relative merits of the cases for and against Spinoza as a proto–Deep Ecologist, I offer no comment. But my reply to the critics of my grounding the land ethic ultimately in a theory of moral sentiments, like that articulated by David Hume and Adam Smith in the eighteenth century, is similar to, although not the same as, Naess's reply to the critics of his appropriation of Spinoza. A theory of moral sentiments provides a moral psychology that makes ethics a matter of the heart as well as the head, a matter of feeling as well as reason. Some such moral psychology is essential for an evolutionary account of the origin and development of ethics, as Darwin clearly recognized. A purely rationalistic theory of ethics puts the cart before the horse.

Reason could not have evolved except in an intensely social circumstance. However, because our prehuman ancestors would have been short on reason—by all accounts one of the most recent and refined human faculties to have evolved—they could have had no ethics, if ethics is grounded exclusively in reason. And if they had no rudimentary ethics, they could not have been members of an intensely social, cooperative community in which reason could have evolved. Thus Darwin drew on the sentiment-based moral philosophies of Hume and Smith in *The Descent of Man*. Aldo Leopold, who probably knew of Hume only as a historian and Smith only as an economist, drew on Darwin in formulating the land ethic. Thus, by the time Hume and Smith inform the land ethic, the particular details of their respective theories are much attenuated. Whether Hume and Smith were or were not hopelessly anthropocentric; whether they believed sympathy to be the most fundamental of the moral sentiments or just one such sentiment among many on a par with, say, patriotism; whether they would or would not have endorsed the land ethic, had they an opportunity to pass judgment on it—all such questions are philosophically interesting and historically relevant, but, in the final analysis, beside the point. The land ethic is grounded in the general approach to ethics that Hume and Smith pioneered. It is not, nor did I ever intend to suggest that it was, a slave to every idiosyncratic nuance of the respective ethical writings of these historical figures. (I return to the issue of selectively borrowing from disparate historical figures in my reply to Hargrove.)

By this I do not essay to demean the splendid contributions of Partridge and Barkdull to this volume. The alternative routes they trace, respectively, to what they consider to be more faithful Humean and Smithian environmental ethics are interesting, and they enrich theoretical and historical research in environmental philosophy. Many of the theoretical problems that Partridge highlights, which a persuasive environmental ethic—Humean or otherwise—must overcome, I have already addressed: the allied problems of symmetry and reciprocity are addressed in my very first published paper in environmental philosophy, "Elements of an Environmental Ethic"; of normative force in "Can a Theory of Moral Sentiments Support a Genuinely Normative Environmental Ethic?"; of fairly balancing (prioritizing) our individualistic duties to members of the various human communities to which we belong with our holistic duties to the various biotic communities of which we are also members (that Barkdull also finds problematic) in "Holistic Environmental Ethics and the Problem of Ecofascism" (about which more in my reply to Shrader-Frechette and Donner). Barkdull argues for a greater separation between the ethics of Smith and Hume than I am wont to recognize. In the course of which, to my pleasant surprise, he shows that Smith himself supports me, against Partridge, in finding a holistic dimension in Hume's moral philosophy. Barkdull writes, "Indeed, Smith argues against Hume's claim that man has a 'natural love for society, and desires that the union of mankind should be preserved for its own sake, although he himself was to derive no benefit from it.'" As we see, Smith attributes to Hume the view that we have a natural love of society per se—a holistically oriented moral sentiment. Interestingly, Barkdull makes his Smith agree with Partridge's Hume that the proper road to environmental ethics runs first through environmental aesthetics.

Of McIntosh and Shrader-Frechette and Ethics and Ecology

I am especially honored, as well as pleased, to have attracted the critical attention of Robert P. McIntosh, who notes that I have been unusual among ecological philosophers in actually giving serious study to ecology. (How much actual ecology do we find, for example, in Deep Ecology, or in the essay here by Hester, McPherson, Booth, and Cheney who speak loosely about an "emergent ecological order"?) Among those from whom I have learned the most about ecology is McIntosh (1985) whose book *Background of Ecology* is the closest thing available to a comprehensive and exhaustive history of the discipline through the mid-1980s. But McIntosh is not only a historian of ecology; he is a maker of ecological history as well. That gives him an axe to grind in that fractious field. As a student and colleague of John Curtis, he is a leading figure in the neo-Gleasonian movement that began in the 1950s and has gained strength ever since. It was Henry Gleason (1926), in opposition to his holistic contemporary F. E. Clements (1916), who insisted that what the latter believed to be self-organized ecological wholes were in fact but fortuitous aggregates of species populations adapted to similar "gradients"—of moisture, soil pH, temperature, and other circumstances of a site—that affect their flourishing. Gleason went unheeded in his own day, but Curtis and McIntosh (1951) and R. H. Whittaker (1951) revived and championed his view. So, although of unparalleled value, McIntosh's history of ideas in ecology, the relevant parts of which his essay for this volume summarizes, is jaundiced. Buyer beware! Golley (1993) and Hagen (1992) provide counterbiased accounts, although far less comprehensive than McIntosh's.

When I began theorizing about environmental ethics, back in the 1970s, I confess I warmly entertained the idea that, in the course of biogeochemical evolution on Planet Earth, complex self-replicating molecules had evolved to form eukaryotic single-celled organisms, enclosing symbiotic organelles, such as mitochondria . . . which in turn had formed associations of symbiotic cells . . . which in turn had evolved to form multicelled organisms with well-articulated tissues and organs . . . which in turn had evolved to form tightly integrated biotic communities and superorganisms. We conscious and reflective multicelled organisms no more noticed these superorganisms, than, were they conscious and reflective, the individual cells of our own bodies would notice the multicelled organism of which they were constituents. That is, we did not notice them until (post)modern science disclosed their existence, as well as the existences of many other wonders of nature that were not disclosed until systematically and methodically searched out. The putative existence of transorganismic entities not only tickles one's fancy and whets one's appetite for mystery and enchantment in the world, it provides, more to the point, a direct object of environmental ethical concern. As McIntosh so authoritatively points out, however, the Clementsian superorganism concept has now been eclipsed by the neo-Gleasonian individualistic view. This gradual but steady paradigm shift in ecology has been under way for fully half a century now.

At first, my knowledge of ecology was limited to a casual study of *Fundamentals of Ecology*, which dominated ecological education well into the 1970s—and its author, Eugene Odum (1971) was a neo-Clementsian, or "systems ecologist" as McIntosh labels it. Then as the 1970s gave way to the 1980s, neo-Gleasonianism began resolutely to over shadow neo-Clementsianism, but I remained at first in ignorance and then in denial. "The Metaphysical Implications of Ecology," written in the mid-1980s represents my attempt to explore the philosophical potential of neo-Clementsianism, which—who knows?—may someday rise again. In one or another guise, it still holds on in various redoubts of the pluralistic science of ecology, as McIntosh admits. So far, it has not gone the way of phlogiston or the luminiferous aether. However, it is certainly not in fashion—a fact I finally faced. As an ecological philosopher, I have been counseled by the old American adage, "you can't buck City Hall," which in our case is ecology. I feel obliged, in accordance with the no-buck-City-Hall principle, to accept the state of the science as it currently exists, however personally unwelcome to me. I am an ecological philosopher, not an ecologist. Who am I to blow against the wind?

Aldo Leopold formulated the land ethic when systems ecology—a more sophisticated reiteration of Clementsian superorganismism—was in ascendancy. However, in "Do Deconstructive Ecology and Sociobiology Undermine the Leopold Land Ethic?" (salient points of which I here summarize shortly), I more or less capitulate and argue, by way of consolation, that even though the philosophically disappointing paradigm shift in ecology that

McIntosh documents has in fact, unfortunately, occurred, with some adjustment and reformulation, the Leopold land ethic is still viable.

In addition to individualism, current thinking in ecology stresses dynamism and disturbance (Pickett & White, 1985). Clements's brand of ecology was, ironically, sometimes called *dynamic ecology*, because Clements focused on ecological succession, the replacement of one plant association by a successor (McIntosh, 1985). The irony is that this process of succession—begun by some "exogenous" disturbance—terminated in a "climax" community, which Clements believed would persist indefinitely. That is, the dynamic process of succession ended in the stasis of the putative climax. Currently, ecologists believe that the process of succession can follow many paths to no particular destination, and that disturbances—fire, wind, flood—are themselves intrinsic parts of the process (Pickett & Ostfeld, 1995). Has environmental ethics then any natural norm by means of which to evaluate human environmental behavior? Aldo Leopold thought we should preserve "the integrity, stability, and beauty of the biotic community." But the ontology of biotic communities is problematic in contemporary ecology, so what does it mean to refer to their "integrity"? Certainly such communities lack "stability." All that remains of Leopold's dictum is "beauty," and everyone believes that that is in the eye of the beholder. Accordingly, in "Deconstructive Ecology," I dared to suggest we emend the "golden rule" of the land ethic thus: "A thing is right when it tends to disturb the biotic community only at normal spatial and temporal scales. It is wrong when it tends otherwise."

The editors of this volume have done as well to pair the McIntosh and Shrader-Frechette essays as to pair those by Partridge and Barkdull. For Shrader-Frechette is as serious a student of ecology as I, and she reiterates McIntosh's attack on the outmoded biotic community concept in ecology. Shrader-Frechette expressly accepts my metaethical principle that, in her words, "the community concept is essential to moral obligation, and that different kinds of communities undergird different moral obligations." In the "Deconstructive Ecology" essay, I argue that however ill-bounded, open to imigration and emigration, and subject to successional change biotic communities may be, human communities are, in these respects, no different. Human communities are no more "robust" than biotic communities. Thus, if the former engender moral obligations, as Shrader-Frechette explicitly affirms, then why shouldn't the latter? Other reasons why biotic communities might not engender moral obligations could be advanced, but state-of-the-science arguments to the effect that they lack robustness should not be among them, because, in this respect, biotic and human communities are in the same boat.

In the same essay, as just noted, I argue for replacing the now scientifically suspect concepts of integrity and stability in the summary moral maxim (or "golden rule") of the Leopold land ethic with the poststructuralist concepts of disturbance and spatiotemporal scale. Human disturbances should not exceed the spatial and temporal scales of natural disturbances. Moreover, our moral obligations—engendered by our community memberships, human and biotic—are delimited by a fairly circumscribed temporal scale. The geological temporal scale is not morally meaningful. The last great extinction event, which occurred 65 million years ago, is not reprehensible. Nor does the expectation that in 50 million years biodiversity will have bounced back from the anthropogenic mass extinction event now in progress let us off the ethical hook. We should not fret over the deep past, nor should we exonerate ourselves by contemplating the deep future. Our duty is to preserve the species populations of the biotic communities that exist now. How to define the specious present indicated by the word *now*? We must build into it room for change; the world is not static. But if not the geological scale, upon what scale is it appropriate morally to evaluate change? An ecological scale, I suggest, calibrated by such ecological processes as disturbance regimes and succession.

Shrader-Frechette's essay presents a unique problem for readers of this volume because in it we find her original critique of my environmental ethics, her summary of my reply to that critique, and her reply to my reply. And now, I here have the latest (if not last) word, but I despair that the reader can keep this multitiered exchange sorted out and make sense of it. Thus, I will answer one of her general complaints—that the land ethic, as I have interpreted it, lacks normative force—in the same way that I answered Partridge's similar complaint by referring the reader to a whole essay of mine devoted to that problem, "Can a Theory of Moral Sentiments Support a Genuinely Normative Environmental Ethic?" and confine myself to only two further comments.

Of Shrader-Frechette and Donner and Ethics and Evolution

First, Shrader-Frechette accuses me of Humpty-Dumptyism, making the word *evolution* mean just what I want it to mean in one place and something else in another. She misses the focus of my earlier disclaimer (which she summarizes). It falls on her supposition that I posit an "*analogy* between evolution and ethics," similar to the way so-called evolutionary epistemologists posit an analogy between genotypes and belief systems. Select memes may be analogous to select genes and survive and flourish in the cultural meme pool in a way analogous to the way select genes survive and flourish in the biological gene pool. I don't know; I have not given the matter enough study to have a well-informed opinion. But I most certainly do not posit a similar analogy between, say, duties and genes. Rather, I posit an evolutionary derivation of ethics, following Darwin's lead in the *Descent of Man*. To derive one thing from another is not the same as to draw an analogy between one thing and another. I argue that like any other normal human psychological characteristic, say a capacity for feeling fear, natural selection has endowed us with moral sentiments. Fear, further, is an open-ended emotion, the objects of which are not rigidly determined by inheritance; fear can be culturally informed, educated. We can learn to fear many things—ghosts and gamma rays, germs and gremlins. Likewise, the moral sentiments are underdetermined and open-ended. We can learn to respect things today, such as universal human rights and animal rights, our nation-states, and our biotic communities, of which our remote ancestors knew nothing when the human moral sentiments were evolving.

Second, which of our community memberships "has primacy," Shrader-Frechette wonders; that is, which of our multiple community-generated duties and obligations should take priority when they conflict—as when, say, our duty to respect human rights conflicts with our duty to preserve an endangered species? As we see, this is a question raised by more than one of my critics. As noted, I have also devoted a whole essay to this problem: "Holistic Environmental Ethics and the Problem of Ecofascism," the heart of which I summarize in my following reply to Wendy Donner.

Donner much better understands and accurately represents my Darwinian evolutionary account of the moral sentiments than does Shrader-Frechette. Furthermore, Donner is right to say that in the absence of "one fundamental principle to resolve conflicts . . . we need principles that set out clearly how we prioritize the interests of or value of these different elements consistently." In the ecofascism essay to which I just referred, I provide two second-order principles (SOPs) to resolve conflicts between the first-order duties and obligations generated by our multiple community memberships. The first, SOP–1, states that duties and obligations generated by membership in nearer and dearer communities take precedence over those generated by membership in more remote and impersonal communities. For example, I feel a greater obligation to contribute to the care of my aged mother because she and I are members of the most venerable and intimate of all communities, the family, than I do to contribute to the care of Donner's mother, who is, with me, a member of a more remote and impersonal community, the global village. The second, SOP–2, states that greater interests prevail over lesser interests. For example, I feel a greater obligation to prevent a murder if I am in a position to do so, than to prevent shoplifting: in the former case, the victim's strongest possible interest in life is at stake; in the latter, the victim's lesser interest in a small amount of money is at stake.

These two SOPs are designed to be employed sequentially, such that the indication of the second either reinforces or countermands that of the first. Thus, in the famous case of the endangered spotted owl and its old growth forest habitat versus the individual loggers, mill workers, and timber company owners in the Pacific Northwest, we reason as follows. Applying SOP–1, we should give preference to the interests of the latter because they are our fellow human beings and, for those of us who are citizens of the United States, our fellow Americans. But SOP–2 countermands SOP–1 because the species in question faces irreversible extinction and its habitat wholesale destruction, while the nonvital interests of the individual human beings in a certain kind of work and a certain kind of lifestyle can be compensated and substituted for. Through local, state, and federal governments, we can and should compensate people who lose jobs because of our collective efforts to conserve endangered species and provide those aggrieved with training for other kinds of employment.

Of Rolston and Ouderkirk and Culture and Nature

Rolston's critique is not only a model of grace—for which I have already expressed my appreciation—it is a model of scholarship and intellectual integrity as well. Unlike some of my critics, Rolston is not content to fasten on to an early essay of mine, quote damagingly from it out of context, and impale a straw man labeled "Callicott" on his well-honed petard. He has been generous in time spent with my work and willing to acknowledge its modifications over a quarter century of reflection and reformulation. Many such modifications, especially on the problem of intrinsic value in nature, I made in the course of a public dialogue with him—which, happily, continues in this volume.

Rolston's title, "Naturalizing Callicott," is ironic because he begins by criticizing my claim that human beings—me included, certainly—and human culture exist as a part of nature, not apart from nature. Therefore, I don't need no naturalizing, thank you very much; I have quite naturalized my own self. My belief that human beings are a part of nature is based on evolutionary considerations. In this belief I follow in the footsteps of both Darwin and Leopold, who notes that:

> It is a century now since Darwin gave us the first glimpse of the origin of species. We know now what was unknown to all the preceding caravan of generations: that men are only fellow voyagers with other creatures in the odyssey of evolution. . . . Above all we should, in the century since Darwin, have come to know that man, while now captain of the adventuring ship, is hardly the sole object of its quest, and that his prior assumptions to this effect arose from the simple necessity of whistling in the dark.

Only on this basis can Leopold argue that we are "plain members and citizens of the biotic community." Thus, the evolutionary naturalization of human beings and human culture is foundational and essential to the land ethic. Note that my Darwinian-Leopoldian naturalization of culture is opposite the stance of other postmodern environmental philosophers who culturize, I suppose we should say, nature (Gare, 1995; Vogel, 1996). They argue that nature is culturally (or socially) constructed, whereas I argue that culture is naturally evolved and remains a part of nature. Rolston hints at this difference here; he is as adamantly opposed to their culturalization of nature as he is to my naturalization of culture (Rolston, 1997). Wayne Ouderkirk in his individual contribution to this volume clearly notes that the nature-culture dichotomy can be reduced in two ways, and that I am guilty of one of them. Following Val Plumwood (1998), he seeks to maintain the dichotomy, but not the dualism—that is, the opposition between them—regarding nature and culture as complementary, like the yin and yang, I suppose, in classical Chinese thought. Rolston, in opposition to every kind of postmodern resolution of the problem, remains a resolute nature-culture dualist. His environmental ethic, thus, is more in the Kantian mode, based on a cold respect for the radical Other, than in the Humean mode, based on love, warm fellowship, and society with our fellow voyagers in the odyssey of evolution, our fellow members of the biotic community.

To his credit, Rolston does not invoke the traditional Judeo-Christian or Greco-Roman grounds—respectively, that human beings are created in the image of God or are exclusively rational—to justify his belief that human culture and its artifacts are sharply distinct from nature and what is so of itself. His grounds are themselves evolutionary. Nature, among human beings, has evolved culture, to be sure; but culture has propelled the species right out of nature's ambit. This argument turns on the very evolutionary analogy that Shrader-Frechette misattributes to me. Both organic nature and emergent culture reproduce themselves by passing "information" from one generation to the next. In organic nature, information is encoded in inheritable genes; in human culture, it is encoded in communicable memes. Evolutionary change in organic nature proceeds at a comparatively slow pace, as genes randomly mutate and the environment remorselessly and blindly culls the far more frequent harmful mutations from the rare beneficial ones. Evolutionary change in culture proceeds at a comparatively rapid pace as memes are often deliberately reshaped and reorganized, and passed on to the next generation. Evolutionary change in culture is Lamarkian, and many orders of magnitude faster than in organic nature, where evolutionary change is Darwinian. This discrepancy in speed—in temporal scale—draws the boundary, which is becoming ever sharper as the speed of cultural change accelerates, between organic nature and human culture. Ouderkirk seconds Rolston's claim that culture is "emergent" from nature.

I am drawn to this argument, but remain unconvinced by its conclusion. It is a matter of emphasis. To me it seems that we human beings are thoroughly primate in our anatomy, physiology, and psychology, except for the more language-dependent cognitive and abstract states of consciousness. Our physical lives, certainly, and the largest part of our conscious lives—our feelings of joy and sorrow, anger and remorse, jealousy and rage; our intense social interactions, negotiated mostly by body language, facial expression, and tone of voice; our pervasive sexuality—all, although culturally shaped around the edges, are utterly animal and therefore natural. The dazzling artifacts of culture—skyscrapers, airplanes, bulldozers, and such—powerful although they may be to transform and destroy nature, seem fragile and ephemeral in comparison with the titanic and persistent forces and processes of nature. We are earthy beings, and remain—culturally, as in every other way—earth bound.

Ancillary to the nature-culture question, is the conservation-preservation question. If human beings are a part of nature, then what rationale have we for preserving pristine nature, wilderness, areas unsullied by (now transcendent) human beings and their culture? Ouderkirk raises this question quite explicitly. Without pristine nature (wilderness) as a standard, "then on what basis can we condemn human environmental destruction," he asks. As noted, I have tried to develop an alternative standard in terms of spatiotemporal scale. People should disturb nature only at normal spatial and temporal scales. Bill McKibben (1989) has convincingly argued, in any case, that no pristine nature remains to be preserved, and William Denevan (1992) adds that there has not been any such on a large scale for a long time. Indeed, I do argue that we can effect conservation in a largely humanly inhabited and economically exploited world, measured by the standard of ecosystem health. In a rare lapse of charitable scholarship, Rolston disses the powerful ecosystem-health standard of conservation by suggesting that "cornfields and wheat fields" might measure up to it. Of course they do not. Neither would single-species tree plantations (or "industrial forests"). A better example of a humanly inhabited and economically exploited healthy ecosystem would be the kind of forestry practiced by the Menominee on their reservation in Wisconsin (Davis, 1999) or the perennial polyculture that Wes Jackson (1980) and his associates are attempting to create at the Land Institute in Salina, Kansas.

Ouderkirk, more thoughtfully and generally, notes that ecosystem health is an ecologically problematic idea. True, but it is certainly no more problematic than the idea of wilderness. Many of our fellow voyagers in the odyssey of evolution, however, do not fit very well into the healthy humanly inhabited and ecologically exploited ecosystems that now exist or may exist in the foreseeable future. Wolves, for example, do not roam the Menominee forest, nor would bison be a welcome part of the Land Institute's perennial polyculture. Such animals, however, need lebensraum—habitat, in a word—free of human interference and conflict of interest. In "Should Wilderness Areas Become Biodiversity Reserves?" I suggest a substitute for the wilderness idea, with which Ouderkirk respectfully disagrees, conservatively preferring to reconfigure the wilderness idea instead. The "biodiversity-reserve" alternative to the wilderness idea provides for the preservation of relatively human-free habitat for those species that need it, without raising the paradoxical matter of the pure "naturalness" or "integrity" of such places. It is also free of the many confounding connotations of the wilderness idea. Wilderness was at first justified—even defined in terms of—virile and manly recreation (Callicott & Nelson, 1998). It was also often venerated, historically, as a resource for human aesthetic and spiritual harvest (Callicott & Nelson, 1998). Are wilderness areas human recreational, spiritual, and aesthetic resources, or are they primarily habitat for nonhuman species? When use of wilderness areas by human backpackers and river runners compromises them as habitat for nonhuman beings, what use takes priority—use by human beings or nonhuman beings? If we call the areas now known as "wilderness" instead "biodiversity reserves," then the priority question is answered before it is even asked.

Of Rolston and Norton and the Intrinsic Value of Nature

Despite our real differences on the nature-culture question, Rolston and I differ less on the question of intrinsic value in nature than he supposes. In my most recent essay on the subject, "The Intrinsic Value of Nature: A Metaethical Analysis," I argue that *value* is first and foremost a verb. To value is an intentional act of a valuing subject. Therefore, something (some object, grammatically speaking) is valuable if and only if it is valued by a valuing subject. Valuing subjects value things in many ways, among them intrinsically, for their own sakes. But without a valuing subject there would be no valuable objects. Rolston actually agrees with this analysis, but he does so only cryptically. We human beings self-consciously value ourselves, as well as other things, intrinsically. But lemurs, he notes, also demonstrably value themselves intrinsically, although perhaps not self-consciously. You see, what Rolston is doing here is finding in nature a wide spectrum of nonhuman reflexively valuing subjects. He begins with human subjects, then moves on to our close relatives, phylogenetically speaking, and on from there, to subjects more distantly related and arguably less acutely conscious than lemurs and other primates—birds, reptiles, insects—all in some sense self-valuing subjects. Finally, Rolston posits the existence of valuing subjects stripped of all subjectivity: "Trees are also valuable in themselves," Rolston writes. But why? How? Because, as he explains, they are "able to value themselves." In what sense? Is Rolston going beyond conventional science and claiming a secret, inner life for plants? Not at all: "Natural selection picks out whatever traits an organism has that are valuable to it, relative to its survival. When natural selection has been at work gathering these traits into an organism, *that organism is able to value* on the basis of those traits. *It is a valuing organism*, even if the organism is not a sentient valuer . . ." (emphasis added). So, clearly, although the valuing subject may lack sentience, indeed consciousness of any kind (i.e., the valuing subject may, paradoxically, lack subjectivity) Rolston agrees with me that the value of any object, a valuee, depends, in the last analysis, on the existence of a valuing subject, a valuer.

For Rolston, the ethical payoff of this analysis is characteristically Kantian. Kant (1959) writes, "Man necessarily thinks of his existence this way [i.e., as an end-in-itself, something of intrinsic value] thus far, it is a subjective principle of human action" (p. 47). Kant is intellectually honest; he is fully aware that value is not objective, in the same sense as a rock is objective, something existing independently of the intentional act of a valuing subject. He goes on, however: "Also every other rational creature thinks of his own existence by means of the same rational ground which holds also for myself, thus it is at the same time an objective principle from which, as a supreme practical ground, it must be possible to derive all laws of the will" (Kant, 1959, p. 47). Rolston's environmental ethic follows this Kantian pattern, but broadens the "subjective principle" to the maximum extent possible. Reflexive self-valuing is not confined to "man," nor to "rational creatures," nor even to sentient or conscious creatures, but to any and all evolved creatures. And, just as Kant, Rolston argues that because they value themselves intrinsically, we should value them intrinsically as well. That makes the principle "objective," but in a different sense of the word, which neither Kant nor Rolston marks. The meaning of *objective*, in the above-quoted fragment from Kant, is "unbiased," not "existing independently of the intentional act of a valuing subject." (Norton here calls it "epistemological" as opposed to "ontological" objectivity.) Each organism should be an unconditional end for all moral agents because for itself it is an unconditional end-in-itself. The problem with basing an environmental ethic on Rolston's particular subjectivist theory of intrinsic value is the same problem vitiating Paul Taylor's environmental ethic: respect for nature is limited to respect for individual organisms. Unlike Taylor, Rolston, however, is not content with an exclusively individualistic environmental ethic. He therefore supplements his theory of intrinsic value in nature with complementary theories of "systemic value" and "projective value" that are more holistic. For me here to discuss critically these supplementary value theories would not be appropriate because Rolston does not mention them in his contribution to this volume.

Bryan Norton takes Rolston at face value and believes that Rolston is defending the view that values in nature are ontologically objective, that is, that they exist independently of a valuing subject. My analysis indicates that Rolston is actually defending, albeit cryptically, a subjectivist theory of intrinsic value in nature. Norton is, however, quite alive to the distinction I just drew in reference to Kant between ontological objectivity and epistemological objectivity, pointing out that the former is not a necessary condition for the latter. Unfortunately, Norton, like most self-styled pragmatists in environmental philosophy, seems to spend his energies critiquing "foundationalist" approaches to environmental ethics, such as mine and Rolston's, then calling for a new post-Cartesian epistemology, and leaving it at that. Norton notes that I have myself actually tried to develop a postmodern value theory—based on epistemological extrapolations from the new physics—that tries to break free of the subject-object bifurcation at the heart of modern philosophy (which is more than he offers us here). But in his judgment, my ventures into a postmodern worldview do not do the trick; they are still confined by the "Cartesian structure" (in a way analogous to the way that Ouderkirk thinks my attempt to overcome the dualism of the wilderness idea is still confined by dualist thinking). All this would seem to lead up to Norton's own stab at "creating a new, post-Cartesian worldview," but alas, all we find is a call for one, along with a call for "breaking down barriers among disciplines."

Norton has conducted a career-long campaign against the idea of intrinsic value of nature. At first, he seems to have believed it simply would not play in Peoria (i.e., it would not be useful to environmental activists and policymakers) and that, therefore, "weak anthropocentrism" would represent a more effective approach to environmental ethics (Norton, 1984). As time has gone on, his anti–intrinsic-value-in-nature campaign seems to have become an obsession with a life of its own (Norton, 1995). The idea that nature possesses intrinsic value has, however, become a staple of mainstream environmentalism out in the real world, as Christopher Preston (1998) massively documents. My own evidence is anecdotal but powerful. Following are three representative anecdotes:

1. Under the leadership of Steven Rockefeller, a proposed United Nations "Earth Charter" has undergone a "consultation" process among a host of constituencies all over the world and has been reiterated in countless drafts. The penultimate draft read, "1. Respect Earth and all life, recognizing the interdependence and *intrinsic value* of all beings, affirming respect for the inherent dignity of every person and faith in the intellectual, ethical, and spiritual potential of humanity" (Earth Council, 1999, emphasis added; Sturm, 1999). The Earth Charter Commission iterated a final version of it in March 2000. The term *intrinsic value*, unfortunately, did not survive the often intensely political consultation process to appear in the final draft. But the concept remains: "1. Respect Earth and all life in its diversity. a. Recognize that all beings are interdependent *and every form of life has value regardless of its worth to human beings*" (Earth Charter Launch, 2000, emphasis added). Norton did not anticipate the capacity of people to embrace the idea of intrinsic value in nature, I believe, precisely because he has underestimated "the intellectual, ethical, and spiritual potential of humanity" to reach beyond anthropocentrism, both strong and weak, for more expansive and inspiring values.
2. I was part of an interdisciplinary research team sponsored by the Great Lakes Fishery Commission charged to chart a course for fishery management in the Great Lakes for the twenty-first century, an example of actual, as opposed to called-for, interdisciplinary work. Part of my task was to discover the values really at play among stakeholders in the Great Lakes. Several stakeholder-generated documents that I assayed for value statements, expressed the intrinsic value of the lakes and their fishes, among them a fishery management plan for the waters of Lake Superior under their jurisdiction crafted by the Minnesota Department of Natural Resources (Callicott et al., 1998). The idea that nature possesses intrinsic value has, as this instance shows, trickled all the way down into a conservative, pedestrian state bureaucracy.
3. Environmental activist par excellence Dave Foreman (1991) notes that "During the 1970s, philosophy professors . . . started

> looking at environmental ethics as a worthy focus for discussion." And he goes on to point out that "by the end of the '80s, few conservation-group staff members or volunteer activists were unaware of the Deep Ecology–Shallow Environmentalism distinction [the anthropocentric-nonanthropocentric distinction, in other words] or of the general discussion about ethics and ecology. At the heart of the discussion was the question of whether other species possessed ***intrinsic value*** or had value solely because of their use to humans" (p. 8, emphasis added).

Foreman's remarks reward close scrutiny. What we environmental "philosophy professors" did was create a useful new discourse first for environmental activists, and eventually for environmental professionals. As the "discussion" spilled out of the "ivory towers" and "dusty academic journals" only its general topic and core issue entered popular awareness and discourse (Foreman, 1991, p. 8). The finer points of theory—Rolstonian objectivism, Callicottian subjectivism, whatever—were filtered out. What remains is the basic idea: nature somehow possesses "intrinsic value." Nor, furthermore, do the finer points of theory make any difference to the pragmatic implications—the power—of the idea. Compare human rights. The discourse of human rights was unknown in the ancient and medieval eras. Early modern philosophers, such as John Locke and Thomas Hobbes, started "discussing" human rights and speculated about how we came by them—a grant either from God or from nature. Thomas Jefferson and other eighteenth-century statespersons institutionalized the new rights discourse in such political documents as the Declaration of Independence and the U.S. Constitution and its Bill of Rights. By now, rights talk is as common as talk of tables and chairs, birds and trees. Few philosophers today, however, believe that human beings possess human rights in the same way that we possess shoes, teeth, kidneys, and thoughts. Rights are, rather, usually theoretically reduced to "justified claims" in contemporary philosophy (Nickel, 1992). However, the fact that most contemporary rights theorists reject an objective God-given or natural rights theory undermines the considerable practical efficacy of rights discourse not one iota. Similarly, the fact that I (and, albeit cryptically, Rolston as well) reject the ontological objectivity of intrinsic value in nature undermines the practical efficacy of the new intrinsic-value-in-nature discourse not one iota.

Comparison of the idea of intrinsic value in nature with the idea of human rights brings us back to the importance of the Earth Charter. The adoption in 1948 of a Universal Declaration of Human Rights by the United Nations represented no international law or treaty binding on member states (Brownlee, 1981). But over the half century since its adoption, the declaration has increasingly set a standard by reference to which member states are criticized and sometimes sanctioned. The efficacy of the Earth Charter is similarly symbolic and rhetorical. After its adoption by the United Nations, we may hope—and, I dare say, expect—that the environmental behavior of member states will also be criticized and sanctioned by reference to its principles, the first of which affirms the intrinsic value of nature. Bryan Norton is one my closest friends in the community of environmental philosophers. He has done some excellent work (which I greatly admire) unrelated to the intrinsic-value-in-nature debate, but, unfortunately, he risks being remembered fifty years after the United Nations adopted the Earth Charter as a naysaying, reactionary twentieth-century anthropocentrist, a self-styled pragmatist who failed to see the pragmatic potential of a grand new idea and the intellectual, ethical, and spiritual potential of humanity to embrace it.

Eugene C. Hargrove, I fear, will share a similar fate, for he too believes that "an emphasis on the revisionary at the expense of the descriptive will . . . simply make environmental ethics completely esoteric and unusable." Quite the contrary, in my opinion; people, especially now in our fast-paced culture, are oriented more to the future than to the past and are receptive to—indeed, have a strong appetite for—new, inspiring ideas. Without providing a scrap of evidence, anecdotal or otherwise, to back his claim, Hargrove flatly asserts as plain fact "the rejection of nonanthropocentric intrinsic value by the general public." The evidence I have offered here, not to mention that which Preston has assembled (1998), indicates just the opposite, a snowballing embrace of the idea by the general public.

Of Hargrove, Larrère, and Palmer and Ecoquantum Metaphysical Foundations

As to Hargrove's charge that I really do not have a metaphysics, not even a metaphysics of morals, as he has rather narrowly defined the terms, I offer no rebuttal. Instead I refer readers to the essays of Catherine Larrère and Clare Palmer to judge for themselves. More important are Hargrove's droll comments on philosophical eclecticism.

Can contemporary environmental philosophers borrow selectively from a wide variety of past philosophical systems in crafting their own? Hargrove interprets me to claim that we cannot; and accuses me of doing so, nevertheless, myself. Certainly I do borrow from earlier philosophers, Hume for example; and quite selectively, accepting Hume's theory of moral sentiments, but ignoring his theory of impressions and ideas, which I would reject were I actively to consider it. I think that selectively borrowing is appropriate, but only if the parts ignored or rejected are not necessary conditions or implications of the part appropriated. I do not think, to continue with the present example, that Hume's theory of impressions and ideas is a necessary condition of his (or any) theory of moral of sentiments, nor does the latter entail the former.

Furthermore, I think that to stitch bits of one philosopher's views together with those of another is appropriate, provided the bits are mutually consistent, even though other, logically independent bits, of the views of the philosophers in question may be mutually inconsistent. Thus, I think that it is appropriate to stitch together Hume's theory of moral sentiments with Darwin's account of the origin of ethics because they are mutually consistent, even though Hume's skepticism about cause-and-effect relationships may be inconsistent with Darwin's attempt to assign causes for differences among species. What I oppose is the attempt to stitch together bits borrowed from one philosopher with bits borrowed from another, when the appropriated bits from disparate sources are mutually inconsistent or rest on mutually inconsistent necessary conditions, assumptions, or implications. Therefore, contrary, it seems, to Angus Taylor, I think that linking animal liberation, à la Peter Singer, with the Leopold land ethic is not appropriate because animal liberation and the land ethic rest on mutually inconsistent assumptions: in the former, the assumption that we have duties only to sentient beings, regardless of their relationship to us; in the latter, the assumption that duties are generated by and oriented to community, regardless of sentience. A better fit with the Leopold land ethic, I think, is Mary Midgley's animal welfare ethic, a key concept of which is the "mixed" human-animal community, because it too rests on a kind of communitarianism (Midgley, 1984).

Larrère's essay is a masterpiece of erudition and penetrating insight, leavened by "the decadent perversity of postmodern [French] philosophy," a vice in which she is happy to indulge. I only wish all my critics combined her virtues—in addition to erudition and insight, a thorough familiarity with the work she is criticizing and generosity of spirit in its interpretation—with so charming (albeit frustrating) a vice. Contrary to Hargrove, Larrère believes that my "main contribution" to environmental ethics is to show that this new field of philosophy "not only require[s] a metaethics, but a metaphysics," and she discusses quite a bit of my work that would tend to support that judgment. I find much in Larrère's discussion illuminating, especially the similarities and differences she sketches between my critique of modernity together with my version of postmodernity and those of such contemporary European "poststructuralist" philosophers as Ilya Prigogine and Isabelle Stengers, Gilles Deleuze and Felix Guattari, René Thom, and Michel Serres.

I might not have fully appreciated Larrère's most general critical point had I not come to a similar realization on my own. In her words, it comes to this: "actual sciences are not identical to natural philosophy." I am—she is right, I admit—wont to conflate "science" and "natural philosophy," as well as to confuse the latter with a "philosophy of nature." I can explain this terminological error, which is responsible for more painful misunderstandings than one might suppose, if not excuse it. As I noted at the beginning of this essay, my first intellectual love and scholarly training was in ancient Greek philosophy. For that reason, I habitually take a rather long perspective, temporally speaking, on intellectual history in the West. From that perspective, science as we know it today, is a rather recent phenomenon that ultimately grew out of Greek natural philosophy. Moreover, the founders of modern science—Galileo, Descartes, Newton—did not call themselves "scientists," nor did they call their work "science"; they called themselves "natural philosophers" or sometimes "natural theologians" (Burtt, 1952). The word *science* did not enter the English language in its contemporary sense until the mid-nineteenth century (Whewell, 1840). Hence science, as we know it, has emerged as something distinct from natural philosophy for only a century and a half. Finally, the work of some of the most celebrated scientists of the twentieth century—such as Albert Einstein, Werner Heisenberg, and Eugene Odum—is indistinguishable from natural philosophy.

In a relatively brief time, however, most normal science has emerged as something quite distinct from natural philosophy. And the word *science* has acquired connotations very different from, even opposed to, the connotations of "natural philosophy." The rigorous, formal (not to say ritualized) hypothetical-deductive-experimental method of science has wed it to manipulative technology. Thus the distinction that is so clear between natural philosophy and technology is not so clear between science and technology. Science, moreover, is not only inherently reductive, as scientific "knowledge" proliferates, science becomes ever more specialized, divided into more finely and sharply partitioned disciplines—various subfields of physics, chemistry, biology, geology—while natural philosophy is synoptic, synthetic, and integrative. Finally, science has become a huge social institution commanding great sums of money, prestige, and power. (Needless to say, natural philosophy has not.) Because of the social status of science, there has also emerged something that Larrère labels "scientism"—a kind of reverence for science, a naive trust in its pronouncements—which is antithetical to the provisional nature of science itself and the falsification-oriented epistemology of the scientific method. Therefore, when I write about the way science informs ethics, no wonder there is confusion, worse frustration, and worse still, irritation, and anger (expressed here especially by Hester et al.)! What I should have all along written about is the way changes in science—the emergence of quantum theory and relativity in physics, of systems theory and later neo-Gleasonianism in ecology—have induced changes in natural philosophy, which in turn informs ethics. Larrèrre's conclusion is worth repeating, nor do I wish to gainsay it: "We must not forget that in 'natural philosophy' there is 'philosophy,' not just science."

Perhaps the distinction between science and natural philosophy that Larrère insists on, and that I now warmly endorse, will serve to deflect Clare Palmer's criticisms that center on my claims about the existence of a "scientific worldview," and especially about the emergence of a new postmodern scientific worldview. I might not be in such trouble with Palmer had I claimed that a mainstream, prevailing, or predominant natural philosophy exists—classical mechanics—and that a postmodern natural philosophy is emerging in its place. Nonetheless, I fear I would still be guilty, in Palmer's opinion, by association with certain figures of New Age notoriety (Fritjof Capra and Rupert Sheldrake are fingered by Palmer) and certain Deep Ecologists (who remain unidentified, but I suspect the culprit Palmer has in mind is Warwick Fox). On this particular point Larrère provides more particular support:

> To argue his thesis of the "metaphysical implications" of contemporary science [that is, as he now better understands, contemporary natural philosophy], Callicott presents the "general conceptual notions" that can be "abstracted" from ecology and what he calls the "new physics" (relativity and quantum theory). Three of them are especially notable:
>
> 1. Quantum physics, and more especially Heisenberg's principle of uncertainty, undermines the subject-object dichotomy, which is at the core of modern metaphysics (from Descartes to Kant).
> 2. Contemporary ontology is relational, "relations are 'prior' to the things related" and entities are not independent objects, but internally related, "knots in the web of life."
> 3. Contemporary ontology is holistic. It is "not possible to conceive of an entity in isolation from its milieu."
>
> Thus summarized, these are the basic assumptions of any constructive alternative to modernity, based on contemporary science. Merleau-Ponty, Serres, Thom, Prigogine, Stengers everybody agrees that nature can no longer be seen, as it was in the modern era, as made of "*partes extra partes*," discrete entities connected by external relations. The emphasis is on internality, upon the interactions between the entities and their milieu. Prigogine and Stengers speak of "entities intrinsically constituted by their irreversible interaction with

the world," Serres criticizes the archaism of those who think that sciences deal with objects, or substances, when relations or interactions prevail.

Are "Merleau-Ponty, Serres, Thom, Prigogine, Stengers everybody" who is anybody in Larrère's estimation, dismissible as New Agers or Deep Ecologists? Are we to prefer the extraordinary archaism of one John Dupre, cited by Palmer, who assumes that the world is such that he can intelligibly speak of its "contents," as if it were a miscellaneous assortment of things in a Euclidian container, *partes extra partes*? Larrère writes from a point of view that is self-consciously French, whereas Palmer seems to exhibit, without noticing it, a bias that seems to me to be characteristically British. Call it the bias of *partes extra partes*. In American English, we say Congress *has* enacted a law, or SUNY Press *has* published this book; in British English, one says Parliament *have* enacted a law or SUNY Press *have* published this book. The very syntax of American English is holistic; governmental bodies and publishing houses are consortiums composed of individual persons, to be sure, but such persons collectively constitute an emergent entity through their relationships with one another and a wider circle of other persons. At least that is the American assumption as revealed in the way subject and verb are conjoined: a singular subject is conjoined with a singular verb. Although *Parliament* and *SUNY Press* are singular proper names, British English, to the contrary, insists on conjoining them with a plural verb. It is as if Parliament were nothing but an aggregate of MPs that do, severally, enact laws (each doing his or her bit by independently casting his or her vote, or so the syntax seems to suggest) and SUNY Press an aggregate of editors who do (each individually the British syntax insists) publish books.

Further to a national difference, contemporary French philosophers are typically also historians of ideas, whereas twentieth-century British philosophy—I am thinking of the ordinary language and analytic schools here—has (or should I say "have"?) tended to break continuity with the past. The rather long view of the emergence of modern science from Greek and eventually modern European natural philosophy and the continuing importance of natural philosophy that I assume may well be unintelligible from an ahistorical perspective on science in which "the fragmentation and specialisation within many scientific disciplines" may loom overwhelmingly large. From a long historical perspective, however, fragmentation and specialization in science is a very recent phenomenon, the emergence of which has not—not yet anyway—compromised the historical unity of science and therefore the possibility of distilling from it some characteristic natural philosophy.

Palmer reiterates a criticism Rolston (1989) leveled at me some time ago and which I answered shortly thereafter in "Rolston on Intrinsic Value: A Deconstruction"—that ontological peculiarities at the quantum scale have nothing to do with the middling scale at which we directly experience the world in which we live. Apparently she is unaware of this exchange. I am not suggesting, in any case, that the uncertainty and indeterminacy at the quantum scale cause something similar at the mesoscale of human experience. Rather, there is an interesting historical correlation—in the fifth century B.C.E. and the seventeenth century C.E.—between the waxing of atomism in natural philosophy and the waxing of egoism and individualism in moral philosophy and in society. In the past, the atomic and mechanical model of nature was, in fact, transposed to moral philosophy and society analogically. Individuals are, as it were, social atoms. History then suggests that the contemporary waning of atomism in natural philosophy might be followed by the waning of egoism and individualism in moral philosophy, and indeed in society. Complementing this retort, Larrère adds one of her own: "To Rolston's objection that the implications of quantum theory are relevant only for the microscopic . . . level, but not for the middle-level," one may reply that, "in both cases, the observer is part of the situation [s]he is observing. With it, [s]he has an internal relationship that knowledge actualizes." Larrère clearly appreciates the relevance of quantum theory to ecological ethics. The science at the basis of the hierarchy of sciences, quantum theory, and the science at the apex, ecology, each reinforce the holism of the other.

Palmer, however, notes that quantum theory is amenable to a wide variety of interpretations—some of which support my natural philosophical conclusions, whereas some do not—and I am not very clear on the one I endorse. Indeed, I am not; I am not even, I confess, familiar with all the available interpretations of quantum theory; it is a very arcane and immense subject of study. But at least this I think we can say with confidence: as science explores the structure of matter ever more finely, we can no longer believe that matter is picturable as ultimately composed of very tiny, solid, externally related substances (like miniature marbles or BBs) extended in Euclidian space, as atomists from Democritus to Newton seem to have supposed; nor can we believe that we are essentially cogitos observing without affecting a world so constituted. From these postmodern features of postmodern natural philosophy, the widely shared conclusions enumerated by Larrère follow. We can also be confident, furthermore, that the dissipation of the modern classical Cartesian-Newtonian natural philosophy will have a ripple effect throughout the rest of philosophy, moral philosophy included, if history is any guide. What that effect will be, I have offered my speculations—tinged, to be sure, with my hopes—nothing more.

Of Wenz, Light, and Bratton and the Moral Pluralism Controversy

The controversy swirling around my critique of moral pluralism can perhaps also be best illuminated by reference to the long history of Western philosophy to which I have always regarded my work as a contemporary contribution. I think that moral philosophy has provided an enormous service to society. Sustaining that service into the contemporary period, requires, in my opinion, a certain loyalty to the methods of our predecessors in the tradition, which have proved to be so fruitful. It also requires, I further opine, establishing continuity with their doctrines, even as we go beyond them, pressing on into new domains of inquiry, such as environmental ethics. Thus I have tried to establish continuity between the novel Leopold land ethic and the moral philosophies of Hume and Smith. In the context of the moral monism versus pluralism debate, however, I am more concerned with methodological than with substantive continuity in the tradition.

The moral philosophers of the past did their damnedest to provide a coherent and comprehensive moral philosophy and sell it, as it were, in a market place of ideas, on its merits in comparison with competing systems. (Doubtless some critics will pounce on this economic simile and my comparison of moral philosophy to competitive sports in the paragraph after next as proof that Callicott uncritically embraces degenerate modern institutions; but these are only similes, and I recognize dissimilarities as well; nor should anyone suppose that I hereby endorse uncritically free-market economics or competitive sports.)

Consider especially the work of two of our greatest predecessors, Jeremy Bentham and Immanuel Kant. Both exhibited the merits of his system in its best light and portrayed the inadequacy of alternative systems in their worst. Bentham (1823, chap. 2, sec. 2) grouped the systems of Hume and Kant under the "principle of sympathy and antipathy," which, in his opinion, was antithetical to the one, true principle of utility. About Hume, he writes, "One man says, he has a thing made on purpose to tell him what is right and what is wrong: and that is called a *moral sense*; and then he goes to work at his ease, and says, such a thing is right and such a thing is wrong—why? 'because my moral sense tells me it is'" (Bentham, 1823, chap. 2, sec. 15, note 7). About Kant he writes, "Another man comes, and says, that as to a moral sense indeed, he cannot find that he has such a thing: that however he has an *understanding*, which will do quite as well. This understanding, he says, is the standard of right and wrong: it tells him so and so. All good and wise men understand as he does: if other men's understandings differ in any point from his, so much the worse for them: it is a sure sign they are either defective or corrupt" (Bentham, 1823, chap. 2, sec. 15, note 7). These characterizations are caricatures, of course. My point is that Bentham, founder of the enormously influential utilitarian school, was not a moral pluralist. He constructed a system of ethics that he believed to be comprehensive and contended that his was the best going. Kant (1959, p. 60), of course, was just as antipluralistic, and equally heavy handed in classifying all the alternatives to his categorical-imperative system under the "principle of heteronomy." And, as we might expect, he reserves his harshest condemnation for utilitarianism: the standard of "happiness is the most objectionable of all . . . for it puts the motives to virtue and those to vice in the same class, teaching us only to make a better calculation while obliterating the specific difference between them" (Kant, 1959, 61).

This is what I thought (and still think) moral philosophers should do. The adversarial, dialectical debate about the "true" nature of morality and the one best system of ethics is (and should be) ongoing. It is like a competitive sport with no end, no last inning, no final buzzer, no Super-Bowl winner. By now the players know that no one, least of all they themselves, will have the final word; no one supposes that everyone will become a Singerian, a Rolstonian, or a Callicottian. But the contestants must play their parts as if there were an end game, a winner; they must give it their best shot. Because in doing so they are making a most valuable social contribution. One aspect of this contribution, I have already mentioned. Moral philosophers create, shape, and infuse powerful ethical discourses: Hobbes and Locke, the discourse of human rights; Bentham and Mill, the discourse of utility; Kant, of duty; Singer, of sentience; Rolston, of intrinsic value in nature. It detracts little from the social services that moral philosophers render that the majority of people who benefit from their services—who, say, insist on their right to drive a car or pursue happiness in a shopping mall—have never heard of any of these philosophers. Louis Pasteur, Thomas Edison, Booker T. Washington, Rachel Carson, and hundreds of other people who have, in one way or another, shaped the cultural environment in which we live are unknown to most of us. "If you asked people worldwide if they are Christian . . . vast numbers would answer in the affirmative," Susan Bratton observes, but "if you asked them if they are Kantian, probably a fraction of 1% (primarily well-educated Westerners) would say 'yes.'" A property-rights zealot defending his acres against the threat of phantom United Nations black helicopters may well call himself a Christian defending Christian values, but property rights, at least as far as I recall, is not an important biblical concern. His worldview is more Hobbesean and Lockean than Christian, although he may never have heard of Thomas Hobbes or John Locke, nor, certainly, identify himself as a "Hobbesean" or "Lockean."

To push the envelope of ethics, in response to new moral problems, and to continually add to and enrich ethical theory is another social contribution of monistic moral philosophers. Ironically, moral pluralism frustrates this process of growth and enrichment. Rather than struggling to construct a system powerful enough to embrace all our expanding interpersonal, interspecies, and environmental moral problems, the moral pluralist throws in the towel and proposes to deal with each ethical domain piecemeal—utilitarianism for this problem, deontology for that, and so forth. We have theory enough, the pluralists seem to suggest; we can deal with any problem by deploying one or another of those already available.

My stance toward pluralism, I admit, is ambiguous for two reasons. The first is that the environmental ethics I construct identifies multiple duties and obligations, generated by our multiple community memberships. In that sense, at what Peter Wenz calls the "level of principle," it is pluralistic. But in another sense, at the level of theory, the system I construct and advocate is monistic, because all our duties and obligations are generated by community membership. Our multiple (pluralistic) duties and obligations are, that is, united by a (monistic) communitarian theory of ethics. Furthermore, I have, in response to pressure from my critics, recently added two second-order principles, as already discussed, to prioritize among the multiple duties and obligations, generated by our multiple community memberships. The second reason that my stance toward pluralism is ambiguous is that I am at once a moral philosopher and a plain member and citizen of various communities. As a moral philosopher, I am, for the reasons just elaborated, committed to monism. I am a part of the Western tradition of moral philosophy, and what moral philosophers traditionally do is construct comprehensive ethical theories and defend the merits of those theories in comparison with the available alternatives; it is our job. But as a plain member and citizen, I am a pluralist. I have just lauded the social value of having a contentious bunch of moral philosophers advancing the front of ethical theory through mutually critical, dialectical discussion, and, in the process, creating a rich diversity of powerful new discourses as instruments of social change.

Andrew Light seems to want to substitute for this social service a much more modest one: we might call it "values reconciliation and conflict resolution." So, it seems, does Susan Bratton who asks rhetorically, "are we searching for 'an ethic'?—the Holy Grail of the environment. Or should we be investigating ways to facilitate environmental problem solving and sensitivity within existing ethical systems?" To which I respond with another rhetorical question: Why not both? And why must I be faulted for pursuing the former, for surely I do not fault anyone for pursuing the latter? I'm an intradisciplinary pluralist; Norton, Bratton, and Light seem to be intradisciplinary monists. If cognitive social work is what they believe to be the best use of their time and energy, so be it. I do not think that it is the best use of mine. I have no objection to their doing the work they think is the most fruitful. Why do they object to me doing mine?

Warming to her theme, Bratton asks, "Is it the task of environmental ethicists to dictate norms, or is it their task to facilitate the greater society's development of norms or values?" Certainly our task is not "to dictate" norms; Plato's dream of philosopher kings (philosopher dictators) has, fortunately, never been realized. As I just tried to explain, for those of us who make it our own, our task is, rather, to propose, theoretically to justify, and to defend new norms, such as the intrinsic value of nature. And the fulfillment of that task contributes profoundly, albeit indirectly, to the task of facilitating "the greater society's development of norms or values." (Once again, I call attention to the Earth Charter as a case in point.) In addition to environmental social science and social work, Bratton also suggests that we need to develop "historic, sociological, or anthropological information" in order better "to relate schools of thought and specific cultural groups to their actions and lifestyles." Certainly I would agree. Environmental philosophy, I repeat, renders an enormous service to society, but it is not the only one to be rendered and may not be the most important. Are philosophers, such as I, however, the best people to develop "historic, sociological, and anthropological information," which would, I am the first to testify, be of great value? Would such information not, however, be better generated by historians, sociologists, and anthropologists? Bratton goes on to enumerate seven sociological criteria for "judging the viability of religious ethical responses to environmental problems." All of them seem reasonable to me; but I am not a sociologist, so I am not in a position authoritatively to evaluate them. Maybe there is just no use for environmental philosophers doing environmental ethics that is continuous with the ethics done by philosophers for 2,500 years, off and on, in the Western tradition. Perhaps we live in a brave, new world that, although a product of that tradition, has no need of those furthering it—except as whipping boys—in the postmodern future. Perhaps, we methodologically old fashioned environmental philosophers should either retrain as historians, sociologists, anthropologists, or pragmatic facilitators, or, better, just retire and fade away into the sunset of the twentieth century. Of course, I vigorously disagree.

Of Gruen and Taylor and Further to the Moral Pluralism Controversy

Lori Gruen's essay continues the conversation about monism and pluralism from an ecofeminist perspective. I am afraid that the ambiguity just noted in my stance toward the issue has led Gruen to interpret my opposition to intrapersonal pluralism as an opposition to interpersonal pluralism. She quotes me as affirming "that we human beings deeply need and mightily strive for consistency, coherency, and closure in our personal and shared outlooks on the world and on ourselves in relationship to the world and to one another . . . [and that] we feel (or at least I feel), that we must maintain a coherent sense of self and world, a unified moral worldview." But my unified moral worldview may be different from that of hers, yours, and a host of others—quite obviously, if the opposition to it variously expressed here is any indication. Both individual diversity and cultural diversity exist. And that is good, for all the reasons I have just enumerated and doubtless for many other reasons as well. But part of the process that refines and enriches that diversity—and above all makes it dynamic—is mutual engagement, debate, and criticism. Thus, although respecting and valuing the way others try to organize their moral experience into a consistent, coherent whole, my social responsibility, as a moral philosopher, is to try to show how my way is better. My responsibility is also to attend to criticisms of my way offered by others, which is precisely what I am doing here, and change my way, which I have done many times (once right before your eyes in my response to Larrère), if I am persuaded that some other way of organizing experience into a comprehensive, self-consistent, and coherent whole is better.

Another misinterpretation that Gruen joins Bratton in making (and, of course, Hester et al.) is that a monist moral philosopher such as I would impose my own moral worldview on others. Perhaps Plato and Karl Marx would impose their personal unified moral worldview on others, if they had the chance, by force of arms or by undemocratic political power, but certainly I would not impose my views on others by any coercive means. Rather, I would hope to win the agreement of others by the persuasive force of argument.

What is especially striking—and, from my point of view, especially welcome—about Gruen's contribution to this volume is her frank and intellectually honest exposition of the monism-pluralism (universalism-multiculturalism) dilemma within ecofeminism. Gruen quotes Isaiah Berlin's particularly strong, unambiguous, and unqualified affirmation of the existence of "true" and "correct" beliefs in the moral and social realm (similar to those he supposes exist in the natural realm) and that ethics should be as certain as scientism supposes science to be. As I read on, I was steeling myself for Gruen's attribution of such views to me, only to be relieved that that was not her purpose in quoting Berlin. For I think that the best we can attain in both the moral and natural domains are not true beliefs, but only more comprehensive, consistent, and coherent beliefs. And I hold out as an ideal but never attainable goal, more or less following Jurgen Habermas (1970), that we could all come to provisional agreement about the best available—not the absolutely truest—set of beliefs, if, as persons of both intellectual and moral good will, we took the time to argue through our differences (Sosa, 1996).

To my relief (and for this I am grateful) Gruen does not join Hester et al. in saddling me with Berlin's absolutistic universalism, but to my astonishment, she goes to admit that to her there "is much appeal to such a view." It would obviate claims that feminists and animal liberationists were trying to impose their culturally spawned values on those members of other cultures for whom the subordination of women and use of animals is culturally institutionalized. (And I think that *impose* here is the right word because animal liberationists and feminists would coercively prevent, if they could, any and all people from engaging in blood sports such as bull fighting, "circumcising" their daughters, and perpetuating other practices that animal liberationists and feminists especially abhor and condemn, although I have no desire to force people against their will to accept the Leopold land ethic, which Bratton as well as Hester et al. falsely accuse me of trying to impose on others or dictate to them.) Gruen joins me, I am happy to note, in finding a middle path between absolutistic universalism and cultural relativism. Although parallel, our middle paths, however, do not seem to be exactly the same.

I certainly agree with Angus Taylor's conclusion that "an alliance of the animal-liberation and environmental movements is not only possible but imperative." His difference with me lies primarily in the theoretical starting point for this project. Do we start with the Leopold land ethic, as I suggest, and try to reach animal liberation through the concept of "mixed" human-domestic animal communities that Midgley (1984) articulates? Or do we start with utilitarianism and deontology as Peter Singer and Tom Regan respectively recommend? Taylor opts for the latter strategy.

With his concluding point (3)—"an end to factory farming of animals"—I not only agree, but add that no policy would be more environment-friendly than a total elimination of industrial animal agriculture. Even environmentalists who are wholly indifferent to animal suffering and think that the idea of animal rights is nonsense on stilts should support a policy of universal vegetarianism. Taylor mentions the reason why: industrial animal agriculture contributes to "habitat destruction." I wish he had elaborated a bit. The removal of cattle, sheep, and other domestic animals from the arid range land in the western United States would return millions of acres to native wildlife (Weurthner, 1992). Much of the plow land in the midwestern United States is devoted to feed crops, such as corn and soybeans, most of which are eaten not by human beings, but fed to factory-farmed cows, pigs, and chickens (Lappé & Collins, 1979). Animals burn about 90 percent of the food they eat and convert the rest (only 10 percent at best) to meat, so Americans would need only about 10 percent of the land now under cultivation to grow food if we consumed grains and legumes directly and altogether eliminated mass-produced meat from our diets (Lappé & Collins 1979). The elimination of industrial animal agriculture would, therefore, make millions of acres available for prairie restoration on a truly grand scale. (The potential for retirement of farmland in parts of the world where people eat less meat would, of course, be more modest.)

Therefore, environmentalists should be even more keen to support policies aimed at the total elimination of industrial animal agriculture than animal liberationists. Why? Because the elimination of industrial animal agriculture would be tantamount to the elimination of all but a few museum specimens of the animals that such policies, from an animal-liberation point of view, are designed to benefit. The environmental benefits of such policies are unambiguous, but the animal-liberation benefits are paradoxical. In saving the millions of specimens of cows, pigs, chickens, and other domestic animals from suffering or having their rights violated, by calling a halt to industrial animal agriculture, their future populations would be reduced to scattered remnants here and there on hobby farms and in zoos. I hope that Taylor and other animal liberationists do no not entertain the environmentally destructive alternative of releasing the legions of domestic animals now confined in factory farms into the wild. Feral populations of domestic animals—mustangs, burrows, razorbacks—are already a scourge in various native wildlife habitats of North America; simply opening the doors of factory farms and literally liberating the inmates would be an environmental disaster.

I do not think that Taylor would support the literal liberation into the environment of the domesticated victims of industrial animal agriculture because he makes a welcome concession to environmental concerns by acknowledging the occasional necessity of "harming individual animals"—the necessity, for example, of reducing irrupting deer populations by lethal means—for the sake of "preserving the integrity of ecosystems." Large populations of feral animals would certainly diminish the integrity of ecosystems. In his concluding point (1), Taylor once again acknowledges that "compelling reasons" of an environmental kind might dictate hunting certain species of wildlife in certain circumstances. He then adds another exception to his proposed "hands-off policy toward wildlife": hunting and fishing are justified when "necessary for subsistence." I can only guess that this is a concession to certain contemporary indigenous peoples, who traditionally subsisted by hunting and fishing (as well as by "gathering" or foraging, we may also suppose). But much turns on the strength of "necessary." In today's global village, few groups of people are so isolated that hunting and fishing are absolutely necessary to live. Rather, for many peoples, hunting and fishing are necessary to maintain continuity with the past and cultural identity. The question arises then, what about the cultural identity of the rural British gentry, which depends, they may claim by parity of reasoning, on maintaining the tradition of fox hunting; what about the cultural identity of American rednecks, among whom blood sports, such as cock fighting, are traditional; and so on and so on? A newsmagazine wag once wrote (I cannot remember who or where I read it) that "multiculturalism ends where feminism begins." Taylor seems to suggest that animal liberation ends where multiculturalism begins. And I ask, if so, how broadly should concessions to cultural identity extend, and by what criterion are legitimate cultural exceptions to be distinguished from illegitimate ones? These are the questions that Gruen confronts directly, but only hints at the way she would answer them.

Of Hester, McPherson, Booth, and Cheney and Multicultural Ecological Ethics

Lee Hester, Dennis McPherson, Annie Booth, and Jim Cheney (hereafter Hester et al.) critique *Earth's Insights*—and, from all one can tell by what they write, the book focuses on capital "I" Indigenous traditions of thought and little else. Most of my book, however, discusses Judeo-Christian cosmogony; Greco-Roman mythology and philosophy; Islam; Hinduism; Jainism; Theravada, Hua-yen, Tendai, Shingon, and Zen Buddhism; Taoism; and Confucianism, as partly witnessed by Bratton. In addition, a number of indigenous traditions of thought are discussed—those of Polynesia, Africa, and Australia, as well as of North and South America. In short, although purportedly a critique of *Earth's Insights*, the essay by Hester et al. touches only on a small portion of the book. Moreover, their habit of capitalizing *Indigenous* essentializes and totalizes the diversity of thought comprehended by the term; it erases the differences internal to indigenous traditions of thought; and permits the self-identified "Indigenous" spokespersons among Hester et al. to speak with a self-conferred authority for all Indigenous traditions of thought. As soon as the particulars of capital "I" Indigenous thought come up, however, the authors narrow their indigenous spatiotemporal horizons to "the North American continent prior to European contact." Are we to believe that what is (allegedly) true of North American indigenous thought is true of all indigenous thought—South American, Polynesian, African, Australian, Asian, European?

In the second endnote, however, Hester et al. write: "This essay was shaped by Western problematics in environmental ethics. Indigenous coauthors [Hester and McPherson] provided *some* Indigenous content, but the problem is Western and in most ways *so is the voice of this essay*" (emphasis added). By process of elimination, that means, in plain English, that Booth and Cheney (or perhaps just one of these two) supplied most of the indigenous content and wrote most of the Hester et al. essay. Surely then, there is some irony in Hester et al.'s charge that "Indigenous thought provides color commentary in the local vernacular" for my project in *Earth's Insights.* In the Hester et al. broadcast studio, apparently Cheney and Booth are calling the game play by play, whereas Hester and McPherson are providing color commentary. We are therefore compelled to wonder on what grounds we are supposed to accept Hester et al.'s pontifications—so confidently asserted—about the indigenous attitudes and values respecting the environing "world" prevailing "on the North American continent prior to European contact." No historical sources are cited for these putative Indigenous "world"-oriented postures of "respect." The justification we are supposed to accept for this appears to be as follows. Two of the coauthors self-identify as indigenous. The approach to ethics, the values, and the epistemological commitments asserted by the other two "Western voices" (or perhaps by just one of them) are, therefore, the same as those of all indigenous peoples everywhere, past as well as present. Apparently, that is, we are supposed to accept undocumented assertions about capital "I" Indigenous thought by the Western voices among Hester et al. because their "indigenous" coauthors provide some of the content and endorse the rest.

We might suspect that Hester et al.'s idyllic and wholly undocumented claims about the environmental attitudes and values indigenous to "the North American continent prior to European contact" are little more than conventional romantic fantasy. They are, rather, projections into the past of the authors' own attitudes and values. Two of the authors do, after all, self-identify as Indigenous. Thus, to know what the attitudes and values of Indigenous peoples on the North American continent were before European contact, all that is necessary is to consult their own attitudes and values and those of their Indigenous friends and relatives. Accordingly, we learn that pre-

Columbian Indigenous North American peoples were "pluralists" just like the authors: they did not fight over "monistic ideologies," but rather lived harmoniously in an "emergent ecological order." I repeat: No historical evidence whatever for this claim is offered; and the evidence that does exist supports an opposite conclusion.

Consider more reliable (that is, actual historical) information about two of the peoples Hester et al. invoke. First, one nineteenth-century Ojibwa author offered this authoritative account of the origins of the name other indigenous peoples called his people: "The word is composed of *O-jib*, 'pucker up,' and *ub-way*, 'to roast,' and it means, 'To roast 'til puckered up.'" As this indigenous author goes on to explain, "because of uncontrolled feeling incited by aggravated wrong," the Ojibwa were widely renowned for the way they "tortured by fire in various ways" their Indigenous enemies (Warren, 1970, p. 36). The name the Ojibwa called themselves, *An-ish-in-aub-ag*, according to the same authority, may be translated as "original man" (Warren, 1970, p. 56). Other indigenous peoples, by implication, are not "original man," that is, not human, at all. Second, Hester et al.'s invocation of both Navaho and Hopi cultural material to illustrate pluralism and harmony among all North American peoples (prior to the corrosive effects of European conquest, that is), is especially ironic. The bitter ethnic conflict, lasting to this day, between the Navajos and the Hopis antedates European contact. Pre-Columbian Hopis called the Navajo "*Tusavutah* (*tu*–person; *savutah*–to pound), because they killed or captured an enemy by pounding his head with a rock" (Waters, 1963, p. 312). The Navajo were, from the Hopi point of view, the Head-Pounding people. The horticultural Hopi felt preyed upon by the nomadic Navajo, who wandered into their world about 1,000 years ago, and whom the Hopi regarded as barbarous murderers and thieves (Kluckhohn & Leighton, 1962; Waters, 1963). The name the Navajos call themselves, *Diné*, means "The People" (Kluckhohn & Leighton, 1962). If you are not *Diné* (i.e., a person among The People) then who (or what) are you? *Diné* connotes the same ethnocentric attitude as *An-ish-in-aub-ag*.

Hester et al. insinuate, again and again, that I believe that there exists "One Truth," that is, that there is one true worldview—that discovered by science—and all the others are false. I do not. In *Earth's Insights* I carefully and assiduously avoid truth claims for the (ever changing) scientific worldview. Indeed, I agree completely with the Carol-Geddes anecdote, quoted by Hester et al., about the Indigenous Ms. Ned who insightfully observed that the scientists she listened to all one day just "tell different stories than we do." All human cognitive constructs are stories, narratives, but all are subject to mutual criticism, as Gruen so forcefully argues. Some stories, I suggest in

Earth's Insights, are more "tenable"—not truer, not even more believable—than others; that is, they may stand up to criticism better than others because they are more comprehensive, consistent, coherent, pragmatic, beautiful, and spiritually inspiring—all qualities I would argue that are exhibited by the new grand narrative of postmodern natural philosophy. Surely this is a modest epistemological claim. For example, I think that the "epic of evolution" is more tenable than Hesiod's epic, the *Theogony*. (I hope this will not outrage any indigenous Greeks.) But when stories from different times and places resonate well with the new grand narrative of postmodern natural philosophy, the two are mutually validated. I thought that some such account as this would lend credibility to the wide variety of such indigenous narratives as those of the pre-Cookian Hawaiians and pre-Columbian Ojibwa, narratives that are usually dismissed as "myth" and "superstition." But Hester, et al. seem to think that we should bring the contemporary scientific narrative to the test of these premodern indigenous narratives. In that they appear to agree in principle with school-board fundamentalists who would bring contemporary natural philosophy to the test of the premodern biblical narrative and prohibit the telling of the scientific story to schoolchildren if it were contradicted by the biblical story. Which indigenous narrative, in other words, do we select to be the standard against which the scientific narrative will be validated? As the Ms.-Ned anecdote testifies, and as Hester et al. remind us, indigenous narratives are quite diverse, and, I add, often mutually inconsistent.

Not only do we learn from Hester et al. that all past as well as present Indigenous peoples were peaceful and tolerant pluralists, just like the contemporary Indigenous coauthors among Hester et al., we also learn that they all subscribe(d) to a form of environmental "respect" just like the one recently developed by Tom Birch (1993) and then publicly endorsed by one of the non-Indigenous authors, namely Cheney (1998). All Indigenous people manifest(ed) "universal consideration," that is, they respect(ed) absolutely everything (except me). This is so, we are informed, because they tell stories and do not theorize. No Indigenous people have an ethic, we learn, environmental or otherwise, nor do they have a metaphysics (such things belong only to Western culture) but they do have an epistemology. It too is exactly like that recently developed by Birch (1993) and Cheney (1998). No less than the scientific method, which places a premium on falsification of hypotheses, the epistemology of Indigenous peoples, we are told, is, and always was, self-critical. We should know this is so because, in the experience of the self-identifying Indigenous authors among Hester et al., when contemporary Indigenous people from different groups get together with one another they "swap stories and come to understand each other better and understand the various ways of understanding the subject matter of the stories better." Because all contemporary Indigenous people, in the experience of the self-identifying Indigenous authors among Hester et al., are so friendly with one another and curious about one another's stories, the accounts of violent and horrible ethnic conflict among some Indigenous peoples, prior to the European conquest of North America, by the nineteenth-century Indigenous historian of the Ojibwa, is a vicious slander, I suppose we must conclude.

The scientific method is infected throughout with Baconian designs on the control of nature and any natural philosophy or worldview informed by science must also be similarly infected, Hester et al. tell us. But they also tell us that Indigenous thought, in sharp contrast to postmodern natural philosophy, is genuinely "ecological." The seeming paradox generated by the fact that ecology is a Western science, and must therefore itself be inherently Baconian, as we are told elsewhere in the essay, should, however, not trouble us. This and a number of other contradictions in the Hester et al. essay does not undermine its credibility, they inform us in an endnote, because the criterion of noncontradiction is a local concern of Western philosophy, inapplicable to Indigenous thought. Hester et al. thus join Bratton in expressing doubts about the usefulness of the noncontradiction criterion in deciding what is tenable and what is not.

As with Lee Hester and many other contemporary Americans, I too have indigenous as well as European ancestry. Heretofore, I have made no public declaration of it because I did not want to exploit something I consider irrelevant to sound scholarship. I do not, that is, think that my indigenous ancestry provides me with any special access whatever to how all Indigenous people—past and present, here, there, and yonder—think. Because Catherine Larrère is French, does that alone qualify her, in the absence of careful study of relevant materials, to pontificate on, say, the fifteenth-century worldview of the French? Furthermore, must one be French to be able to understand and credibly characterize the fifteenth-century French worldview? My attempt to characterize the traditions of thought of various, and very different, indigenous peoples in *Earth's Insights* was based on a careful study of a wide variety of cultural materials. In the case of the Polynesian tradition, for example, I relied in part on the *Kumulipo*, a seventeenth-century royal Hawaiian genealogical chant; in that of the Ojibwa, I relied in part on a body of narratives collected and roughly translated by William Jones, an anthropologist of indigenous ancestry, and recorded in the original language. Hester et al. challenge my interpretation of indigenous traditions globally, that is, without respect to the differences among them and with no reference whatever to any sources, except their own, apparently innate knowledge of things Indigenous.

Frankly, I do not know to defend myself against such papal bull, except to baldly state its preposterous assumptions.

And when I say, "defend myself," I choose my words carefully because, unlike all the other authors in this volume, Hester et al. mount a mean-spirited personal attack on me. Their title rhetorically associates me with George Armstrong Custer. I consider this to be as tasteless and offensive as if they titled their essay "Callicott's Mein Kampf." The essay is full of the rhetoric of political violence. I am accused of an "attempted intellectual coup d'état of Indigenous thought"; of employing an "imperialist conquest model" of "heart(less)" analysis. They call me "arrogant," "misguided," out to "export" and "impose" the Leopold land ethic throughout the world. (Actually, in *Earth's Insights*, I find only one indigenous ecological ethic to be similar to the Leopold land ethic; others are more or less "ecological" in different ways.) They accuse me of hubris. This is not only disheartening, it defies understanding. In a document coauthored by one of the indigenous coauthors of the Hester et al. essay, McPherson, the same account of the Ojibwa land ethic that Hester et al. characterize as "distorted" is warmly endorsed for including an analysis of the very thing it is now accused of failing to appreciate, namely, Indigenous respect for things natural:

> One of the conclusions which Callicott reaches which has particular importance for The Native Philosophy Project here in North-Western Ontario is, for example, that "the Ojibwa regarded animals, plants, and assorted other natural things and phenomena as persons with whom it was possible to enter into complex social intercourse." . . . In Callicott's original research on the Ojibwa narrative-tradition, . . . he came to realize that "Ojibwa narratives consistently represent the natural world as a world of other-than-human persons organized into congeries of societies." . . . This concept of "other-than-human persons" is an extremely important one in moral philosophy. A person indeed is someone who has our respect. (Rabb & McPherson 1994, p. 5)

What Hester et al. reject with particular disdain, Rabb and McPherson (1994, p. 7) go on warmly to endorse: "Callicott and others have suggested that the North American Indian attitude to nature constitutes a land ethic very close to that proposed by the famous American environmentalist Aldo Leopold, [who] argues that what is important in making decisions affecting the environment is not the right of the individual, human or otherwise, but the good of the biotic community." In many other passages in the same document, my work is extensively cited as a precedent for taking indigenous thought seriously as philosophy, as a study in "ethno-metaphysics." Furthermore, the material upon which I base my interpretation of an Ojibwa worldview (i.e., the narratives collected by Jones) is, in the opinion of Rabb and McPherson (1994, p. 10), so authentic that "many of our [indigenous] students have found these narratives very familiar and are able to tell the entire story with astonishing accuracy after reading only the first few sentences." As of this writing, McPherson still uses the text I coauthored and that he and Rabb cite, *Clothed-in-Fur and Other Tales: An Introduction to an Ojibwa World View*, in his courses. And Rabb and McPherson warmly endorse my methods of ethno-metaphysical analysis. How are we to account for this reversal of judgment? Is this more evidence that the principal nonindigenous coauthor is using the indigenous identity of two of his collaborators as a mantle to hide his own naked speculations and otherwise groundless assertions? As noted, Hester et al. claim that consistency is not a concern in Indigenous thought. Neither, it seems, is gratitude to a fellow philosopher for work that helped make an academic program in Indigenous philosophy at Lakehead University possible.

Conclusion

Let me conclude "My Reply" by returning to the point with which I began—thanksgiving. By way of transition from the immediately preceding discussion, I am grateful that the two versions of the attack on *Earth's Insights* (and its author) by Hester et al. are both accompanied by my retort—here and, and more fully, in *Environmental Ethics* (the journal). Again, I thank all the authors represented in this volume for their engagement with my work, and the great majority of them for the graceful and appreciative way in which they do it. Reading these essays and mounting a reply has been both an exhilarating and humbling experience. I wish I could have devoted a reply of equal length to each (as I did to that by Hester et al. in the pages of *Environmental Ethics*). Short of that, my first challenge was to select the central points in each of their essays and reply to those. The second challenge was to craft a reply that had a unity and integrity of its own. It is a Herculean task, to which only a true philosophical Hercules would be adequate. A mere mortal, I have here done the best I could.

Works Cited

Bentham, J. (1823). *An introduction to principles of morals and legislation.* Oxford: Clarendon Press.

Birch, T. H. (1993). Moral considerability and universal consideration. *Environmental Ethics, 15,* 313–332.

Brownlee, I. (1981). *Basic documents on human rights* (2nd ed.). Oxford: Clarendon Press.

Burtt, E. A. (1952). *The metaphysical foundations of modern science.* New York: Humanities Press.

Callicott, J. B., Crossman, E. J., Crowder, L. B., Cudmore, B., Eby, L., & Mumford, K. (1998). Great Lakes Fishery Commission biodiversity workshop for citizens: Conference proceedings. Ann Arbor, Mich.: Great Lakes Fishery Commission.

Callicott, J. B., & Nelson, M. P. (1998). *The great new wilderness debate.* Athens: University of Georgia Press.

Cheney, J. (1998). Universal consideration, an epistemological map of the terrain. *Environmental Ethics, 20,* 265–277.

Clements, F. E. (1916). *Plant succession: An analysis of the development of vegetation* (Publication no. 242). Washington, D.C.: Carnegie Institution of Washington.

Curtis, J. T., & McIntosh, R. P. (1951). An upland forest continuum in the prairie-forest border region of Wisconsin. *Ecology, 32,* 476–496.

Davis, T. (1999). *Sustaining the forest, the people, the spirit.* Albany, NY: SUNY Press.

Denevan, W. M. (1992). The pristine myth: The landscape in the Americas in 1492. *Annals of the Association of American Geographers, 82,* 369–385.

Earth Council. (1999). *Introduction to the Earth Charter initiative and benchmark draft II.* Middlebury, VT: Earth Charter International Secretariat.

Earth Charter Launch. (2000). *Earth Charter briefing book.* The Hague: Peace Palace.

Foreman, D. (1991). The new conservation movement. *Wild Earth, 1,* 6–12.

Gare, A. E. (1995). *Postmodernism and environmental crisis.* London: Routledge.

Gleason, H. A. (1926). The individualistic concept of the plant formation. *Bulletin of the Tory Botanical Garden, 53,* 1–20.

Golley, F. B. (1993). *A history of the ecosystem concept in ecology: More than the sum of the parts.* New Haven, CT: Yale University Press.

Habermas, J. (1970). Toward a theory of communicative competence. *Inquiry, 13,* 347–364.

Hagen, J. B. (1992). *An entangled bank: The origins of ecosystem ecology.* New Brunswick, NJ: Rutgers University Press.

Jackson, W. (1980). *New roots for agriculture.* Lincoln: University of Nebraska Press.

Kant, I. (1959). *Foundations of the metaphysics of morals* (L. W. Beck, Trans.). New York: Bobbs-Merrill.

Kluckhohn, C., & Leighton, D. (1962). *The Navaho.* Garden City, NY: Doubleday.

Lappé, F. M., & Collins, J. (1979). *Food first: Beyond the myth of scarcity.* New York: Ballantine.

Lloyd, G. (1980). Spinoza's environmental ethics. *Inquiry, 23,* 293–311.

McIntosh, R. P. (1985). *The background of ecology: Concept and theory.* Cambridge, England: Cambridge University Press.

McKibben, B. (1989). *The end of nature.* New York: Random House.

Midgley, M. (1984). *Animals and why they matter.* Athens: University of Georgia Press.

Naess, A. (1977). Spinoza and deep ecology. In S. Hessing (Ed.), *Speculum spinozanum* (pp. 418–425). Boston: Routledge and Kegan Paul.

Naess, A. (1980). Environmental ethics and Spinoza's ethics. *Inquiry, 23,* 313–325.

Nickel, J. (1992). Human rights. In L. C. Becker & C. B. Becker (Eds.). *Encyclopedia of ethics* (vol. 1). New York: Garland.

Norton, B. G. (1984). Environmental ethics and weak anthropocentrism. *Environmental Ethics, 6,* 131–148.

Norton, B. G. (1995). Why I am not a nonanthropocentrist: Callicott and the failure of monistic inherentism. *Environmental Ethics, 17,* 341–358.

Odum, E. P. (1971). *Fundamentals of ecology* (3rd ed.). Philadelphia: W. B. Saunders.

Pickett, S. T. A., & R. S. Ostfeld, (1995). The shifting paradigm in ecology. In R. L. Knight & S. F. Bates (Eds.), *A new century for natural resources management* (pp. 261–278). Washington, D.C.: Island Press.

Pickett, S. T. A., & White, P. S. (1985). *The ecology of natural disturbance and patch dynamics.* San Diego, CA: Academic Press.

Plumwood, V. (1998). Wilderness skepticism and wilderness dualism. J. B. Callicott & M. P. Nelson (Eds.), *The great new wilderness debate* (pp. 652–690). Athens: University of Georgia Press.

Preston, C. J. (1998). Epistemology and intrinsic values: Norton and Callicott's critique of Rolston. *Environmental Ethics, 20,* 409–428.

Rabb, J. D., & McPherson, D. H. (1964). Native philosophy project: Lakehead University. Thunder Bay, ON: Lakehead University.

Rolston, H. (1989). Review of Andrew Brennan, Thinking about nature: An investigation of nature, value, and ecology. *Environmental Ethics, 11*, 259–267.

Rolston, H. (1997). Nature for real: Is nature a social construct? In T. D. J. Chappell (Ed.), *The philosophy of the environment* (pp. 38–64). Edinburgh: Edinburgh University Press.

Sosa, N. M. (1996). The ethics of dialogue and the environment: Solidarity as a foundation for environmental ethics. In J. B. Callicott & F. J. R. da Rocha (Eds.), *Earth summit ethics: Toward a reconstructive postmodern philosophy of environmental education* (pp. 47–70). Albany, NY: SUNY Press.

Sturm, D. (1999). *Identity and alterity: Summons to a new axial age (Perspectives on the Earth Charter movement)*. Lewisburg, PA: Bucknell University.

Vogel, S. (1996). *Against nature: The concept of nature in critical theory*. Albany, NY: SUNY Press.

Warren, W. (1970). *History of the Ojibway nation*. Minneapolis, MN: Ross and Haines.

Waters, F. (1963). *Book of the Hopi*. New York: Ballantine.

Whewell, W. (1840). *The philosophy of the inductive sciences, founded upon their history*. London: Parker.

Whittaker, R. H. (1951). A criticism of the plant association and climatic climax concepts. *Northwest Science, 25*, 17–31.

Weurthner, G. (1992). Some ecological costs of livestock. *Wild Earth, 2*, 10–14.

Bibliography

Those interested can find several sources of bibliographies for the field of environmental philosophy. First, there is the Master Bibliography provided through the International Society of Environmental Ethics, at its Web site: http://www.cep.unt.edu/ISEE.html. This bibliography is also available on disk; purchasing information is available on the Web site.

Two quite similar bibliographical essays on the field, both by Wayne Ouderkirk, are: "Mindful of the Earth: A Bibliographical Essay on Environmental Philosophy," *Centennial Review*, 47 (1998): 353–392, and "Earthly Thoughts," *Choice: Current Reviews for Academic Libraries*, 35/3 (Nov., 1997): 421–434

Select Bibliography of J. Baird Callicott's Writings

This partial listing of Callicott's extensive writings is intended to give sufficient information about most of his works. By combining it with the citations in the essays collected in this volume and with the bibliographies cited above, the reader will have access to the full range of environmental philosophy literature.

Books Authored

Beyond the Land Ethic: More Essays in Environmental Philosophy. Albany, N.Y.: SUNY Press, 1999.

Earth's Insights: A Survey of Ecological Ethics from the Mediterranean Basin to the Australian Outback. Berkeley: University of California Press, 1994.

In Defense of the Land Ethic: Essays in Environmental Philosophy. Albany. N.Y.: SUNY Press, 1989.

Clothed-in-Fur and Other Tales: An Introduction to an Ojibwa World View, with Thomas W. Overholt. Washington, D.C.: University Press of America, 1982.

"Plato's Aesthetics: An Introduction to the Theory of Forms." Ph.D. diss., Syracuse University, 1971.

Books Edited

For the Health of the Land: Previously Unpublished Essays and Other Writings on Conservation by Aldo Leopold, with Eric T. Freyfogle. Washington, D.C.: Island Press, 1999.

The Great New Wilderness Debate, with Michael P. Nelson. Athens: University of Georgia Press, 1998.

Environmental Philosophy: From Animal Rights to Social Ecology, 2d ed., with Michael E. Zimmerman, George Sessions, Karen J. Warren, and John Clark. Upper Saddle River, N.J.: Prentice-Hall, 1998.

Earth Summit Ethics: Toward a Reconstructive Postmodern Philosophy of Environmental Education, with Fernando J. R. da Rocha. Albany, N.Y.: SUNY Press, 1996.

Environmental Philosophy: From Animal Rights to Social Ecology, with Michael Zimmerman, George Sessions, Karen Warren, and John Clark. (Englewood Cliffs, N.J.: Prentice-Hall, 1993.

The River of the Mother of God and Other Essays by Aldo Leopold, with Susan L. Flader. Madison: University of Wisconsin Press, 1991.

Nature in Asian Traditions of Thought: Essays in Environmental Philosophy, with Roger T. Ames. Albany: SUNY Press, 1989.

Companion to A Sand County Almanac: Interpretive and Critical Essays. Madison: University of Wisconsin Press, 1987.

Miscellaneous Stand-Alone Items Authored or Edited

Proceedings of the Great Lakes Fishery Commission Biodiversity Workshop for Citizens: Biodiversity Task Presentations and Discussion Summaries, with Larry B. Crowder, E. J. Crossman, Becky Cudmore, Lisa A. Eby, and Karen Mumford. Ann Arbor, Mich.: Great Lakes Fishery Commission, 1998.

Monist 75, no. 2 (1992). This is a theme issue on the intrinsic value of nonhuman natural entities and nature as a whole.

Book Chapters

"A Hierarchical Theory of Value Applied to the Great Lakes and Their Fishes," with Karen Mumford. In Dorinda G. Dallmeyer, ed., *Values at Sea: Ethics for the Marine Environment.* Athens: University of Georgia Press, forthcoming.

"Contemporary Criticisms of the Received Wilderness Idea." In David N. Cole and Stephen F. McCool, Wayne A. Freimund, and Jennifer O'Loughlin, eds., *Wilderness Science in a Time of Change*, vol 1: *Changing Perspectioves and Future Directions.* Ogden, Utah: U. S. Department of Agriculture, Forest Service, Rocky Mountain Research Station, 2000, 24–31.

"Diversité culturelle." In Catherine Larrère, ed., *Beautés de la nature.* Paris: Muséum National d'Histoire Naturelle, 2000.

"The Land Ethic." In Dale Jamieson, ed., *A Companion to Environmental Philosophy.* (Oxford: Blackwell, 2000, 204–217.

"Conservation Values and Ethics" [substantially revised]. In Gary K. Meffe and C. Ronald Carroll, eds., *Principles of Conservation Biology,* 2d ed. Sunderland, Mass.: Sinauer Associates, 1997, 24–49.

"Après le paradigme industriel." In Raphaël Larrère and Catherine Larrère, eds., *La crise environnementale.* Paris: Institut Nationale de la Recherche Agronomique Press, 1997, 205–219.

"Whaling in Sand County: A Dialectical Hunt for Land-Ethical Answers about the Morality of Minke-Whale Catching." In Jon Wetlesen, ed., *Likeverd og Forskjell: En Etisk Intuisjon og dens Grenser.* Oslo: Lobo Grafisk, 1997, 61–96.

"Benevolent Symbiosis: The Philosophy of Conservation Reconstructed." In J. Baird Callicott and Fernando J. R. da Rocha, eds. *Earth Summit Ethics: Toward a Reconstructive Postmodern Philosophy of Environmental Education,* 139–159.

"Conservation Ethics at the Crossroads." In Jennifer L. Nielsen and Dennis A. Powers, eds., *Evolution and the Aquatic Ecosystem: Defining Unique Units in Population Conservation.* Bethesda, Md.: American Fisheries Society, 1995, 3–7.

"Paul Shepard and the Development of Mid-Twentieth-Century Ecological Metaphysics." In Max Oelschlaeger, ed., *The Company of Others: Essays in Celebration of Paul Shepard.* Durango, Colo.: Kivakí Press, 1995, 101–108.

"Environmental Philosophy Is Environmental Activism: The Most Radical and Effective Kind." In Don E. Marietta and Lester Embree, eds., *Environmental Philosophy and Environmental Activism.* Lanham, Md.: Rowman and Littlefield, 1995, 19–35.

"American Indian Environmental Philosophy," with Thomas W. Overholt. In Robert Solomon and Kathleen Higgins, eds., *Readings in World Philosophy.* New York: McGraw Hill, 1995, 188–218.

"Il Ruolo Della Tecnologia Nel Concetto Mutevole di Natura" ["The Role of Technology in the Evolving Concept of Nature"]. In Corrado Poli, ed., *Etica Ambietale: Teoria e Pratica*, Eleonora Cussini, Silvia Bertolo, and Gabriella De Boni, trans. Milano: Angelo Guerini e Associati, 1994, 99–131.

"Toward a Global Environmental Ethic." In Mary Evelyn Tucker and John Grim, eds., *World Views and Ecology: Religion, Philosophy and the Environment.* (Maryknoll, N.Y.: Orbis Books, 1994.

"The Role of Technology in the Evolving Concept of Nature." In Frederick Ferré, ed., *Ethics and Environmental Policy: Theory Meets Practice.* Athens: University of Georgia Press, 1994, 58–83 (English original).

"A Brief History of American Conservation Philosophy." In W. Wallace Covington and Leonard F. Debano, eds., *Sustainable Ecological Systems: Implementing an Ecological Approach to Land Management.* Fort Collins, Colo.: Rocky Mountain Forest and Range Experiment Station, U. S. Department of Agriculture, 1994, 10–14.

"Conservation Values and Ethics." In Gary K. Meffe and C. Ronald Carroll, eds., *Principles of Conservation Biology.* Sunderland, Mass.: Sinauer Associates, 1994, 24–49.

"Traditional American Indian Attitudes Toward Nature," with Thomas W. Overholt. In Robert Solomon and Kathleen Higgins, eds., *From Africa to Zen: An Introduction to World Philosophy* (New York: Rowman and Littlefield, 1993, 55–80.

"The Search for an Environmental Ethic" [substantially revised]. In Tom Regan, ed., *Matters of Life and Death*, 3d ed. New York: McGraw Hill, 1992, 322–382.

"Aldo Leopold's Metaphor." In Robert Costanza, Bryan Norton, and Ben Haskell, eds., *Ecosystem Health: New Goals for Environmental Management.* Washington, D.C.: Island Press, 1992, 42–56.

"Individual and Community in Society and Nature," with Frances Moore Lappé. In Michael Zweig, ed., *Religion and Economic Justice.* Philadelphia: Temple University Press, 1992, 245–252.

"The Wilderness Idea Revisited." In John J. Stuhr and Robin M. Cochran, eds., *Ethics and Free Enterprise: The Social Responsibility of Business.* Eugene: University of Oregon Books, 1992, 35–52, The 1990 Autzen Lectures in the Humanities.

"Conservation of Biological Resources: Responsibility to Nature and Future Generations." in D. J. Decker, M. E. Krasny, G. R. Goff, and C. R. Smith, eds., *Conserving Biological Resources: A Practitioner's Guide.* Boulder, Colo.: Westview Press, 1991, 33–42.

"Genesis and John Muir." In Carol Robb and Carl Casebolt, ed., *Covenant for a New Creation.* Maryknoll, N.Y.: Orbis Books, 1991, 106–138.

"Introduction" to J. Baird Callicott and Susan Flader, eds., *The River of the Mother of God, and Other Essays by Aldo Leopold*. Madison: University of Wisconsin Press, 1991, 3–31.

"American Indian Land Wisdom." In Paul A. Olson, ed., *Struggle for the Land: Indigenous Insights and Industrial Empire in Semi-Arid Lands*. Lincoln: University of Nebraska Press, 1990, 255–272.

"On the Relation of Idea and Action." Epilogue to J. Baird Callicott and Roger T. Ames, eds., *Nature in Asian Traditions of Thought*. Albany: State University of New York Press, 1989, 279–289.

"Marx Meets Muir: Toward a Synthesis of the Progressive Political and Ecological Visions," with Frances Moore Lappé. In P. Allen and D. Van Dusen, eds., *Global Perspectives on Agroecology and Sustainable Agricultural Systems: Proceedings of the Sixth International Scientific Conference of the International Federation of Organic Agriculture Movements*. Santa Cruz: University of California Agroecology Program, 1988, vol. 1, 21–30.

"Agroecology in Context." In *Global Perspectives on Agroecology and Sustainable Agricultural Systems: Proceedings of the Sixth International Scientific Conference of the International Federation of Organic Agricultural Movements*. Santa Cruz: University of California Agroecology Program, 1988, vol 2, 39–43.

"The Scientific Substance of the Land Ethic." In Tom Tanner, ed., *Aldo Leopold: The Man and His Legacy*. Ankeny, Iowa: Society for Soil and Water Conservation, 1987, 87–104.

"The Philosophical Value of Wildlife." In Daniel J. Decker and Gary Goff, eds., *Economic and Social Values of Wildlife*. (Boulder, Colo.: Westview Press, 1987, 214–221.

"The Land Aesthetic." In J. Baird Callicott, ed., *Companion to A Sand County Almanac: Critical and Interpretive Essays*. Madison: University of Wisconsin Press, 1987, 157–171.

"The Conceptual Foundations of the Land Ethic." In J. Baird Callicott, ed., *Companion to A Sand County Almanac: Critical and Interpretive Essays*. Madison: University of Wisconsin Press, 1987, 186–217.

"On the Intrinsic Value of Non-Human Species." In Bryan G. Norton, ed., *The Preservation of Species*. Princeton, N.J.: Princeton University Press, 1986, 138–172.

"Moral Considerability and Extraterrestrial Life." in Eugene C. Hargrove, ed., *Beyond Space Ship Earth: Environmental Ethics and the Solar System*. San Francisco: Sierra Club Books, 1986, 227–259.

"Aldo Leopold's Land Aesthetic and Agrarian Land Use Values." In Richard Haynes, ed., *Agriculture, Change, and Human Values, Proceedings of a Multidisciplinary Conference, October 18–21, 1982*. Gainesville: University of Florida, 1984, 475–486.

Refereed Journal Articles

"Choosing Appropriate Spatial and Temporal Scales for Ecological Restoration," *Journal of Biosciences*: forthcoming.

"Indigenous Worlds or the Indigenous World: A Reply to My 'Indigenous' Critics." *Environmental Ethics* 22 (2000): 291–310.

"Aldo Leopold and the Foundations of Ecosystem Management." *Journal of Forestry* 98 (2000): 5–13.

"Silencing Philosophers: Minteer and Foundations of Anti-Foundationalism." *Environmental Values* 8 (1999): 499–516.

"'The Arboretum and the University': The Speech and the Essay." *Transactions of the Wisconsin Academy of Sciences, Arts, and Letters* 87 (1999): 5–21.

"After the Industrial Paradigm, What?" *Research in Philosophy and Technology* 18 (1999): 13–25.

"Current Normative Concepts in Conservation," with Larry B. Crowder and Karen Mumford. *Conservation Biology* 13 (1999): 22–35.

"'Back Together Again' Again." *Environmental Values* 7 (1998): 461–475.

"In Defense of *Earth's Insights*." *Worldviews: Environment, Culture and Religion* 1 (1997): 167–182.

"Ecological Sustainability as a Conservation Concept." *Conservation Biology* 11 (1997): 32–40.

"Whaling in Sand County: A Dialectical Hunt for Land-Ethical Answers to Questions about the Morality of Norwegian Minke-Whale Catching." *Colorado Journal of International Environmental Law and Policy* 8 (1997): 1–30 [North American version].

"The Challenge of a World Environmental Ethic." *American Journal of Theology and Philosophy* 18 (1997): 65–79

"Do Deconstructive Ecology and Sociobiology Undermine the Leopold Land Ethic?" *Environmental Ethics* 18 (1996): 353–371.

"Should Wilderness Areas Become Biodiversity Reserves?" *George Wright Forum* 13 (1996): 32–38.

"How Environmental Ethical Theory May Be Put into Practice." *Ethics and the Environment* 1 (1996): 3–14.

"Environmental Wellness." *Literature and Medicine* 15 (1996): 148–162.

"Whaling in Sand County: A Dialectical Hunt for Land-Ethical Answers to Questions about the Morality of Norwegian Minke-Whale Catching." *Ecology of Industrial Regions* 1 (1995): 83–98 [European version].

"The Value of Ecosystem Health." *Environmental Values* 5 (1995): 345–361.

"Intrinsic Value in Nature: A Metaethical Analysis." Special issue: Justifying Value in Nature. *Electronic Journal of Analytic Philosophy* 3 (1995): http://www.phil.indiana.edu/ejap/1995.spring/callicott.abs.html.

"A Review of Some Problems with the Concept of Ecosystem Health." *Ecosystem Health* 1 (1995): 101–112.

"Moral Monism in Environmental Ethics Defended." *Journal of Philosophical Research* 19 (1994): 51–60.

"International Environmental Ethics." *National Geographic Journal of India* 39 (1993): 25–32.

"Sustainability in Historical-Philosophical Context." *George Wright Forum* 10 (Winter 1993): 26–33.

"Toward a Global Environmental Ethic." *Bucknell Review* 37 (November 1993): 30–38.

"American Conservation Philosophy—A Brief History." *Ecos: A Review of Conservation* 14 (1993): 41–46.

"La Nature est Morte, Vive la Nature!" Translated by Marc Saint-Upéry. *Écologie Politique* 7 (Summer 1993): 73–90 (French translation). *Hastings Center Report* 22 (September/October 1992): 16–23 (English original).

"The Role of Technology in the Evolving Concept of Nature." *Research in Philosophy and Technology* 13 (1993): 201–222.

"The Land Ethic Today." *Topoi: An International Review of Philosophy* 12 (1993): 41–51.

"The Land Aesthetic." *Renewable Resources Journal* 10, no. 4 (Winter 1992): 12–17.

"Can a Theory of Moral Sentiments Support a Genuinely Normative Environmental Ethic?" *Inquiry* 35 (1992): 183–198.

"Rolston on Intrinsic Value: A Deconstruction," *Environmental Ethics* 14 (1992): 129–143.

"Principal Traditions in American Environmental Ethics: A Survey of Moral Values for Framing an American Ocean Policy." *Ocean and Shoreline Management* 17 (1992): 299–308.

"The Wilderness Idea Revisited: The Sustainable Development Alternative." *Environmental Professional* 13 (1991): 235–247.

"Conservation Ethics and Fishery Management." *Fisheries* 16, no. 2 (March–April 1991): 22–28.

"Leopold's 'Means and Ends in Wild Life Management,'" with Eugene C. Hargrove. *Environmental Ethics* 12 (1990): 333–337.

"Standards of Conservation: Then and Now." *Conservation Biology* 4 (1990): 229–232.

"Genesis Revisited: Muirian Musings on the Lynn White, Jr. Debate." *Environmental History Review* 14 (1990): 65–90.

"The Metaphysical Transition in Farming: From the Newtonian-Mechanical to the Eltonian-Ecological." *Journal of Agricultural Ethics* 3 (1990): 36–49.

"Whither Conservation Ethics?" *Conservation Biology* 4 (1990): 15–20.

"The Case against Moral Pluralism." *Environmental Ethics* 12 (1990): 99–124.

"Genesis and John Muir." *ReVision* 13, no. 3 (1990): 31–47.

"American Indian Land Wisdom?: Sorting Out the Issues." *Journal of Forest History* 33 (January 1989): 35–42.

"Animal Liberation and Environmental Ethics: Back Together Again." *Between the Species* 4 (1988): 163–169.

"Agroecology in Context." *Journal of Agricultural Ethics* 1 (1988): 3–9.

"Just the Facts, Ma'am." *The Environmental Professional* 9 (1987): 279–288.

"Tertium Organum and Mankind's Role in Future Evolution." *Philosophica* 39 (1987): 101–112.

"Conceptual Resources for Environmental Ethics in Asian Traditions of Thought: A Propaedeutic," *Philosophy East and West* 31 (1987): 115–130.

"The Metaphysical Implications of Ecology." *Environmental Ethics* 9 (1986): 300–315.

"Intrinsic Value, Quantum Theory, and Environmental Ethics," *Environmental Ethics* 7 (1985): 257–275.

"Non-Anthropocentric Value Theory and Environmental Ethics." *American Philosophical Quarterly* 21 (1984): 299–309.

"The Land Aesthetic." *Environmental Review* 7 (1983): 345–358.

"Leopold's Land Aesthetic." *Journal of Soil and Water Conservation* 38 (1983): 329–332.

"Aldo Leopold on Education, as Educator, and His Land Ethic in the Context of Environmental Education." *Journal of Environmental Education* 14 (1982): 34–41.

"Traditional Western European and American Indian Attitudes toward Nature: An Overview." *Environmental Ethics* 4 (1982): 293–318.

"Hume's Is/Ought Dichotomy and the Relation of Ecology to Leopold's Land Ethic." *Environmental Ethics* 4 (1982): 163–74.

"Animal Liberation: A Triangular Affair." *Environmental Ethics* 2 (1980): 311–228.

"Elements of an Environmental Ethic: Moral Considerability and the Biotic Community." *Environmental Ethics* 1 (1979): 71–81.

Anthologized Articles

"A Critique of and an Alternative to the Wilderness Idea." In Tom Butler, ed., *Wild Earth Reader.* Milkweed Editions: Minneapolis, forthcoming.

"Toward a Global Environmental Ethic." Excerpt in Darrell Addison Posey, ed., *Cultural and Spiritual Values of Biodiversity.* Nairobi: United Nations Environment Program, 1999, 447

"La Éthica de la Tierra Hoy" ["The Land Ethic Today"]. In Margarita Valdéz and Alejandro Herrera, eds., *Éthica Ambiental.* Mexico City: Instituto de Investigaciones Filosoficas, Universidad Nacional Autónoma de México y Fondo de Cultura Económica, forthcoming.

"Rolston on Intrinsic Value: A Deconstruction." In *Zhexue Yicong: Journal of the Chinese Academy of Social Sciences.* Beijing: Institute of Philosophy, 1999, 19–26.

"Teoria Non-Anthropocentrica del Valore ed Ettica Ambientale" ["Non-Anthropocentric Value Theory and Environmental Ethics"]. In Mariachiara Tallacchini, *Etiche della Terra: Antologia di Filosofia dell'Ambiente.* Milan: Vita en Pensiera, 1998.

"A Critical Examination of 'Another Look at Aldo Leopold's Land Ethic.'" In Society of American Foresters, ed., *The Land Ethic: Meeting Human Needs for the Land and Its Resources.* Bethesda, Md.: Society of American Foresters, 1998, 107–121.

"Ecological Sustainability as a Conservation Concept," with Karen Mumford. In John Lemons, Laura Westra, and Robert Goodland, eds., *Ecological Sustainability and Integrity: Concepts and Approaches.* Dordrecht: Kluwer, 1998, 31–45.

"Do Deconstructive Ecology and Sociobiology Undermine Leopold's Land Ethic?" In Michael E. Zimmerman, J. Baird Callicott, George Sessions, Karen J. Warren, and John Clark, eds., *Environmental Philosophy: From Animal Rights to Radical Ecology*, 2d Ed. Upper Saddle River, N.J.: Prentice-Hall, 1998, 145–164.

"Whaling in Sand County." In T. D. J. Chappel, ed., *The Philosophy of the Environment.* Edinburgh: Edinburgh University Press, 1997, 156–179.

"The Wilderness Idea Revisited: The Sustainable Development Alternative." In J. Baird Callicott and Michael P. Nelson, eds., *The Great New Wilderness Debate.* Athens: University of Georgia Press, 1998, 337–366; in John Lemons, ed., *Readings from the Environmental Professional: Natural Resources.* Oxford: Blackwell Science, 1995, 59-71; in Andrew Brennan, ed., *The International Research Library of Philosophy*, vol. 11, *Ethics of the Environment.* Aldershot, England: Dartmouth Publishing, 1994, 431–443; in Christopher Key Chapple, ed.,

Ecological Prospects: Scientific, Religious, and Aesthetic Perspectives. Albany, N.Y.: SUNY Press, 1994, 37–63; excerpt in Lori Gruen and Dale Jamieson, eds., *Reflecting on Nature: Readings in Environmental Philosophy.* New York: Oxford University Press, 1994, 252–265; and "The Ethnocentricity of Wilderness Values" [excerpted from "The Wilderness Idea Revisited"]. In Carolyn Merchant, ed., *Major Problems in Environmental History: Documents and Essays.* Lexington, Mass.: D. C. Heath, 1992, 409–412.

"Should Wilderness Areas become Biodiversity Reserves?" In J. Baird Callicott and Michael P. Nelson, eds., *The Great New Wilderness Debate.* Athens: University of Georgia Press, 1998, 585–594.

"That Good Old-Time Wilderness Religion." In J. Baird Callicott and Michael P. Nelson, eds., *The Great New Wilderness Debate.* Athens: University of Georgia Press, 1998, 387–394; and Brennan, *The Ethics of the Environment*, 453-454

"The Case against Moral Pluralism." In Brennan, *The Ethics of the Environment*, 527–552.

"On the Intrinsic Value of Nonhuman Species." In Susan Armstrong and Richard Botzler, eds., *Environmental Ethics: Divergence and Convergence.* (New York: McGraw-Hill, 1993, 66–70

"The Land Aesthetic." In Armstrong and Botzler, *Environmental Ethics*, 148–157; and in Chapple, *Ecological Prospects*, 169–183.

"The Conceptual Foundations of the Land Ethic." In Joseph Des Jardins, ed., *Environmental Ethics: Concepts, Policy, Theory.* Mountain View, Calif.: Mayfield, 1998, 227–237; Louis Pojman, ed., *Environmental Ethics: Readings in Theory and Application*, 2d ed. Belmont, Calif.: Wadsworth, 1998, 124–134; in *Environmental Philosophy: From Animal Rights to Radical Ecology*, 2d ed., 101–123; as "Die begrifflichen Grundlagen der *land ethic.*" In Angelika Krebs, ed., *Naturethik: Grundtextes der gegenwärtgen tier—und ökoethichen Diskussion.* Frankfurt: Suhrkamp Tashenbuch Wissenschaft, 1997, 211–246; in Donald VanDeVeer and Christine Pierce, eds., *The Environmental Ethics and Policy Book*, 2d ed. Belmont, Calif.: Wadsworth, 1997, 184–198; in Christine Pierce and Donald VanDeVeer, eds., *People, Penguins, and Plastic Trees: Basic Issues in Environmental Ethics*, 2d ed. Belmont, Calif.: Wadsworth, 1995, 151–162; in Louis P. Pojman, *Environmental Ethics: Readings in Theory and Application.* Boston: Jones and Bartlett Publishers, 1994, 92–102; in Donald VanDeVeer and Christine Pierce, eds., *The Environmental Ethics and Policy Book.* Belmont, Calif.: Wadsworth, 1994, 147–161; in Fred Westphal, ed., *Planet in Peril: Can Its Life be Saved?* New York: Harcourt Brace, 1994, 77–102; in Armstrong and Botzler, eds., *Environmental Ethics*, 386–397; and in Michael Zimmerman et al., eds, *Environmental Philosophy: From Animal Rights to Radical Ecology* (Englewood Cliffs, N.J.: Prentice-Hall, 1993, 110–134.

"Animal Liberation and Environmental Ethics: Back Together Again." In Eugene C. Hargrove, ed., *The Animal Rights/Environmental Ethics Debate: The Environmental Perspective.* (Albany, N.Y.: SUNY Press, 1992, 249–261.

"Moral Considerability and Extraterrestrial Life." in Hargrove, *The Animal Rights/Environmental Ethics Debate*, 137–150.

"The Search for an Environmental Ethic" ["En Busca de una Ética Ambiental," traducción por Jorge Issa] in Teresa Kwiatkowska y Jorge Issa, compiladores, *Los Caminos de la Ética Ambiental: Una Antología de Textos Contemporáneos* (México City: Plaza y Valdéz, 1998, 85–159; excerpt in William H. Shaw, ed., *Social and Personal Ethics.* Belmont, Calif.: Wadsworth, 1993, 186–198 [2d ed., 1995, 193-205]; and in Jeffrey Olen and Vincent Barry, eds., *Applying Ethics: A Text with Readings*, 4th ed. Belmont, Calif.: Wadsworth, 1992, 413–418.

"Aziatskaya Traditsiya i Perspektivy Ekologicheskoy Etiki: Propedivtika" ["Conceptual Resources for Environmental Ethics in Asian Traditions of Thought: A Propaedeutic"]. In L. I. Vasilenko and V. E. Yermolayeva, eds., *Global'nye Problemy i Obshchelovecheskie Tsennosti* [*Global Problems and Human Values*], trans. by Y. A. Shreydera Moscow: Progress Publisher, 1990.

"The Metaphysical Implications of Ecology." In J. Baird Callicot and Roger T. Ames, eds., *Nature in Asian Traditions of Thought: Essays in Environmental Philosophy* Albany, N.Y.: SUNY Press, 1989, 51–64; and in *Ziranexue Zhexuewenti* [*Philosophical Problems in Natural Science*] no. 4 (1988): 66–74, translated into Chinese by Yu Hui.

"Traditional Western European and American Indian Attitudes toward Nature: An Overview" In Larry May and Shari Collins Sharatt, eds., *Applied Ethics: A Multicultural Approach* Englewood Cliffs, N.J.: Prentice-Hall, 1994, 95–105; and in Robert Elliot and Arran Gare, eds., *Environmental Philosophy: A Collection of Readings*, St. Lucia: University of Queensland Press, 1983, 231–259.

"Animal Liberation: A Triangular Affair." In Pojman, ed., *Environmental Ethics*, 2d ed., Belmont, Calif.: Wadsworth, 1998, 57–68; "La Liberazione animale: una questione triangulare." In Mariachiara Tallacchini, *Etiche della Terra: Antologia di Filosofia dell'Ambiente.* Milan: Vita en Pensiera, 1998; in John Arthur, ed., *Morality and Moral Controversies*, Fourth Edition, Englewood Cliffs, N.J.: Prentice-Hall, 1996, 236–252; in Robert Elliot, ed., *Environmental Ethics*, Oxford Readings in Philosophy series. Oxford: Oxford University Press, 1995, 29–59; in Christine Pierce and Donald VanDeVere, ed., *People, Penguins and Plastic Trees: Readings in Environmental Ethics*, 2d ed. Belmont, Calif.: Wadsworth, 1995, 237–254; in Pojman, ed., *Environmental Ethics*, 52–63; in Westphal, ed., *Planet in Peril*, 221–248; in John Arthur, ed., *Morality and Moral Controversies*, 3d ed. Englewood Cliffs, N.J.: Prentice-Hall, 1993, 236–252; in Hargrove, *Animal Rights/Environmental Ethics Debate*, 37–69; in Christine Pierce and Donald VanDeVere, ed., *People, Penguins and Plastic Trees: Readings in Environmental Ethics* Belmont, Calif.: Wadsworth, 1986, 184–203; and in Donald Scherer and Thomas Attig, eds., *Ethics and the Environment*, Englewood Cliffs, N.J.: Prentice-Hall, 1983, 54–72.

Invited Journal Articles

"A Critical Examination of 'Another Look at Leopold's Land Ethic.'" *Journal of Forestry* 95, no. 12 (1997): 20–26.

"In Eintracht mit der Natur leben." *Natur* (December 1996): 18–19.

"A Critique of and Alternative to the Wilderness Idea." *Wild Earth* 4 (1994): 54–59.

"The Wilderness Idea: A Critique and an Alternative." *Inner Voice* 6, no. 4 (July 1994): 12–13.

"Does the Iowa Prairie Have Intrinsic Value?" *Ag Bioethics Forum* 6, no. 1 (1994): 1–5.

"A Brief History of the American Land Ethic Since 1492." *Forest History Today* (1995): 16–20, reprinted from *Inner Voice* 6, no. 1 (January 1994). 5–7.

"Anthropocentrism: The Original Sin." *Earth Ethics* 4, no. 1 (Fall 1992): 12–15.

"Reading the Land." *Orion Nature Quarterly* 9, no. 3 (Winter 1990): 16–17.

"Marx Meets Muir: Toward a Synthesis of the Progressive Political and Ecological Visions," with Frances Moore Lappé. *Tikkun* 2, no. 3 (September/October 1987): 16–21.

"The Land Aesthetic" *Orion Nature Quarterly* 3, no. 3 (Summer 1984): 16–23.

"Leopold's Land Aesthetic." *Illinois Audubon* (Spring 1985): 5–9.

About the Contributors

John Barkdull (Ph.D., University of Wisconsin-Madison, 1993) is Associate Professor of Political Science at Texas Tech University. His research interests include U.S. foreign policy and the environment, international relations theory, and globalization. He has published in the *American Political Science Review*, *Environmental Ethics*, *Ethics and International Affairs*, and *Presidential Studies Quarterly*.

Annie Booth is an Associate Professor in Environmental Studies at the University of Northern British Columbia in Canada. Her research and teaching interests are in environmental philosophy, aboriginal philosophy and resource management, environmental policy, and community sustainability and resource management. She can be contacted at the University of Northern British Columbia, 3333 University Way, Prince George, BC V2N4Z9 Canada.

Susan P. Bratton, Lindaman Chair of Science, Technology and Society at Whitworth College, Spokane, Washington, is the author of two books on Christianity and environmental ethics: *Six Billion and More: Human Population Regulation and Christian Ethics* (Louisville, Kentucky: Westminster, 1992) and *Christianity, Wilderness and Wildlife: The Original Desert Solitaire* (Scranton, Pa.: University of Scranton Press, 1993). She has produced recent book chapters on the ethics of toxic chemical exposure, western religion in U.S. environments, historic roots of environmental anti-semitism, and the ethics of commercial fishing. Dr. Bratton has also published recent scientific articles on biodiversity, eastern old-growth forest, and the effects of forest fragmentation. She received a Ph.D. in ecology from Cornell University, an M.A. in theology from Fuller Seminary, and a graduate certificate in environmental ethics from the University of Georgia, and a Ph.D. in interdisciplinary arts and humanities at the University of Texas at Dallas. Her previous professional positions include Coordinator of Uplands Field Research Laboratory in Great Smoky Mountains National Park, and Coordinator of the National Park Service Cooperative at the Institute of Ecology, University of Georgia.

J. Baird Callicott is professor of philosophy and religion studies at the University of North Texas. He was visiting professor or scholar in residence at the following institutions: University of Florida; University of Hawai'i; University of California, Berkeley; University of California, Santa Barbara; University of Kansas; Presbyterian College; and James Cook University of North Queensland. In 1971 he offered the world's first course in environmental ethics at the University of Wisconsin–Stevens Point. From 1997–2000 he was president of the International Society of Environmental Ethics.

Jim Cheney teaches Environmental Ethics and American Indian Philosophy at the University of Wisconsin–Waukesha. He was a Rockefeller Foundation Visiting Humanities Fellow with the Native Philosophy Project at Lakehead University, Thunder Bay, Ontario during 1996–1997, and the first Visiting Scholar in Residence in Ecophilosophy and Earth Education at Murdoch University in Western Australia in 2000.

Wendy Donner is an associate professor of philosophy at Carleton University, Ottawa. She is author of *The Liberal Self: John Stuart Mill's Moral and Political Philosophy* (Ithaca, N.Y.: Cornell, 1991) and has published articles in environmental ethics, political philosophy, and ethics.

Lori Gruen has taught environmental philosophy across North America, including at the University of British Columbia–Vancouver, University of North Carolina–Chapel Hill, and Stanford University. She is now calling Connecticut home, where she is Assistant Professor of Philosophy at Wesleyan University. She has published on topics in ecofeminist theory, human obligations to nonhuman animals, and environmental justice. She is currently working on a book on animal minds and animal ethics.

Eugene C. Hargrove is the founding editor of the journal *Environmental Ethics*, the author of *Foundations of Environmental Ethics* (Englewood Cliffs, N.J.: Prentice Hall, 1989; reprint, Denton, Texas: Environmental Ethics Books, 1996), the editor of *Beyond Spaceship Earth: Environmental Ethics and the Solar System* (Denton, Texas: Environmental Ethics Books, 1988) and

The Animal Rights/Environmental Ethics Debate: The Environmental Perspective (Albany, N.Y.: SUNY Press, 1992), the editor of Environmental Ethics Books, the director of the Center for Environmental Philosophy, and the Chair of the Department of Philosophy and Religion Studies at University of North Texas, where he developed a master's program devoted entirely to environmental philosophy and environmental ethics.

Thurman Lee Hester Jr. is a citizen of the Choctaw Nation. He is active in the Indian community of central Oklahoma and has served as Chairman of the Board of the Oklahoma Choctaw Tribal Alliance. Lee has taught Indigenous Philosophy and Native American Law and Policy at universities in Canada and the United States. Along with Dennis McPherson, he edits *Ayaangwaamizin: The International Journal of Indigenous Philosophy.* He currently serves as Director of American Indian Studies at the University of Science and Arts of Oklahoma.

Jim Hill, friend to alligator and snake, spends a lot of time hiding in the Okefenokee Swamp, where he is rumored to keep a still. He also goes on extended "backpacking" trips in the Big Bend area of Texas, where he is well known by the locals and the Border Patrol. When not getting "lost" on Billy's Island (in the Okefenokee), he teaches environmental ethics and philosophy at Valdosta State University (Georgia). He did his graduate work in philosophy at the University of Chicago and the University of Georgia, has published and presented numerous papers in environmental ethics at various conferences, has directed several conferences at Valdosta State University in environmental ethics, and currently is planning an environmental ethics conference in the Big Bend area.

Catherine Larrère, a specialist in political philosophy, has published three books in the field of environmental philosophy: *Les philosophies de l'environnement* (Paris: PUF Collection Philosophies, 1997); *Du bon usage de la nature. Pour une philosophie pour l'environnement,* with R. Larrère (Paris: Aubier, collection Alto, 1997); *La crise environnementale,* edited by C. et R. Larrère (Paris: INRA éditions, 1997); and numerous articles in French and English, such as "Nature, naturalisme" in *Dictionnaire de philosophie morale,* M. Canto (ed.) (Paris: PUF, 1996); "Environnement," with R. Larrèrre, in *Dictionnaire d'histoire et de philosophie des sciences,* Dominique Lecourt (ed.) (Paris: PUF, 1999); "Ethics, politics, science and the environment" in *Earth Summit Ethics: Toward a Reconstructive Postmodern Philosophy of Environmental Education,* edited by J. Baird Callicott and Fernando da Rocha

(Albany, N.Y.: SUNY Press, 1996); "Animal rearing as a contract," with R. Larrère, *Journal of Agriculture and Environmental Ethics*, (The Netherlands: Kluwer, 1999.

Andrew Light is Assistant Professor of Environmental Philosophy and Director of the Environmental Conservation Education Program at New York University and Research Fellow at the Center for Environmental Philosophy at Lancaster University (England). His primary areas of interest are environmental ethics and policy, philosophy of technology, and philosophy of film. Light is author of more than forty articles, reviews, and book chapters, and has edited or coedited twelve books, in print and forthcoming, including, with Eric Higgs and David Strong, *Technology and the Good Life?* (Chicago: University of Chicago Press, 2000); with Eric Katz and David Rothenberg, *Beneath the Surface: Critical Essays on the Philosophy of Deep Ecology* (Cambridge, Mass.: MIT Press, 2000); *Social Ecology After Bookchin* (New York: Guilford, 1998), and again with Eric Katz, *Environmental Pragmatism* (London: Routledge, 1996). Forthcoming titles include *Environmental Ethics: The Big Questions*, with Holmes Rolston (Blackwell). He also coedits the journal *Philosophy and Geography* (Carfax Publishers) with Jonathan Smith. In 1994 he cofounded the Society for Philosophy and Geography and is now serving as President of the Society for Philosophy and Technology.

Robert P. McIntosh is Emeritus Professor of Biological Science at the University of Notre Dame. He started as a plant ecologist at the University of Wisconsin in the post–World War II 1940s working on forest communities. Before coming to Notre Dame he taught at Middlebury College (Vermont), and Vassar College. After some twenty years as a field ecologist in Wisconsin and New York he became editor of the *American Midland Naturalist* in 1970. He switched his emphasis to the history and philosophy of ecology, and he has published numerous articles on history of ecology, in addition to a book, *The Background of Ecology: Concept and Theory*, along with essays on various aspects of ecology heterogeneity, competition, theory, and citation classics. Apart from teaching, he was involved for some years at the Argonne National Laboratory in the 1960s–1970s and served as program director for ecology at the National Science Foundation. His current interests in ecology are now largely in its history, its idiosyncrasies as a science, and its extensions into philosophy.

Dennis McPherson is an Ojibwa and a member of the Couchiching First Nation at Fort Frances, Ontario. To understand the issues concerning First

Nations communities better he has earned degrees in the arts (B.A.), in social work (H.B.S.W.), in philosophy (H.B.A., Philosophy), and in law (LL.B. and LL.M.). He is the coordinator and codirector of the Native Philosophy Project at Lakehead University and coauthor with Dr. Douglas Rabb of *Indian from the Inside: A Study in Ethno-Metaphysics.* As Chair he was instrumental in the development of the Department of Indigenous Learning within Lakehead University and the master's program in Native Philosophy. He is currently a professor in the Department of Indigenous Learning.

Bryan Norton received his Ph.D. in philosophy from the University of Michigan in 1970. Currently Professor of Philosophy in the School of Public Policy, Georgia Institute of Technology, he writes on intergenerational equity, sustainability theory, biodiversity policy and on valuation methods. His current research includes work on intergenerational impacts of policy choices (sponsored by U.S. Environmental Protection Agency) and on cultural aspects of environmental protection. He is author of *Why Preserve Natural Variety?* (Princeton, N.J.: Princeton University Press, 1987); *Toward Unity among Environmentalists* (Oxford: Oxford University Press, 1991), editor of *The Preservation of Species* (Princeton, N.J.: Princeton University Press); and coeditor of *Ecosystem Health: New Goals for Environmental Management* (Washington, D.C.: Island Press, 1992); *Ethics on the Ark* (Washington, D.C.: Smithsonian Press 1995); and *Wolves and Human Communities* (Washington, D.C.: Island Press, forthcoming). He has contributed to journals in several fields, including philosophy, biology, and environmental management. Norton has served on numerous panels, including the Ecosystem Valuation Forum, the Risk Assessment Forum (U.S. Environmental Protection Agency) and he was a charter member of the Environmental Economics Advisory Committee of the EPA Science Advisory Board. He is a member of the Board of Directors of Defenders of Wildlife.

Wayne Ouderkirk, a hiker who recently completed his climbs of the forty-six high(est) peaks in New York's Adirondack Mountains, is occasionally seen also on the dance floor. Most of the time he teaches environmental philosophy and philosophy at Empire State College, the State University of New York's nontraditional college. He also manages a small, rural office of the college, serving adult students. One of the faculty coordinators for Empire State College's interdisciplinary Adirondack/Environmental Studies Residency Program, he has published and presented papers on several aspects of environmental philosophy. His current research interests are wilderness issues and the epistemological dimensions of environmental ethics. He intends to continue both his dancing and his Adirondack exploration, as well as to climb the 111 peaks higher than 4,000 feet in the Northeastern United States. He earned his Ph.D. in philosophy at the State University of New York at Albany.

Clare Palmer lectures in the Centre for Philosophy, Institute for the Environment, Philosophy and Public Policy at Lancaster University (UK). She is editor of the journal *Worldviews: Environment, Culture, Religion* and has published two books: *Environmental Ethics and Process Thinking* (Oxford: Oxford University Press, 1998) and *Environmental Ethics* (Santa Barbara, California: ABC–Clio, 1997).

Ernest Partridge is a research associate in the Department of Philosophy, University of California, Riverside where he is currently studying disequilibrium ecology under a research grant from the National Science Foundation. His published papers deal with environmental ethics, policy analysis, moral philosophy, and applied ethics. He is the editor of the anthology *Responsibilities to Future Generations* (Buffalo, N.Y.: Prometheus, 1981). In recent and continuing visits to Russia, Dr. Partridge has established productive and ongoing communication and cooperation with international scholars and scientists involved in global environmental issues. Partridge maintains a Web site, "The Online Gadfly" (www.igc.org/gadfly), which contains his recent publications as well as news and opinion regarding environmental ethics and policy.

Holmes Rolston III is University Distinguished Professor and Professor of Philosophy at Colorado State University. He has written seven books, most recently *Genes, Genesis and God* (Cambridge); *Philosophy Gone Wild* (Prometheus); *Environmental Ethics: Duties to and Values in the Natural World* (Temple); *Science and Religion: A Critical Survey* (Random House, McGraw Hill, Harcourt Brace); and *Conserving Natural Value* (Columbia University). He gave the Gifford Lectures, University of Edinburgh, 1997–1998, has lectured on six continents, and is past president of the International Society for Environmental Ethics.

Kristin Shrader-Frechette is associate editor of *Bioscience* and editor-in-chief of the Oxford University Press monograph series on Environmental Ethics and Science Policy. She serves on the editorial boards of 17 professional journals. Shrader-Frechette has been the first woman president of three international organizations: the Society for Philosophy and Technology (1985–1987); the Risk Assessment and Policy Association (1997–1999); and the

International Society for Environmental Ethics (1997–1999). She has published more than 250 articles and 14 books/monographs. Currently she is O'Neill Professor of Philosophy and Philosophy of Science and concurrent Professor of Biology at the University of Notre Dame.

Angus Taylor teaches philosophy at the University of Victoria in British Columbia. He is the author of *Magpies, Monkeys, and Morals: What Philosophers Say about Animal Liberation* (Peterborough, Ontario: Broadview, 1999).

Peter S. Wenz, Professor of Philosophy and Legal Studies at the University of Illinois at Springfield and Adjunct Professor of Medical Humanities at Southern Illinois University School of Medicine, is one of those aging, leftist, feminist, vegetarian environmentalists who one often sees jogging near universities. He teaches a variety of courses in practical philosophy, including moral issues in the law, and environmental values. His books include *Environmental Justice* (Albany, N.Y.: SUNY Press, 1988); *Abortion Rights as Religious Freedom* (Philadelphia: Temple University Press, 1992), *Nature's Keeper* (Philadelphia: Temple University Press, 1996), and *Environmental Ethics Today* (Oxford: Oxford University Press, 2001).

Index